THE SPARROWS

THE SPARROWS

A study of the genus *Passer*

by J. Denis Summers-Smith

Illustrations and colour plates by

ROBERT GILLMOR

T & AD POYSER

Calton

© *J. Denis Summers-Smith 1988*

ISBN 0 85661 048 8

First published in 1988 by T & A D Poyser Ltd
Town Head House, Calton, Waterhouses, Staffordshire, England

All rights reserved. No part of this book may be reproduced, stored in a retrieval system, or transmitted in any form or by any means, electrical, mechanical, photocopying or otherwise, without the permission of the publisher

British Library Cataloguing in Publication Data

Summers-Smith, J. Denis
 The sparrows: a study of the genus Passer.
 1. Passer
 I. Title
 598.8'73   QL696.P264

ISBN 0-85661-048-8

Text set in Monophoto Times, printed and bound in Great Britain by Butler & Tanner, Frome and London
Colour plates by Lawrence-Allen Ltd, Weston-super-Mare

Contents

Preface and acknowledgements		11
Introduction		13
1	The Afrotropical grey-headed sparrows	17
	Grey-headed Sparrow	19
	Swainson's Sparrow	27
	Parrot-billed Sparrow	31
	Swahili Sparrow	35
	Southern Grey-headed Sparrow	38
2	Golden Sparrow	46
3	Chestnut Sparrow	61
4	Cape Sparrow	67
5	Rufous Sparrow	78
6	Iago Sparrow	93
7	Somali Sparrow	101
8	Desert Sparrow	106
9	House Sparrow	114
10	Willow (Spanish) Sparrow	162
11	Dead Sea Sparrow	180
12	Sind Jungle Sparrow	194
13	Pegu Sparrow	199
14	Cinnamon Sparrow	205
15	Tree Sparrow	216
16	Saxaul Sparrow	245
17	Characteristics and interrelationships	253
18	Origins and evolution	276
19	The systematic position of the sparrows	297
Appendix A	A Key to the names of the sparrows	307
Appendix B	Gazetteer	314
References		316
Index		337

List of Figures

1	Range of Grey-headed Sparrow	23
2	Breeding season data for Grey-headed Sparrow	25
3	Range of Swainson's Sparrow	29
4	Breeding season data for Swainson's Sparrow	29
5	Range of Parrot-billed Sparrow	33
6	Breeding season data for Parrot-billed Sparrow	34
7	Range of Swahili Sparrow	36
8	Breeding season data for Swahili Sparrow	37
9	Spread of Southern Grey-headed Sparrow in Cape Province, since 1940	40
10	Range of Southern Grey-headed Sparrow	41
11	Breeding season data for Southern Grey-headed Sparrow	43
12	Ranges of grey-headed sparrows and regions of overlap	44
13	Seasonal variation in mean weights of Golden Sparrow	48
14	Range of Golden Sparrow	50
15	Main vegetation types occurring in Africa	51
16	Breeding season data for Golden Sparrow	53
17	Zone of breeding of Golden Sparrow in Senegal, 1975–1984	57
18	Range of Chestnut Sparrow	63
19	Breeding season data for Chestnut Sparrow	65
20	Range of Cape Sparrow	69
21	Breeding season data for Cape Sparrow	74
22	Seasonal distribution of nests of Cape Sparrow in two localities	75
23	Distribution of African rufous sparrows	79
24	Range of Great Sparrow	82
25	Breeding season data for southern African populations of Rufous Sparrow	84
26	Range of Kenya Rufous Sparrow	86
27	Breeding season data for Kenya Rufous Sparrow	87
28	Range of White Nile Rufous Sparrow	88
29	Breeding season data for White Nile Rufous Sparrow	89
30	Range of Kordofan Rufous Sparrow	90
31	Breeding season data for Kordofan Rufous Sparrow	91
32	Wing lengths of male rufous sparrows	95
33	Male wing lengths for Iago Sparrow and four races of Rufous Sparrow as a function of mean monthly temperature	96
34	First records for Iago Sparrow on different islands of the Cape Verde Archipelago	97
35	Breeding season data for Iago Sparrow	99
36	Range of Somali Sparrow	103
37	Breeding season data for Somali Sparrow	104
38	Distribution of Desert Sparrow in North Africa	108
39	Distribution of Desert Sparrow in Asia	109

List of Figures

40	Breeding season data for Desert Sparrow	112
41	Variation of wing length of male House Sparrows with longitude	118
42	Seasonal pattern of mean weights of House Sparrows	120
43	Zone of intergradation between *P. d. domesticus* and *P. hispaniolensis italiae*	124
44	Range of House Sparrow *P. d. tingitanus* and the zone of hybridisation of House and Willow Sparrows in northwest Africa	126
45	Distribution of races of House Sparrow in the Palaearctic and Oriental regions	128
46	Distribution and relative abundance of House Sparrow in North America during breeding season and in mid winter	130
47	Spread of House Sparrow in northeastern Brazil south of the Amazon	132
48	Dates of introduction of House Sparrow in Afrotropical region and timing of spread in southern Africa	134
49	Distribution of House Sparrow	136
50	Relationship between start of breeding season and latitude in House Sparrow	147
51	Duration of breeding season of House Sparrow as a function of latitude	147
52	Number of clutches laid per year by House Sparrows as a function of latitude	148
53	Mean clutch size of House Sparrow in North America as a function of latitude	149
54	Breeding success of House Sparrow as a function of latitude	151
55	Number of young House Sparrows per pair per year as a function of latitude	151
56	Percentage mortality curve for young House Sparrows	154
57	Seasonal composition of two hypothetical House Sparrow populations	155
58	Seasonal variation in type of food taken by House Sparrows in North America	160
59	Seasonal variation in mean weight of Willow Sparrow	165
60	Distribution of Willow Sparrow in northwest Africa	166
61	Distribution of Willow Sparrow in northwest Africa and first-recorded dates on Madeira and the Canary Islands	167
62	First-recorded dates of Willow Sparrow on different islands of Cape Verde Archipelago	168
63	Distribution of Willow Sparrow	171
64	Breeding season data for Willow Sparrow	176
65	Spread of Dead Sea Sparrow in Israel	183
66	Distribution of Dead Sea Sparrow in Middle East	185
67	Range of eastern race of Dead Sea Sparrow	187
68	Male Dead Sea Sparrow on nest	191
69	Breeding season data for Dead Sea Sparrow	192
70	Range of Sind Jungle Sparrow	196
71	Breeding season data for Sind Jungle Sparrow	197
72	Range of Pegu Sparrow	201
73	Breeding season data for Pegu Sparrow	203
74	Range of Cinnamon Sparrow *P. r. rutilans*	208
75	Range of Cinnamon Sparrow *P. r. intensior*	209

8 *List of Figures*

76	Range of Cinnamon Sparrow *P. r. cinnamomeus*	210
77	Altitudinal breeding limits of Cinnamon Sparrow	211
78	Distribution of Cinnamon and House Sparrow nests in Mussooree	212
79	Breeding season data for Cinnamon Sparrow	213
80	Range of Tree Sparrow in Sardinia	220
81	Range of Tree Sparrow in North America	221
82	Distribution of Tree Sparrow in Palaearctic and Oriental regions	225
83	Habitat utilisation by House and Tree Sparrows in Afghanistan	227
84	Breeding season data for Tree Sparrow	236
85	Changes in Tree Sparrow population in British Isles, 1945–1985	241
86	Comparison of the effect of food availability on timing of breeding and moult in two populations of Tree Sparrows in Singapore	242
87	Range of Saxaul Sparrow *P. a. ammodendri* (western population)	247
88	Range of Saxaul Sparrow *P. a. ammodendri* (eastern population)	248
89	Range of Saxaul Sparrow *P. a. nigricans*	248
90	Range of Saxaul Sparrow *P. a. korijewi*	249
91	Range of Saxaul Sparrow *P. a. stolickzae*	249
92	Range of Saxaul Sparrow *P. a. timidus*	250
93	Distribution of Saxaul Sparrow	251
94	Breeding season data for Saxaul Sparrow	252
95	Zone of overlap between Grey-headed and Swainson's Sparrows	256
96	Zone of overlap between Grey-headed and Parrot-billed Sparrows	257
97	Zone of overlap between Grey-headed and Swahili Sparrows	258
98	Zone of overlap between Grey-headed and Southern Grey-headed Sparrows in central coastal Angola	259
99	Zone of overlap of the isolated *P. diffuses luangwae* population of Southern Grey-headed with Grey-headed Sparrow	260
100	Zone of overlap between Grey-headed and Southern Grey-headed Sparrows in Zambia, Zimbabwe and Mozambique	260
101	Zone of overlap between Swainson's and Parrot-billed Sparrows	261
102	Zone of overlap between Parrot-billed and Swahili Sparrows	262
103	Proposed spread of sparrows from Afrotropical region to the Levant and subsequent expansion in Palaearctic and Oriental regions during the Pleistocene	280
104	The 'Fertile Crescent', proposed centre for the origin of the House Sparrow	281
105	Distribution of agricultural communities in Europe, approximately 7,000 B.P.	282
106	Scheme for evolution of Willow, Italian and House sparrows	284
107	A tentative scheme for subspeciation of House Sparrow	285
108	Distribution of six Asian sparrows	286
109	Proposed evolution centres for northern hemisphere sparrows	289
110	Proposed vegetation map for Africa during a dry climate phase	292
111	Proposed vegetation map for Africa during a wet climate phase	293
112	Wing of House Sparrow (dorsal view)	299
113	Proposed classification for passerine-finches, family Passeridae	302
114	Proposed classification for sparrow-weavers, subfamily Plocepasserinae	303

List of Tables

1	Bill colour scores for Grey-headed Sparrow	20
2	Biometric data for Grey-headed Sparrow	21
3	Bill colour scores for Swainson's Sparrow	27
4	Biometric data for Swainson's Sparrow	28
5	Bill colour scores for Parrot-billed Sparrow	32
6	Biometric data for Parrot-billed Sparrow	32
7	Bill colour scores for Swahili Sparrow	35
8	Biometric data for Swahili Sparrow	35
9	Bill colour scores for Southern Grey-headed Sparrow	39
10	Biometric data for Southern Grey-headed Sparrow	39
11	Biometric data for Golden Sparrow	47
12	Golden Sparrow breeding statistics	58
13	Moult of primaries by Golden Sparrow	59
14	Biometric data for Chestnut Sparrow	62
15	Biometric data for Cape Sparrow	68
16	Variation in average clutch size of Cape Sparrow with latitude	74
17	Variation in average clutch size of Cape Sparrow with season	74
18	Comparison of key features of subspecies of Rufous Sparrow	80
19	Biometric data for Rufous Sparrow	81
20	Biometric data for Iago Sparrow	95
21	Biometric data for Somali Sparrow	102
22	Biometric data for Desert Sparrow	108
23	Activity of Desert Sparrow at nest during incubation period	112
24	Biometric data for House Sparrow	119
25	Seasonal changes in the bill length of House Sparrow	121
26	Limits of the transition zone between *P. d. domesticus* and *P. hispaniolensis italiae*	125
27	House Sparrow breeding statistics	152
28	House Sparrow densities in different habitat types	157
29	Percentage of animal food given to nestling House Sparrows as a function of their age	160
30	Biometric data for Willow Sparrow	165
31	Stomach contents of adult Willow Sparrows	178
32	Stomach contents of nestling and fledgling Willow Sparrows	178
33	Biometric data for Dead Sea Sparrow	182
34	Seasonal changes in wing and tail length of Dead Sea Sparrow	182
35	Seasonal change in bill length of Dead Sea Sparrow	182
36	Biometric data for Sind Jungle Sparrow	195
37	Biometric data for Pegu Sparrow	200
38	Biometric data for Cinnamon Sparrow	207
39	Biometric data for Tree Sparrow	219
40	Situation of Tree Sparrow nests in the Balkans	234
41	Tree Sparrow breeding statistics	238
42	Tree Sparrow densities in different habitat types	240
43	Stomach analyses of Tree Sparrows in Yugoslavia	243
44	Biometric data for Saxaul Sparrow	246

45	Comparison of grey-headed sparrows	256
46	Types of nest sites used by different sparrow species	269
47	Regional distribution of sparrow nest-site types	277
48	Use of houses for nesting by sparrows	278
49	Comparison of Golden and Chestnut Sparrows	290
50	Classification of the genus *Passer*	304

Colour plates

between pages 128 and 129

Preface and Acknowledgements

The sparrows, and most particularly the House Sparrow, get a bad press:

> 'The sparrow, meanest of the feathered race'
> *William Cowper*

> 'How I hate the sparrows, the sparrows, the sparrows'
> *Dora Sigerson*

But not all are against them:

> 'Many people speak ill of sparrows. I can understand dislike of them in the country, but I cannot understand it in the towns. In the country they are invaders driving out better birds than themselves. In towns on the other hand the sparrow is at home. He has no music for the traffic to drown – no bright plumage for the smoke to blacken. He is a little parasite, who can pick up a living where a more sensitive bird would starve. He may not, as an individual, be so confiding as the robin. But the robins do not come crowding round a human being in families like the family of the old woman who lived in a shoe.'
>
> *Robert Lynd*

and, judging by the list of references at the end of this book, professional ornithologists would appear to consider them eminently worthy of study, with probably more written about sparrows than any other genus of birds.

In 1949 I began to make a special study of the House Sparrow. This culminated in a monograph on the species in 1963. An over-riding impression that came out of this study was the undoubted success of the House Sparrow as an animal species. It is one of the most widespread passerine birds in the world, both with its natural range from North Africa through Europe and Asia, and also through its successful introductions to many other parts of the world: in North and South America, Africa south of the equator and Australasia, occurring from sea level (even almost 400 m below it near the Dead Sea) to elevations well above 3,000 m (3,100 m at Leadville, Colorado, USA, 3,750 m at La Paz, Bolivia, and up to 4,500 m in the Himalayas).

Moreover, over most of this range it is also a very common bird, probably more familiar to more people than any other bird species.

This posed the question: 'Wherein lies the reason for its success'. The immediate answer is, of course, the close association of the House Sparrow with that most successful of all animals, man. This, however, is a trivial answer and merely exposes the more fundamental problem of how this association came about. The question now becomes an evolutionary one and I resolved to study as many of the other closely related sparrows as possible in an attempt to find an answer. There are nineteen other sparrows in the genus *Passer,* to which the House Sparrow belongs, and this book is based on the study of the twenty *Passer* species. It has involved me in much travel over Europe, Africa and Asia, together with an extensive correspondence. I am indebted to a very large number of people for unstinting help. I should like to pay special tribute to a few of these and apologise to any I have overlooked. For particular help during travels I want specially to thank Mrs Janet Barnes (Botswana), Sugato Chaudhuri (India), Dr Upen de Zylva (Sri Lanka), Professors H. Mendelssohn and Yoram Yom-Tov (Israel), Dr Gèrard Morel and Dr Marie-Yvonne Morel (Senegal), David Skead and Warwick Tarboton (South Africa), Dr Schwann Tunhikorn (Thailand).

I should like to record my gratitude to a succession of Heads of the Sub-department of Ornithology, British Museum (Natural History) – J. D. Macdonald, Dr D. W. Snow, I. C. J. Galbraith – and the current Keeper of the Bird Collection, G. Cowles, for permission to examine the sparrow skins in the collection, and to members of the staff for their help.

A number of friends have read chapters in draft and have suggested many improvements; any remaining weaknesses and omissions are mine. In particular, I wish to acknowledge the help given by L. R. Lewis during many discussions held over the years since I started working with House Sparrows and for his comments on Chapters 17 and 18; Dr Gèrard Morel for his comments on Chapters 1 and 2; Professor Yoram Yom-Tov on Chapter 11, Dr David Parkin on Chapter 18, Dr John Coulson on Chapter 19 and most particularly Dr Peter Evans (Guisborough) who read the complete manuscript in draft.

I am grateful to the following for permission to reproduce figures from their publications: T. M. Crowe – Figs 110 and 111; the American Ornithologists' Union – Figs 46a and 81; the United States Department of Agriculture – Fig. 55; E. K. Urban and the Academic Press (*The Birds of Africa*, Vol. 1) – Fig. 15; J. A. Wiens and R. F. Johnston – Fig. 46b – and R. F. Johnston and W. J. Klitz – Fig. 105 – both from *Granivorous Birds in Ecosystems*, Cambridge University Press.

John Turner prepared the maps and graphs. I should like to thank him most particularly for his skill in interpreting my requirements.

Finally, I wish to pay tribute to my wife for her company and tolerance during the more trying times that inevitably occur during foreign field expeditions, and even more for having put up with my lack of attention during the long period of gestation and writing, much of which occurred when I was already involved in a full-time and demanding job in industry.

Introduction

The modern word sparrow is derived from the Old English speerwa (from the Aryan root 'little flutterer' – spar = to flutter). Probably, this was applied originally to many sorts of small birds, but came to be used particularly for the House Sparrow and its congeners (genus *Passer*) and the related Rock and Bush Sparrows (genus *Petronia*). It was given, as well, to other similar, small, dull-coloured birds that bear no family relationship, such as the Hedge Sparrow *Prunella modularis* (Family Prunellidae) – now usually called the Dunnock to avoid confusion. Later, as birds from other parts of the world began to receive English names, it was used for such diverse birds as the Java Sparrow *Padda oryzivora* (Family Estrildidae) and a large group of Nearctic and Neotropical seed-eating birds, including, for example, the Song Sparrow *Melospiza melodia*, which are more properly buntings (Family Emberizidae).

Although the different species that have been called sparrows are clearly separated by their scientific names, the vernacular names are still widely used and it would be more satisfactory to restrict the name sparrow to the House Sparrow and species closely related to it. This would, however, be flying in the face of common usage and the best that can be hoped for is to restrict its use to three groups of birds: the Old World Sparrows, including the sparrows of the genus *Passer*, which can be distinguished by the epithet 'true', the Rock and Bush Sparrows of the related genus *Petronia* that already have distinguishing epithets, and the American sparrows that are best described as the New World Sparrows.

Misunderstanding is most likely to arise when two birds share the same English name, Tree Sparrow: *Passer montanus* of the True Sparrows and *Spizella arborea* of the New World Sparrows, particularly as the former species was introduced to the USA in the middle of the 19th century and the two birds now overlap in Illinois. For the sake of clarity it might be better to refer to them as the Eurasian Tree Sparrow and the American Tree Sparrow respectively, though the name Tree Sparrow is now well entrenched for each.

I am concerned in this book solely with the genus *Passer*, the True Sparrows. I consider that there are twenty species in this genus. These birds have an extensive natural distribution in the Afrotropical, Palaearctic and Oriental regions, and members of the genus have been successfully introduced to the Nearctic, Neotropical and Australasian regions, so that the True Sparrows have now virtually a cosmopolitan distribution.

In thirteen of the twenty species, the male has a black bib with a characteristic facial pattern of light cheeks and a darker crown, usually grey or chestnut. Twelve of these species are sexually dimorphic, with the female a nondescript brown colour; the thirteenth, the Tree Sparrow, is sexually monomorphic with the female having the characteristic male plumage pattern with a black bib. As a convenient shorthand I shall refer to them as the 'black-bibbed sparrows'. The remaining seven species, which are basically African, except for one, the Golden Sparrow *Passer luteus* that has a small extension of range into Saudi Arabia, lack the black bib and are either sexually monomorphic or sesquimorphic. (I use the latter term to describe those species in which both sexes have basically the same plumage pattern, though the female is clearly distinguishable by reason of her paler or washed out colour.) Again, for convenience, I shall refer to these seven species as the 'plain sparrows'. I have made observations on nineteen of the twenty species in the field. The one exception is the Saxaul Sparrow *Passer ammodendri* that occurs in remote parts of Russian and Chinese Turkestan and has so far eluded me.

The most significant published literature has been consulted up to the end of 1985, though with the rapidly increasing number of new journals in the developing world it is inevitable that some relevant papers will have been overlooked. The sources of original information are listed in the References towards the end of the book.

The first part of the book is concerned with detailed accounts of the twenty species, each species being treated separately apart from the closely related grey-headed sparrows that are more conveniently considered in one chapter. A standardised format is used in the species accounts for ease of comparison and to provide a data base that will enable a detailed examination of the relationships between the different species and allow some speculation on the evolution of the genus as a whole.

Each species is given one or more distribution maps. These maps are based on all available published records, supplemented by correspondence with experienced ornithologists in cases where the published data appeared confusing or inadequate. The maps have been drawn to enclose all the records; no attempt has been made to exclude areas of unsuitable habitat lying within this range. The identification of place names is a problem: some are not marked in readily available maps, spellings vary from time to time and source to source, and, even worse, changes have occurred. In general, I have adhered to the name used in the original source, but, where necessary as an aid to identification, I have included the spellings given in *The Times Atlas of the World: Comprehensive Edition*, Times Books, 5th Edition 1975. In a few cases I have retained the more familiar anglicised spellings, eg Naples for Napoli, Turkmenistan instead of Turkmenskaya, where these make it easier for the English reader without causing confusion. Many of the spelling problems arise through the Romanisation of non-Roman scripts. In particular, the official pinyin Romanisation of Chinese characters results in considerable differences in the current spellings of Chinese place names; where relevant the pinyin spellings given in *Zhonghua Renmin Gongheguo Fen Sheng Dituji*, Zhongguo, Beijing: Ditu Chubanshe, 1980 (Map of the Peoples' Republic of China, Beijing 1980) are shown. (This atlas is readily available in bookshops and at airports in China.)

Use of these two atlases will give the locations of the majority of the places mentioned in the text, though it should be pointed out that in some cases place names appear in the maps in the Times Atlas without being listed in the Index. A gazetteer of the places that do not appear in either of these atlases is given in Appendix B. In addition, it has been convenient at times to make use of less precise geographical names of common usage, such as Middle East, Sahel and so on, that do not appear in the Times Atlas. The precise sense in which I use these terms is given in Appendix B.

At times it has been desirable to distinguish between the age classes of the birds. I have used three terms: *juvenile* refers to the stage from fledging to the first moult, which for the sparrows occurs within a few weeks of fledging; *immature* is used for the first-year birds after the post-juvenile moult and *adult* for older birds.

In a study of this sort, biometric information is particularly important for comparison of the sexes, identification of regional variations within a species and differentiation between races and between species. The treatment of published biometric data has caused particular problems and it has not been easy to summarise the data in a meaningful way. This is partly because of the variability in quality of the published data available: in some cases only small numbers of specimens have been measured, in others ranges are given without any indication of the numbers of measurements involved or mean values; only in a few cases have adequate samples been analysed to justify the calculation of statistical values: means, standard deviations, standard errors. In addition, however, there are genuine seasonal differences in some of the characters. Abrasion of feathers takes place continuously. This is the way that the breeding plumage is developed as the tips of the feathers are worn away to reveal the underlying colours, but it also means that measurements based on feathers change during the year. For example, wear of remiges and rectrices results in decreases of wing and tail lengths by up to 3% between the newly-grown condition and the time when they are moulted a year later. The rate of wear is not, however, uniform, being influenced among other things by the nesting and roosting behaviour – birds using holes are subject to greater abrasion. This makes it virtually impossible to make sensible seasonal allowances. Bill length is a balance between the rate of growth and the rate of wear, the latter being related to the type of food that is eaten. In three of the species in which this has been studied, House, Tree and Dead Sea Sparrow, the bill was 5–15% shorter in the non-breeding season, when the birds feed exclusively on a diet of seeds, compared with the breeding season when the diet has a large, softer, invertebrate component; moreover, this effect varies with different populations depending on their precise dietary regime. A further problem with bill length is that, with many workers, it is not entirely clear how the measurements have been taken, whether from the cranio-facial hinge, the end of the feathering on the skull or even from the nostril. Weight, one of the most useful characteristics when making comparisons, varies not only seasonally, but also diurnally as well. Hence, without information on the time of year that the specimens were measured, and even the time of day for weight, caution must be exercised when using the values for comparative purposes.

A different source of variability is apparent from studies of species for which large samples of data are available. It is apparent from several extensive studies on the House Sparrow, for example, that significant size differences can exist between neighbouring populations independently of any larger-scale clinal variations.

Extensive tables have been prepared covering the published data for weight, overall length, wing and tail length, tarsus and culmen. The tables given in the text are my interpretation of the complete tables. Copies of the latter have been deposited in the libraries of the British Museum (Natural History), Tring, and the Edward Grey Institute, Oxford, and are available for consultation by courtesy of the Librarians for those requiring to make use of the full data.

In an attempt to present the available information in as useful a way as possible, the results are given in a standardised tabular format for each species, or subspecies where relevant, together with localities in the case of widely distributed populations to show clinal effects. Separate data are given for each sex, though where these are not available for some of the sexually monomorphic species the combined data are presented. As an optimum a range and mean are given for each character, with a median where the mean is not available. The median is a less satisfactory statistic, though as the characters tend to have normal distributions it gives a reasonable approximation of the mean. One advantage of dealing with such a homogeneous group as the sparrows is that the variability of the measured characters seems to be very similar between species, typically $\pm 7\%$ (3 standard deviations) of the mean (or median) value for the linear dimensions, up to 25% of the mean (or median) value for the weight. Throughout, linear dimensions are given in millimetres, weights in grammes. To highlight the quality of the data, entries based on samples of more than 50 individuals are printed in bold type.

Similar problems arise with other numerate data such as clutch size, incubation and fledging periods, number of broods per year, etc, as these vary not only within the range, but in one locality from year to year. Comparison between the performances of one species in different parts of its range and between different species are thus fraught with uncertainty because of the variable nature of the data, making even statistical analyses somewhat suspect.

The comparisons of population densities, particularly of colonial or semi-colonial species is another possible pitfall, depending on whether the area chosen is biased towards a colony or an attempt has been made to obtain a more representative value covering a wider area.

The above caveats have always to be borne in mind when reading the following chapters and too much should not be read into the absolute figures, though it is hoped that the figures given have sufficient validity for comparative purposes.

While the English language vernacular names are more easily read and thus more convenient to use in the text, the scientific names are essential to avoid any misunderstanding. The nomenclature of each species is dealt with briefly at the beginning of the species accounts to show any recent changes, listing names that are still used in current literature, and giving my own ideas with regard to the species and their recognisable races. Over the years a very large number of names has been given to these birds. This sometimes makes it difficult when reading the old literature to be sure exactly which species or race is being discussed; to simplify the situation for future workers on sparrows an extended list of names that have appeared in the literature is given in Appendix A showing their synonymy. Although this Appendix contains over 200 names it is still likely to be incomplete, though it is hoped that the major usages have been covered.

This is basically an evolutionary study and thus I have considered it essential to deal with each species as fully as possible, even if it necessarily means that the treatment of different species is very unequal because of the variable amount of information available. In order to draw comparisons it is necessary to make generalisations, but these tend to be somewhat suspect in the more widespread species, where genuine differences in behaviour occur in different parts of the range. Again, with more well-studied species it is inevitable that observers have drawn attention to variations from the normal pattern. While there is a risk that such variations tend to obscure the generalisations, nevertheless it is important that they are treated fully as they form the very stuff of evolution. Equally, there is a risk of over-simplification with the less well-studied species, giving the impression that they are more rigid in their behaviour patterns. Due allowance must be given to these effects when reading the text.

1: The Afrotropical grey-headed sparrows

Grey-headed sparrows occur over almost the whole of Africa south of the Sahara from about 15°N, with the exception of the Horn of Africa and parts of western Cape Province, South Africa. The grey-headed sparrows are monomorphic, the sexes being indistinguishable in the field; over the whole of Africa differences are small and ever since the birds were first described in 1817 opinions have differed on how far a number of separate species should be recognised, or whether only one polymorphic species is involved.

A recent comparative study by Hall and Moreau (1970) has shown that five different species can be recognised on the grounds that small areas of overlap occur where the two taxa live side by side without interbreeding, though in other parts of their range these two taxa may apparently intergrade. The situation is an extremely complicated one taxonomically and it seems best to treat these five closely related species as allospecies* (Amadon 1966) that are members of a superspecies *Passer* [*griseus*], in which differences have developed sufficiently to prevent interbreeding in certain areas in which they overlap, whereas in others they intergrade to some extent. The five species are the Grey-headed Sparrow *Passer griseus* occurring from Senegal to the Sudan and south to Angola; Swainson's Sparrow *P. swainsonii* almost entirely confined to Ethiopia: the Parrot-billed Sparrow *P. gongonensis* largely restricted to Kenya; the Swahili Sparrow *P. suahelicus* occurring south of the Parrot-billed Sparrow in Tanzania, and the Southern Grey-headed Sparrow *P. diffusus* that occurs from Angola to Mozambique, south to the Republic of South Africa. In addition to slight morphological differences, the species also tend to differ ecologically, these differences becoming more accentuated in those regions of overlap in which there is sufficient ecological diversity in the habitat.

---

* The use of the term allospecies is perhaps not strictly correct, suggesting as it does no sympatry between the different taxa. Amadon in his definition, however, allows 'or nearly allopatric taxa', which would appear to be applicable in this case.

The grey-headed sparrows were probably originally birds of mesic grassland savanna; they are associated with trees, do not occur in completely open country and tend to be absent from the more arid parts; on the other hand, they do not penetrate forested areas. As is typical of most of the sparrows they spend much of the time in the trees, but feed mostly on the ground on the seeds of grasses and small herbs. They are social species, living and breeding in small groups, though not usually in the large associations found with some sparrow species, particularly outside the breeding season.

Grey-headed Sparrow *Passer griseus*

NOMENCLATURE
Fringilla grisea Vieillot, N. Dict. d'Hist. Nat. 1817 12: 198.
    United States [*error* = Senegal, Lafresnaye, Rev. Zool. 1839: 95].
Passer griseus (Vieill.) Reichenow, Vög. Afr. 1904 3: 230.
  Subspecies: *Passer griseus griseus* (Vieill.) 1817.
    Synonyms: *Pyrgita gularis* Lesson 1839.
        Senegal.
    Passer occidentalis Shelley 1893
        West Africa [Lokoja, south Nigeria according to Lynes Ibis 1926: 346]
    Passer diffusus thierryi Reichenow 1899
        Mangu [Sausanné Mango], Togo.
    Passer griseus kleinschmidti Grote 1923
        Ngaundere [N'Gaoundéré], [French] Cameroons.
Passer griseus ugandae Reichenow 1904
    Uganda [Manjonga].
  Synonyms: *Passer albiventris* Madarasz 1911
        Sudan.
    Passer nikersoni Madarasz 1911
        Chor-em-Dul, Sennar district, Sudan.
    Passer griseus zedlitzi Gyldenstolpe 1922
        near Benguella town, Portugese West Africa [Angola].
Passer griseus laeneni Niethammer 1955
    Bol, east bank of lake Chad.

This last race was named by Niethammer after the collector, Mr Laenen.

DESCRIPTION
The Grey-headed Sparrow is a medium-sized sparrow; it is one of the 'plain sparrow' group. The sexes are monomorphic.

Passer griseus griseus
The sexes are alike. Head and neck ash grey with darker grey mask through eye, cheeks paler than rest of head; mantle and upper back grey-brown, unstreaked; lower back and rump rather brighter, reddish brown; tail brown; wings brown with a cinnamon-red patch on scapulars, tips of median coverts forming a small, rather variable, white patch; underparts pale grey, becoming white on belly, with a striking, sharply defined, white patch on throat and chin. Legs greyish brown. Eyes light brown. length about 150 mm.
Immature. Similar, but has dusky streaks on mantle and lacks white wing patch.

P. griseus laeneni
Similar to nominate race, but much paler. Head and nape light grey; back fawn; rump sand-coloured; underparts white so that white bib patch is not differentiated.

P. griseus ugandae
Again, similar to nominate race but is darker above and below, with head tending towards grey-brown, upperparts more rufous and white bib less clearly defined.

Most authorities state that bill colour in *Passer griseus* is horn with that of male becoming black when it comes into breeding condition. White and Moreau (1958), on the other hand, imply that, in contrast to other grey-headed sparrows, bill of *griseus* is black in both sexes throughout year. I have examined the specimens in the British Museum collection, scoring bill colour as follows: black 8, blackish horn 6, dark horn 4, horn 2, pale horn 0. Unfortunately, with the wide variation in time of breeding over the range of the Grey-headed Sparrow (see later), it was not possible to relate these specimens to the breeding season. The results given in Table 1 are divided into three groups: adult males, adult females and immatures as identified by the collector. Allowing for difficulties in identifying the age of the bird without examining the state of ossification of the skull, these results tend to support the view that the bill is horn in the immature, but black in adult male and in the majority of adult females; one female collected from a nest with eggs was noted as having a black bill.

Table 1: Bill colour scores for Grey-headed Sparrow

Population	Males		Females		Immatures	
	No.	Mean score	No.	Mean score	No.	Mean score
griseus	31	7.0	16	6.1	—	—
laeneni	10	7.6	9	5.9	—	—
ugandae	108	7.1	59	5.3	13	3.3

BIOMETRICS
A summary of typical body measurements for the three recognised subspecies is given in Table 2. Because of the small amount of information and differences of opinion by different authors on the distribution of the different races, it is possible to draw only tentative conclusions. There is a tendency for the largest birds to occur in the northeast, with a cline of decreasing size in *griseus* from east to west and in *ugandae* from northeast to southwest, with *ugandae* generally larger; *laeneni* is larger than *griseus* in the area in which they come into contact. females have similar weights to males, but on average their wing length is about 3% shorter.

Table 2: Biometric data for Grey-headed Sparrow

Feature	Subspecies	Locality	Males Range	Males Mean (Median)	Females Range	Females Mean (Median)
weight	griseus	—	20–26	23	20–28	24
	laeneni	—	24–30*	(27)	26–31*	(28.5)
wing	griseus	W. Africa	77–85	(81)	75–86	(80.5)
		Cameroun/Chad	79–88	(83.5)	75–87	(81)
		Sudan	84–90	(87)		
	laeneni	—	80–87	(83.5)	79–85	(82)
	ugandae	Sudan/Eritrea	85–91	(88)		
		Uganda	**82–88**	**85**		
		Zambia	70–87	84.1	79–83	80.2
		Angola	77–87	82	78–83	(81)
tail	griseus	W. Africa	55–65	(60)	53–61	(57)
		Cameroun	55–65	(60)	57–67	(62)
tarsus	griseus	—	20–21	(20.5)	20–21	(20.5)
culmen	griseus	—	14.0–16.0	14.8	14.1–16.0	15.0
	ugandae	—	14.5–16.5	15.7	14.0–17.0	15.8

* very small samples

DISTRIBUTION

The Grey-headed Sparrow is the most widely distributed of the members of the [griseus] superspecies.

Passer g. griseus

Nominate race occurs in west central Africa from Gabon, where it is present in the north at least (information on central and southern Gabon is lacking), northwards to Cameroun and the extreme north of the Central African Republic, where it has been recorded from the Bamingui-Bangoran National Park. It extends west of Cameroun through southern Nigeria to Senegal and into southern Mauritania north of the Senegal river to between 17–18°N, and occasionally as far as Nouakchott, 18°07N (Gee 1984). The situation in Guinea, Sierra Leone, Liberia and the Ivory Coast is not clear, though in the latter it is said to be abundant in all the villages and sometimes found very far from any human settlement (Brunel & Thiollay 1969; Thiollay 1985); it appears to be common along the coastal strip, but of more sporadic occurrence inland, where its distribution is restricted by the hinterland forest zone and is only to be found there in villages and forest clearings. Although present throughout Nigeria up to about 10°N, again it is commonest on the coast and less common in the bush and the more sparsely populated country inland, particularly in the west. The transition zone between *griseus* and the northern race *laeneni* has not been clearly established. According to Lamarche (1981) the birds in Mali vary considerably, suggesting, in the absence of better information, that the boundary may lie there.

Passer g. laeneni

The paler birds of the subspecies *laeneni* occur in the more arid regions of the southern Sahel to the east of nominate *griseus*, from Tombouctou in Mali, north to the Aïr ou Azbine in Niger and Ennedi in Chad, east to the Darfur Province of western Sudan and south to Kano and the north of Nigeria (Niethammer 1955; Vaurie 1956; White 1963). It is described as very common in southern Niger, north to about 15°N, and occurs at a density of five pairs per hectare in the southern Aïr; a few extralimital birds were seen north of Bilma on 8 December 1970 (P. Giraudoux, *in litt.*).

Passer g. ugandae
The third subspecies, *ugandae*, reaches its northern limit in the Sudan about Atbara on the Nile, extends eastwards into western Eritrea up to the Ethiopian plateau, and west in Sudan to Kordofan Province. from there it ranges south through Zaire to reach the west coast of Africa in northern Angola, and in the east through Uganda, south through most of Zambia to the Zambezi and into western Malawi.

The situation is much less clear in Kenya and Tanzania. Birds showing the characteristics of *P. griseus* occur to the northeast of Lake Victoria: I have seen white-bibbed birds at the Mara Serena Lodge in the Masai Mara Game Reserve, Clancey (1959) collected some at Lake Magadi in the extreme south of Kenya, and G. R. Cunningham-van Someren (*in litt.*) informs me that small, slender-billed birds (*ie* contrasting with the larger, heavier-billed Parrot-billed Sparrow that is the common Kenyan grey-headed sparrow) occur in the Nairobi area, together with many intergrades. There is a report, dating from the 1970s, of grey-headed sparrows breeding in holes in houses in the centre of Nairobi (J. Reynolds, pers. comm.). Although these birds have not been specifically identified, it seems to me more likely that they are Grey-headed or Swahili Sparrows than the Parrot-billed Sparrow, the common grey-headed sparrow of Kenya. This last species is the least associated with man of the five grey-headed sparrows and does not breed in any other Kenyan town (see later). The Grey-headed Sparrow seems to be the most likely as it is a species that regularly nests in towns and, moreover, appears to be extending its range in east Africa.

According to Chapin (1954) the Sudanese race (*viz ugandae*) reaches the border of the Congo (Zaire) on the Oubangui river and the southern Bahr el Ghazal province of Sudan that borders Zaire.

To the north and west of this the situation, once again, is not clear. The Grey-headed Sparrow has been recorded from Bangui and the Manovo-Goundi-Saint Floris national park in the extreme south of the Central African Republic, but I have not found any records for a zone extending from western Zaire and south Gabon through the middle of the Central African Republic. I have been unable to establish whether this means that there is a gap here between the range of *ugandae* and the southern limit of the nominate race in northern Gabon and the north of the Central African Republic, or merely that this is a lacuna through the lack of published records. Much of this gap consists of forested country, unsuitable for Grey-headed Sparrows, and the bird is uncommon at the distribution limits given above, so a gap in range is quite possible. Again, the lack of records to the northeast suggests that *ugandae* and *laeneni* may also be separated.

The distribution of the three subspecies is shown in Fig. 1.

HABITAT
The Grey-headed Sparrow is commonly associated with cultivated land and human habitations, being found in villages and even towns, where it takes over the role of a 'house sparrow', though in parts of its range it also occurs away from human settlements and activities. For example, it is found in light woodland in western Ethiopia and Kenya, and, according to Drs G. and M.-Y. Morel (pers. comm.), it is common – sometimes in flocks of up to a hundred birds – in dry deciduous woodland

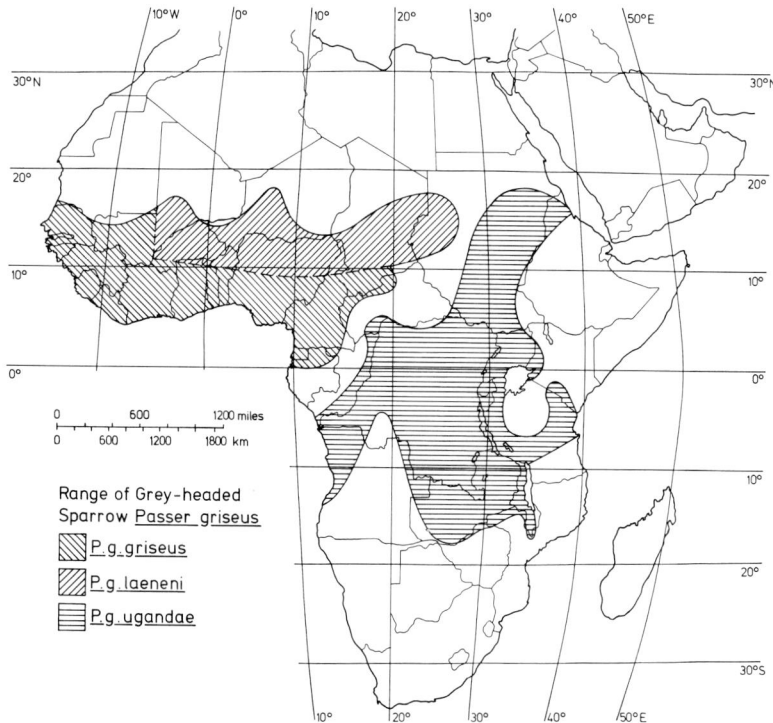

Fig. 1   Range of Grey-headed Sparrow

in southern Senegal. It was probably originally a bird of bush country and open grassland savanna with trees, that has secondarily become associated with man and is now most commonly found near the activities of the latter, being particularly common where there are horses. This no doubt has allowed an extension of distribution as more country becomes cleared, the bird advancing along railway lines and penetrating into forest clearings. Apart from the extreme north of its range, where the Grey-headed Sparrow reaches into arid country on the fringes of the desert, it tends to be more a bird of the relatively humid areas.

The northern race *laeneni* is much more adapted to an arid environment and is equally at home in open country away from human habitations, provided there are bushes and trees, as it is in the villages and towns.

BEHAVIOUR

The Grey-headed Sparrow is not shy and is frequently to be seen hopping around streets in the villages and perching on the roofs of huts, and, as well described by Serle (1940), shows a combination of familiarity, impudence and wariness, resembling in this respect the House Sparrow *Passer domesticus**. A. Brossett (in press) describes

---

* The House Sparrow is described in Chapter 9, but as it is likely to be the best known of the sparrows to most readers it is useful to make it the base species against which other, less well known members of the genus can be compared.

it as noisy and quarrelsome. Although the bird can be described as a social species, generally breeding in loose colonies and forming small parties or flocks of 20–50 birds outside the breeding season (in the more arid parts of its range the flocks can be much larger and cause damage to crops), it is not infrequently to be found in isolated pairs. According to Dr G. Morel (*in litt.*) the bird behaves differently in different parts of Senegal: in the rather lusher south of the country it lives, as throughout most of the rest of its range, as a gregarious species, whereas in the arid north it is territorial, breeding in isolated pairs. According to Brosset (in press) this is also the case in the Ivindo basin, northeast Gabon, the pairs defending an area of some hundreds of square metres round the nest.

Where the birds occur in flocks, they stay together searching for food on the ground or resting in trees, and they roost together in a village or in thick bushes on its outskirts.

Out of the breeding season the birds may move away from the normal village habitat, wandering over more open country, including cultivated land and marshy areas. Lynes (1926), for example, remarks that they appeared commonly in the Nuba region of the Sudan in winter, away from the breeding areas, and the bird is an occasional visitor to Nouakchott in Mauritania, north of the normal breeding range (Gee 1984).

BREEDING BIOLOGY

Little has been published about display; in the pre-breeding period, I have watched them indulging in headlong chases and when the leading bird lands the pursuer (presumably the male) lands nearby and postures in front of the other, chirping excitedly with neck and head stretched upwards and tail elevated.

The Grey-headed Sparrow normally breeds in, or close to, villages. The nest is an untidy accumulation, mainly of grasses, lined with feathers, frequently from chickens and domestic ducks, domed over and with an entrance on the side. The species is very catholic in its choice of sites; the nest can be placed in a wide variety of situations, both in holes and openly in the branches of trees. It will nest in houses in holes in the thatch or under projecting roofs, in hollow pipes on electricity pylons, holes in trees (frequently using the old nests of barbets and woodpeckers), and in old nests of swifts and swallows (*eg Apus affinis, Hirundo semirufa*). According to Serle (1940) both sexes visit the nest during building, though only one of the pair actually carries nesting material.

Reported data on the breeding season are summarised in Fig. 2, together with such information on the rains as I have been able to obtain (mostly from Kendrew 1961). This figure greatly oversimplifies the situation and has to be interpreted with some care for a number of reasons. The information has been organised in terms of recognisable political boundaries and it is by no means certain that the breeding and rainfall data relate to the same areas; secondly the breeding periods shown depend on published records and may not reflect the complete breeding season in some of the less well observed areas. As might be expected with a species that is widely distributed between 15°N and 18°S, covering a wide range of climatic zones, breeding occurs somewhere at all times of the year. Allowing for the caveats given above, it is apparent that the breeding season of the Grey-headed Sparrow is closely associated with the rains. This is shown very clearly for east Africa by Brown and Britton (1980), is the case in northern Senegal (G. and M.-Y. Morel, *in litt.*) and is supported by the verbal statements of a number of authors, *eg* Chapin (1954), Lippens and Willie (1976) for Zaire.

In some areas breeding can take place at any time of the year with little relation to rainfall. This is most likely in irrigated cultivated parts where the presence of water throughout the year ensures an adequate availability of the food necessary for rearing

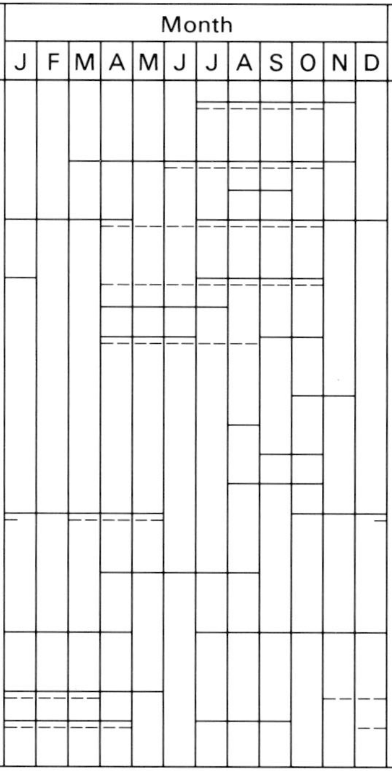

Subspecies	Locality	Month J F M A M J J A S O N D	Reference
griseus	Senegal 14–16°N		Mackworth-Praed & Grant 1973, Morel & Morel 1973b, 1982
	The Gambia 13°N		Jensen & Kirkeby 1980, Gore 1981
	Senegal 12–13°N		Morel & Morel 1982, *in litt.*
	Nigeria 4°N		Bannerman 1948, Elgood 1982, Mackworth-Praed & Grant 1960, 1973
	Cameroun 5°N		Serle 1950, 1981
	Gabon 0°		Mackworth-Praed & Grant 1973
	Niger 15°N		Hartert 1921, Bannerman 1948, Mackworth-Praed & Grant 1973
laeneni	Mali 13°N		Guichard 1947
	Nigeria 13°N		Serle 1940
	Sudan 14°N		Mackworth-Praed & Grant 1960
	Sudan 4–18°N		Mackworth-Praed & Grant 1960, Brown & Britton 1980
	Kenya 0–4°N		Mackworth-Praed & Grant 1960, Betts 1966, Brown & Britton 1980
ugandae	Uganda 2°N		Seth-Smith 1913, Mackworth-Praed & Grant 1960, Brown & Britton 1980
	Zaire 4°N 6–10°S		Chapin 1954, Mackworth-Praed & Grant 1963, Lippens & Willie 1976
	Zambia 8–18°S		Bannerman 1948, Winterbottom 1936
	Malawi 13°S		Benson 1953, Mackworth-Praed & Grant 1960, 1963, Benson *et al.* 1973

Fig. 2   Breeding season data for Grey-headed Sparrow (breeding ——— rains -----)

the young independent of the season, and probably accounts for the protracted breeding season in Nigeria and The Gambia.

The nesting cycle of the Grey-headed Sparrow has received little detailed attention. The clutch size is normally 3–4 eggs (occasionally 2) with apparently little variation over the extensive range of the species. The nestlings are fed mainly on insects. At a nest that I watched in Senegal, both sexes took an almost equal share in feeding the young (the adults were distinguishable by minor plumage differences); the feeding rate ranged from about 20 visits per hour in the early morning and late afternoon to about 10 visits per hour in the middle of the day. Neither bird appeared to be dominant over the other at the nest. One bird, presumably the female, roosted overnight in the nest cavity. Observations at a nest in Zambia by Winterbottom (1936) gave typical feeding rates of 6–8 visits per hour, though when winged termites were available the rate increased to 25 visits per hour.

SURVIVAL

There are no statistics on the life span of the Grey-headed Sparrow in the wild, but two individuals have survived for over 11 years in captivity (Flower 1925).

MOULT

Lynes (1926) gives the date for the annual moult in western Sudan, where the birds breed in September–October, as March.

VOICE

The basic call is a monosyllabic *chip, chirp* or *cheerp*; the individual chirps can be strung together to form a song (using the term in its functional sense rather than suggesting tunefulness!) *cheep chirp cheerp chirp*. There is also an alarm churr (*churr, chur-it-it*) similar to that used by many of the sparrow species.

FOOD

The adult birds are largely granivorous, feeding on the seeds both of grasses and of cultivated cereals, though they also take small fruits. Where the Grey-headed Sparrow is a town bird, it scavenges the streets and is almost omnivorous, eating bread, fruit and other household scraps. The nestlings are mainly fed on insects; representatives of the following orders have been reported: orthoptera (locusts), isoptera (termites, in their aerial phase) and hymenoptera, particularly formicoidae (ants). According to Brosset (in press) it comes to the houses in the mornings to collect insects that have been attracted to the lights at night and lie littered on the ground below.

Swainson's Sparrow *Passer swainsonii*

NOMENCLATURE
Pyrgita swainsonii Rüppell, Neue Wirbelt. Vög. 1840: 94
    northern Abyssinia [Ethiopia].
Passer swainsonii (Rüppell) 1840.
  Subspecies: none.
    Synonyms: *Passer griseus abyssinicus* Neumann 1908.
        Ghadi-Saati, Mareb River, Eritrea
        Intergrades with *gongonensis*.
Passer griseus neumanni Zedlitz 1908.
        Salamona about 16 miles west of Massawa, Eritrea.

DESCRIPTION
Swainson's Sparrow is slightly larger (length *ca* 160 mm), but very similar to Grey-headed Sparrow. Sexes are alike. Generally darker than Grey-headed, head darker grey and mantle and back duller brown, contrasting with chestnut rump; underparts are darker and, more particularly, it does not show white bib and belly of Grey-headed. Differences sufficient to allow species to be distinguished in the field.

The birds that have been described as *neumanni* show characters somewhat intermediate between *swainsonii* and *griseus*.

Most authors state that bill is horn-coloured and becomes black in breeding season. Bill colour scores for the British Museum specimens, obtained as described for Grey-headed Sparrow, are given in Table 3. The results are not very different from those obtained for Grey-headed Sparrow and once more are not consistent with a seasonal change in bill colour.

Table 3: Bill colour scores for Swainson's Sparrow

Males		Females		Immatures	
No.	Mean score	No.	Mean score	No.	Mean score
46	7.5	35	6.6	4	2.0

BIOMETRICS

Typical biometric data for Swainson's Sparrow are given in Table 4. Swainson's Sparrow is very similar in size to the more northerly Grey-headed Sparrows, though the limited data on weight suggest that it is somewhat heavier; in contrast the bill appears to be smaller than that of the race of the Grey-headed Sparrow which it abuts on the western border of Ethiopia.

Table 4: Biometric data for Swainson's Sparrow

Feature	Locality	Males		Females	
		Range	Mean (Median)	Range	Mean (Median)
weight	Ethiopia	27.3–35.2*	31.6*		
wing		**80–92**	**86**	80–88	(84)
tail		**64–72**	(68)		
tarsus		19–20*	(19.5)		
culmen		13.0–15.5	14.5	13.4–15.8	14.7

* both sexes

DISTRIBUTION

Swainson's Sparrow is present over much of Ethiopia, north into extreme coastal Sudan, as far north as Port Sudan, and south to north-central Kenya as far as Marsabit, where I saw them in February. It is typically a bird of the highlands, and extends into the high interior plateau of northwest Somalia, though it is also present in the maritime plain from the southeast of Sudan to northern Somalia as far as 49°E. it is absent from northwest Eritrea below 1,200 m and the lower lying Ogaden region of eastern Ethiopia. The distribution is given in Fig. 3.

HABITAT

This is a common bird of the towns and villages throughout its range; on the high Ethiopian plateau, above 1,400–1,500 m, it is also to be found in open country with bush and scrub.

BEHAVIOUR

In the towns and villages Swainson's Sparrow is very similar in most respects to the Grey-headed Sparrow, behaving as a 'house sparrow'. Outside the breeding season, however, the birds tend to form larger flocks, at times consisting of several hundred birds, which undertake seasonal wandering rather than a true migration, and cause damage to crops. For example, Erlanger (1907) found it breeding at Harar (Harer) from March to May, but absent in October, turning up once again in December.

BREEDING BIOLOGY

Nothing has been published about the display of this species. In northern Kenya, in February, I watched one bird posturing in front of another with the wings held out and slightly drooped, exposing the chestnut of lower back and rump; the feathers of the upper back were ruffled, giving the bird a hunched appearance.

Swainson's Sparrow breeds in much the same situations as the Grey-headed Sparrow: in holes in buildings and trees, crowns of palm trees, openly in the branches

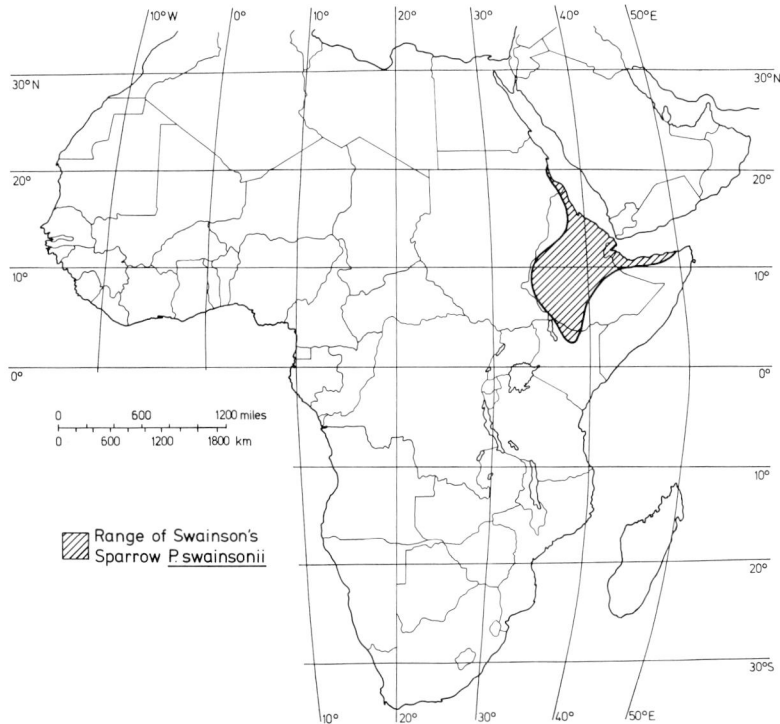

Fig. 3   Range of Swainson's Sparrow

of trees and in other birds' nests, which it relines; for example, I have watched a pair at the nest of an Ethiopian Swallow *Hirundo aethiopica* under the eaves of a house in the centre of Addis Ababa.

The tree nest is the usual untidy bundle of grass that is typical of many of the sparrows, domed over with the entrance hole in the side, and lined with feathers. Clutch size ranges from 3–6 eggs.

Data on the breeding season for four localities in Ethiopia, together with data on rainfall, are summarised in Fig. 4. It is evident that breeding is closely associated with the rains, occurring during and slightly after their start.

Locality	Month												Reference
	J	F	M	A	M	J	J	A	S	O	N	D	
Coastal Eritrea													Smith 1955b, Archer & Godman 1961
Central Ethiopia (plateau)													Erlanger 1907, Friedmann 1930, Smith 1957
South Ethiopia													Friedmann 1930, Benson 1947
East Ethiopia & Somalia													Erlanger 1907, Archer & Godman 1961

Fig. 4   Breeding season data for Swainson's Sparrow (breeding ——— rains - - - -)

VOICE

I have transcribed the call as *chirri-up*.

FOOD

No detailed analysis has been made of the food taken by Swainson's Sparrow, though Cheesman and Sclater (1936) reported that they do damage in corn fields and gardens. I have watched the birds searching for insects under house eaves.

Parrot-billed Sparrow *Passer gongonensis*

NOMENCLATURE
Pseudostruthus gongonensis Oustalet, Naturaliste 1890: 274.
    Gongoni, near Mombasa.
Passer gongonensis (Oustalet) 1890.
  Subspecies: none
    Synonyms: *Passer griseus turkanae* Granvik 1934.
              Lotonok, Turkana, northwestern Kenya.
    Passer griseus tertale Benson 1942.
              30 miles east of Yavello, South Abyssinia [Yabelo, Ethiopia].
              Intergrades with *gongonensis*.
    Passer griseus jubaensis Benson 1942.
              Mandera, Juba [Giuba] River valley, southern Italian Somaliland
              (Mandera is in Kenya according to the Times Atlas)
              Intergrades with *gongonensis*.

DESCRIPTION
Parrot-billed Sparrow, length 180 mm and a massive bill, is quite the most distinctive of grey-headed sparrows and has long been considered a separate species, even by authorities who lumped the other grey-headed sparrows in one species.

Sexes are alike and bird is generally similar to Grey-headed and Swainson's Sparrows, but is separable in field on account of larger size and heavier, more conical bill that gives Roman-nosed appearance. Head and underparts dark grey, the latter uniform without white bib or belly. Wings dark grey with chestnut patch at scapulars and striking white wing patch. Mantle is browner than in previous two species and lower back and rump brighter chestnut, with chestnut extending to lower flanks. In northern and western extremities of range, bill noticeably less heavy and belly paler than rest of underparts.

Immature. Young birds lack chestnut and white on wings; sides of rump white instead of chestnut.

Bill colour scores for the British Museum specimens are given in Table 5. These results are closely similar to those for preceding two species.

Table 5: Bill colour scores for Parrot-billed Sparrow

	Males		Females
No.	Mean score	No.	Mean score
25	7.4	21	7.2

BIOMETRICS

Biometric data are given for the Parrot-billed Sparrow in Table 6. These measurements shown how much larger this species is than either of the two preceeding species (and, as will be seen below, than the remaining two members of the [*griseus*] superspecies). Several authors suggest that there is a cline of decreasing wing length and bill size, westwards from the coast, though published data are lacking to confirm this. In the field the bill of the Parrot-billed Sparrow appears massive, justifying its English name, though this is merely a reflection of the overall size of the birds as, in fact, it is proportionately no larger than the bills of the other grey-headed sparrows.

Table 6: Biometric data for Parrot-billed Sparrow

Feature	Locality	Males		Females	
		Range	Mean (Median)	Range	Mean (Median)
wing	Kenya coast	95–102	97.8	91–96	93.5
	Ethiopia*	89–98	(93.5)	85–96	(90.5)
tail	Kenya	**63–69**	**66**		
	Ethiopia*	59–61	(60)		
culmen	Kenya	16.1–18.6	17.3	16.7–19.6	17.5
	Ethiopia*	16–17	(16.5)		

*intergrades between *gongonensis* and *swainsonii*

DISTRIBUTION

The Parrot-billed Sparrow occurs all over Kenya, extending in the north to the Shoa (Shewa) Province of Ethiopia and eastwards to southern Somalia, where it overlaps with Swainson's Sparrow, with limits of about 7°N in Ethiopia and 5°N in coastal Somalia, where the intergrades *tertale* and *jubaensis* respectively have been described. In the west it just reaches into the extreme southeast of the Sudan and eastern Uganda to about 33°30′E, overlapping there with the Grey-headed Sparrow *P. g. ugandae*. The southern limit occurs in extreme northern Tanzania. The distribution is shown in Fig. 5.

HABITAT

The Parrot-billed Sparrow is more a bird of open country and bush than the other members of the [*griseus*] superspecies, being found in grass savanna with scattered trees and more arid country than Swainson's Sparrow and the Grey-headed Sparrow, apart from the arid-loving race of the latter, *laeneni*. It is much less associated with human habitations than any of the other members of the superspecies, though Dr J. S. Ash informs me (*in litt.*) that it behaves as a 'house sparrow' in Mogadiscio towards the northern limit of its range in Somalia. It appears to be absent from most Kenyan towns and villages; I searched and failed to find any in Isiolo and Nanyuki in February 1982.

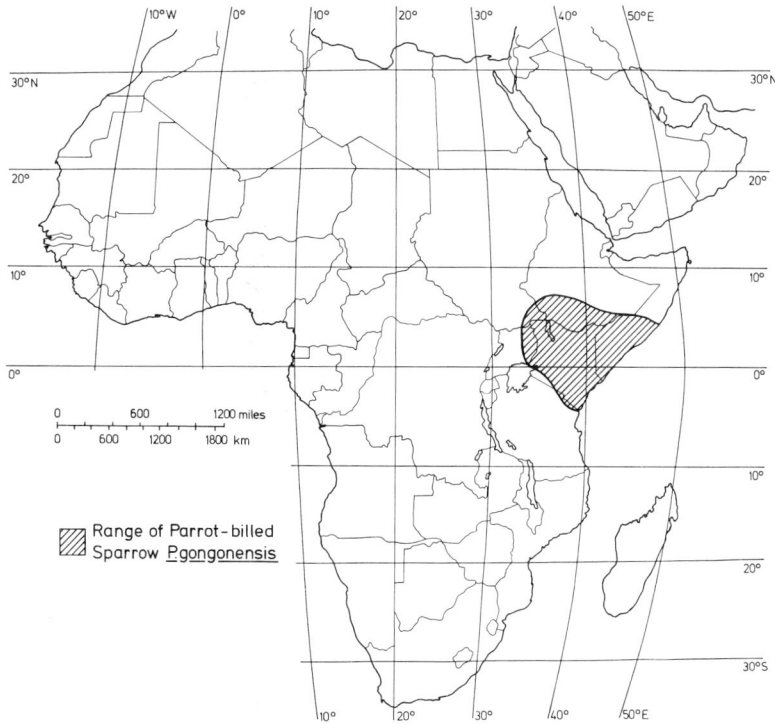

Fig. 5   Range of Parrot-billed Sparrow

BEHAVIOUR

This is a shyer species than the other grey-headed sparrows; although it comes into the outskirts of villages, it is less frequently to be seen feeding around the village streets than the other members of the superspecies. It has, however, now learned to recognise the potential of the lodges and picnic sites in the game reserves as feeding places, and flocks of 20–30 will assemble to feed on the crumbs and other scraps provided. More usually, however, they are to be seen in small parties of only 4–6 birds.

BREEDING BIOLOGY

In courtship display, one bird chases another in headlong flight. When they land the pursuing bird (presumably the male) hops in front of the other with wings held out and slightly drooped.

This species nests openly in trees and also uses the old nests of other species. Nesting in holes occurs only occasionally, if at all. The nest is similar to that of the other grey-headed sparrows, being built of grass and lined with feathers.

Published data on the breeding season are summarised in Fig. 6. Breeding occurs in south Ethiopia from December, probably through into May (though there are no records for April), following the October–November rains and coinciding with the

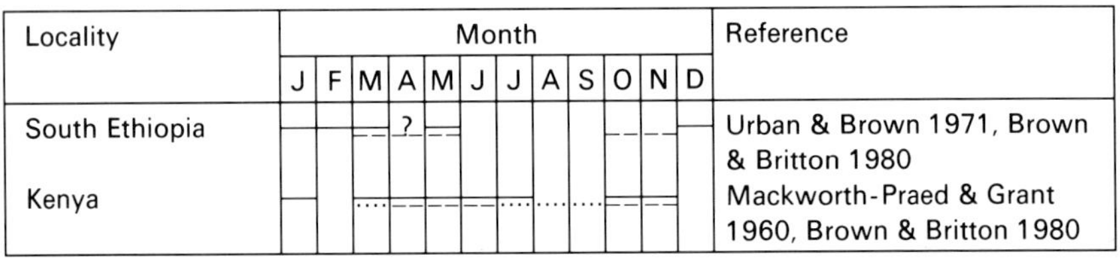

Locality	Month												Reference
	J	F	M	A	M	J	J	A	S	O	N	D	
South Ethiopia				?									Urban & Brown 1971, Brown & Britton 1980
Kenya													Mackworth-Praed & Grant 1960, Brown & Britton 1980

Fig. 6 Breeding season data for Parrot-billed Sparrow (breeding ——— rains: main –––→ minor ....)

March–April rains. Again, in Kenya, breeding usually follows or takes place during the rains, though it also occurs in the dry season. Brown and Britton (1980) suggest that dry-season breeding is associated with cultivated land with irrigation, where there is much less seasonal variation in vegetation, and hence food availability, than in more natural circumstances. The clutch consists of 2 eggs.

VOICE

The Parrot-billed Sparrow has a similar chirp to the other grey-headed sparrows, though somewhat deeper in tone; I have transcribed it as *choop*.

FOOD

While principally seed eaters, they will also feed on crumbs and other scraps as mentioned above. Food is mostly taken on the ground.

Swahili Sparrow *Passer suahelicus*

NOMENCLATURE
Passer griseus suahelicus Reichenow, Vog. Afr. 1904 3: 231.
  Bussissi, [Mwanza district, northern Tanzania].
Passer suahelicus (Reichenow) 1904.
  Subspecies: none.
    Synonyms: none.

DESCRIPTION
Sexes alike. Swahili Sparrow very similar to Grey-headed, though possibly slightly larger, length *ca* 160 mm. Differs in that back and mantle are greyish brown (paler than *P. g. ugandae*), contrasting sharply with red-brown rump, but scarcely at all with grey of head. Underparts are uniform mid grey except for traces of white throat, though this much less distinct than in Grey-headed and is doubtfully visible in field. Grey of underparts lighter than in Parrot-billed, but like that species does not become paler towards belly as does Grey-headed Sparrow. Chestnut shoulder patch paler than in *P. g. ugandae*.

Bill colour scores for the British Museum specimens are given in Table 7. Unfortunately, the number of specimens is too small for the results to be significant, though there is a suggestion that there may be a seasonal change in both sexes.

Table 7: Bill colour scores for Swahili Sparrow

Males		Females	
No.	mean score	No.	mean score
5	6.0	5	6.2

BIOMETRICS
Very few measurements are available for the Swahili Sparrow. Typical values for wing and bill length are given in Table 8. As far as can be judged these suggest that it is similar in size to the Grey-headed and Swainson's Sparrow.

Table 8: Biometric data for Swahili Sparrow

Feature	Males		Females	
	Range	Mean	Range	Mean
wing	83.91*	85.6	81–89*	85
culmen	14.9–15.4*	15.2	14.6–15.8*	15.0

* very small sample

DISTRIBUTION
The Swahili Sparrow extends from the border of Tanzania and Kenya at Lake Victoria, north into southern Kenya – Loito Plains, Lebetero Hills, Lake Magadi, and, according to Clancey (1959), as far north as Lake Elmenteita. Small fine-billed birds are also found in the Nairobi area and have recently begun to breed in the capital. These could be the present species, though, as already mentioned under the Grey-headed Sparrow, I think the latter species more probable. Birds I watched

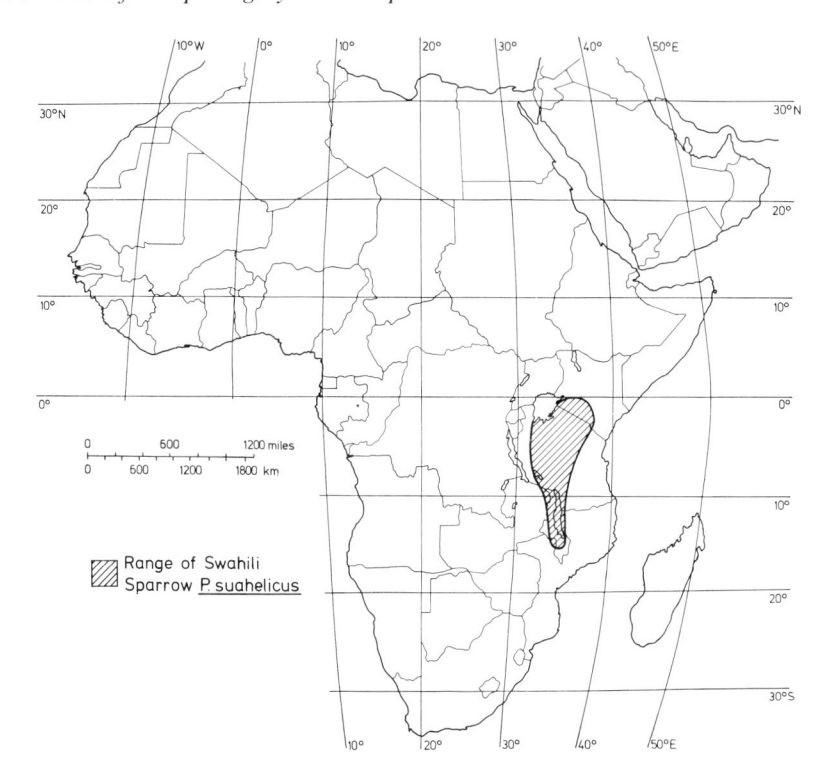

Fig. 7 Range of Swahili Sparrow

beside the Mara river, in the Masai Mara Game Reserve in southwest Kenya, were small, uniform grey underneath and had fine bills; I identified them as Swahili Sparrows. To the south the range extends through central Tanzania, most of Malawi except for the south, and northwestern Mozambique. The distribution is shown in Fig. 7.

HABITAT

This species is predominantly a bird of grassland with scattered trees, usually near human habitations; it extends, also, into cultivated land and villages as well as into light woodland. It is much more a village bird than the Parrot-billed Sparrow, but less so than the Grey-headed Sparrow.

BEHAVIOUR

Nothing significant has been recorded, though I have seen small parties feeding together on the ground in February. At the same time they were visiting holes in fig trees, but there was no sign of any breeding activity.

BREEDING BIOLOGY

Nests are built in holes, both in trees and in houses, but open nests in the branches

Swahili Sparrow *Passer suahelicus*

NOMENCLATURE
Passer griseus suahelicus Reichenow, Vog. Afr. 1904 3: 231.
      Bussissi, [Mwanza district, northern Tanzania].
Passer suahelicus (Reichenow) 1904.
   Subspecies: none.
   Synonyms: none.

DESCRIPTION
Sexes alike. Swahili Sparrow very similar to Grey-headed, though possibly slightly larger, length *ca* 160 mm. Differs in that back and mantle are greyish brown (paler than *P. g. ugandae*), contrasting sharply with red-brown rump, but scarcely at all with grey of head. Underparts are uniform mid grey except for traces of white throat, though this much less distinct than in Grey-headed and is doubtfully visible in field. Grey of underparts lighter than in Parrot-billed, but like that species does not become paler towards belly as does Grey-headed Sparrow. Chestnut shoulder patch paler than in *P. g. ugandae*.

Bill colour scores for the British Museum specimens are given in Table 7. Unfortunately, the number of specimens is too small for the results to be significant, though there is a suggestion that there may be a seasonal change in both sexes.

Table 7: Bill colour scores for Swahili Sparrow

	Males			Females	
No.		mean score	No.		mean score
5		6.0	5		6.2

BIOMETRICS
Very few measurements are available for the Swahili Sparrow. Typical values for wing and bill length are given in Table 8. As far as can be judged these suggest that it is similar in size to the Grey-headed and Swainson's Sparrow.

Table 8: Biometric data for Swahili Sparrow

Feature	Males		Females	
	Range	Mean	Range	Mean
wing	83.91*	85.6	81–89*	85
culmen	14.9–15.4*	15.2	14.6–15.8*	15.0

*very small sample

DISTRIBUTION
The Swahili Sparrow extends from the border of Tanzania and Kenya at Lake Victoria, north into southern Kenya – Loito Plains, Lebetero Hills, Lake Magadi, and, according to Clancey (1959), as far north as Lake Elmenteita. Small fine-billed birds are also found in the Nairobi area and have recently begun to breed in the capital. These could be the present species, though, as already mentioned under the Grey-headed Sparrow, I think the latter species more probable. Birds I watched

36 *The Afrotropical grey-headed sparrows*

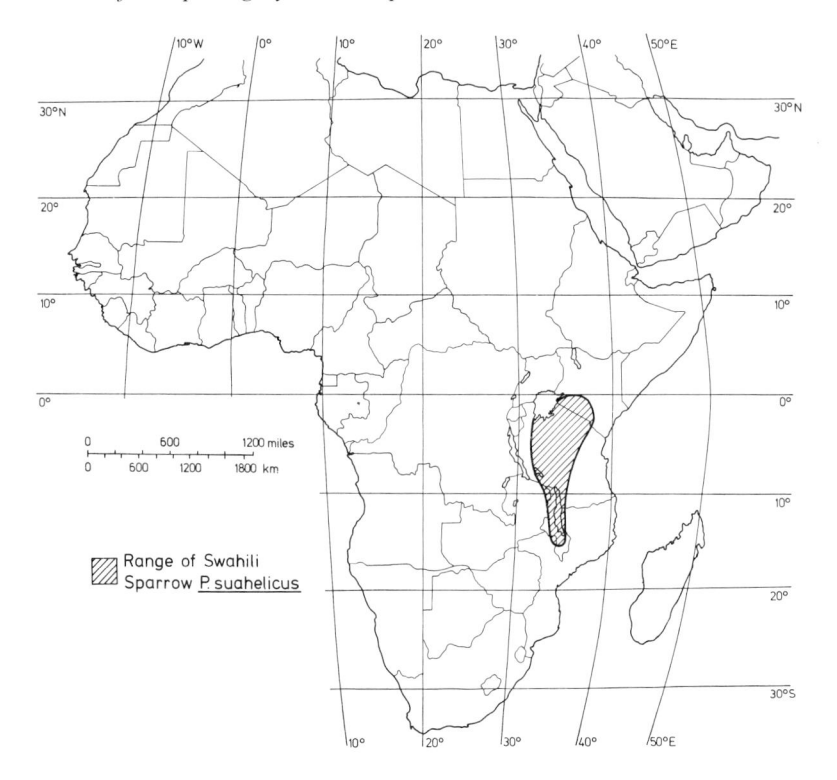

Fig. 7   Range of Swahili Sparrow

beside the Mara river, in the Masai Mara Game Reserve in southwest Kenya, were small, uniform grey underneath and had fine bills; I identified them as Swahili Sparrows. To the south the range extends through central Tanzania, most of Malawi except for the south, and northwestern Mozambique. The distribution is shown in Fig. 7.

HABITAT

This species is predominantly a bird of grassland with scattered trees, usually near human habitations; it extends, also, into cultivated land and villages as well as into light woodland. It is much more a village bird than the Parrot-billed Sparrow, but less so than the Grey-headed Sparrow.

BEHAVIOUR

Nothing significant has been recorded, though I have seen small parties feeding together on the ground in February. At the same time they were visiting holes in fig trees, but there was no sign of any breeding activity.

BREEDING BIOLOGY

Nests are built in holes, both in trees and in houses, but open nests in the branches

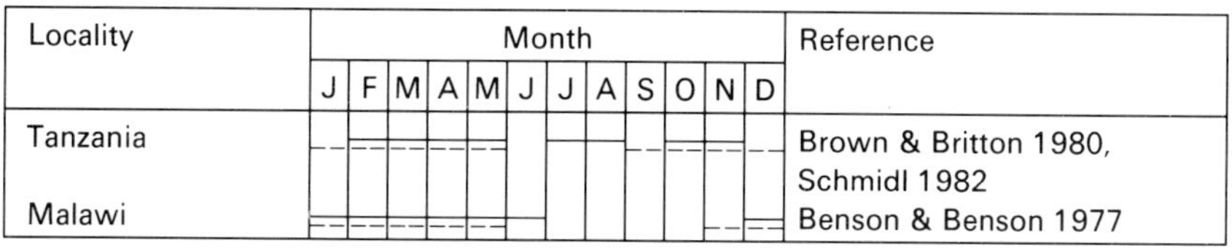

Locality	Month												Reference
	J	F	M	A	M	J	J	A	S	O	N	D	
Tanzania													Brown & Britton 1980, Schmidl 1982
Malawi													Benson & Benson 1977

Fig. 8 Breeding season data for Swahili Sparrow (breeding ——— rains ----)

of trees have not been reported.

Published data on the breeding season are summarised in Fig. 8. This is very protracted and apparently coincides with the rains.

VOICE

Nothing has been published about the voice of this rather neglected species. The birds I have watched have been silent.

FOOD

Nothing significant has been reported about the food taken by the Swahili Sparrow, though I have seen them feeding on crumbs put out at game lodges.

Southern Grey-headed Sparrow *Passer diffusus*

NOMENCLATURE
Pyrgita diffusus Smith, Report Exped. Centr. Afr. 1836: Appendix 50.
    'north of the Orange River': restricted to near Kuruman, Cape Province, by Macdonald and Hall, Ann. Transvaal Mus. 1957 23: 35.
Passer diffusus (Smith) 1836.
  Subspecies: *Passer diffusus diffusus* (Smith) 1836.
    Synonyms: *Passer diffusus georgicus* Reichenow 1904
        Damaraland, South West Africa [Otimbingwe, Namibia].
    Passer diffusus stygiceps Clancey 1954
        Umzinyati Falls, Inanda, near Durban, Natal.
    Passer diffusus mosambicus van Someren 1921
        Lumbo, Portuguese West Africa [Mozambique].
    Passer diffusus luangwae Benson 1956
        Mupamadzi River, Mpika area, Luangwa Valley, Northern Rhodesia [Zambia].

DESCRIPTION

Passer diffusus diffusus
Sexes alike. Very similar to Grey-headed, but generally a little larger, length 155–160 mm, overall much paler than *P. griseus ugandae,* race of Grey-headed Sparrow with which it comes into contact. Underparts are uniform pale grey, almost tending

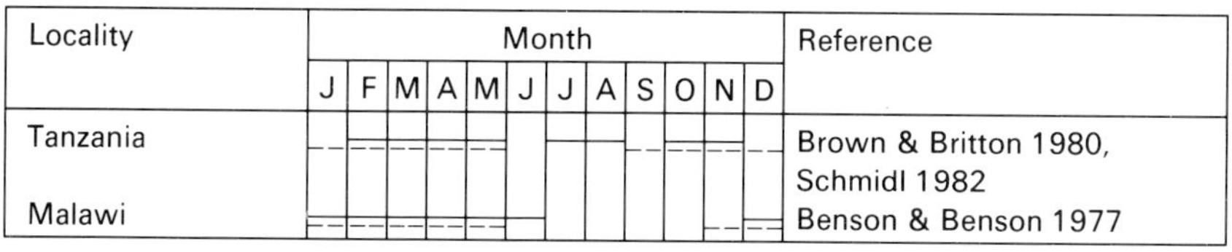

Locality	Month												Reference
	J	F	M	A	M	J	J	A	S	O	N	D	
Tanzania													Brown & Britton 1980, Schmidl 1982
Malawi													Benson & Benson 1977

Fig. 8 Breeding season data for Swahili Sparrow (breeding ——— rains ----)

of trees have not been reported.

Published data on the breeding season are summarised in Fig. 8. This is very protracted and apparently coincides with the rains.

VOICE

Nothing has been published about the voice of this rather neglected species. The birds I have watched have been silent.

FOOD

Nothing significant has been reported about the food taken by the Swahili Sparrow, though I have seen them feeding on crumbs put out at game lodges.

Southern Grey-headed Sparrow *Passer diffusus*

NOMENCLATURE
Pyrgita diffusus Smith, Report Exped. Centr. Afr. 1836: Appendix 50.
       'north of the Orange River': restricted to near Kuruman, Cape Province,
       by Macdonald and Hall, Ann. Transvaal Mus. 1957 23: 35.
Passer diffusus (Smith) 1836.
  Subspecies: *Passer diffusus diffusus* (Smith) 1836.
     Synonyms: *Passer diffusus georgicus* Reichenow 1904
          Damaraland, South West Africa [Otimbingwe, Namibia].
     Passer diffusus stygiceps Clancey 1954
          Umzinyati Falls, Inanda, near Durban, Natal.
     Passer diffusus mosambicus van Someren 1921
          Lumbo, Portuguese West Africa [Mozambique].
     Passer diffusus luangwae Benson 1956
          Mupamadzi River, Mpika area, Luangwa Valley,
          Northern Rhodesia [Zambia].

DESCRIPTION

Passer diffusus diffusus
Sexes alike. Very similar to Grey-headed, but generally a little larger, length 155–160 mm, overall much paler than *P. griseus ugandae,* race of Grey-headed Sparrow with which it comes into contact. Underparts are uniform pale grey, almost tending

to white, but lacking contrasting white bib that is a conspicuous feature of latter. White wing bar more prominent than in Grey-headed.

There is a cline from the palest birds in the west, '*georgicus*' to the darker ones in the more humid east, '*stygiceps*', in conformation with Gloger's ecogeographical rule.

Immature. Similar to adult except that back is mottled.

Passer diffusus mosambicus
Darker and more richly coloured than nominate *diffusus*, approaching more to *P. griseus*, but this race is smaller than *griseus* and has a finer bill.

Passer diffusus luangwae
Intermediate in colouring between nominate *diffusus* and *P. griseus*, but is definitely smaller, weighing on average over 6% less than nominate race, and with a very fine bill.

Bill colour scores for the British Museum specimens of the Southern Grey-headed Sparrow are given in Table 9. These are significantly smaller than for the other grey-headed sparrow species and confirm, strongly, the view of Dowsett and Dowsett-Lamaire (1980) that the bill colour changes seasonally, becoming black in breeding season (September–March).

Table 9: Bill colour scores for Southern Grey-headed Sparrow

Males		Females		Immatures	
No.	Mean score	No.	Mean score	No.	Mean score
46	4.5	34	4.4	1	3.0

BIOMETRICS
Body measurements for the Southern Grey-headed Sparrow are given in Table 10.

Table 10: Biometric data for Southern Grey-headed Sparrow

Feature	Subspecies	Locality	Males		Females	
			Range	Mean (Median)	Range	Mean (Median)
weight	*diffusus*	Transvaal	20.4–27.5	24.9	19.5–25.4	23.5
		—	20.0–97.6*	24.2*		
	luangwae	Luangwa Valley	22.7–23.2*	22.9*		
wing	*diffusus*	Namibia/Angola	79–86	(82.5)	73–82	(77.5)
		Natal	83–88	(85.5)	79–82	(80.5)
	luangwae	Luangwa Valley	74–82	77.5	72–79	75.5
	mosambicus	—	76.5–78.5*	81.6*		
tail	*diffusus*	—	55–65*	(60)*		
tarsus	*diffusus*	—	16–19*	(17.5)*		
culmen	*diffusus*	—	13.5–15.8	14.2	13.8–15.0	14.3
	luangwae	—	12.5–13.8†	13.2†	12.9–14.0†	13.6†
	mosambicus	—	12–13*	12.5*		

* not sexed   † very small sample

Skead (1977) weighed 130 adults (not sexed) and 272 juveniles trapped in southwest Transvaal. According to his results the weight of the adults was fairly constant throughout the year (mean 24.2 ± 0.6 g); whereas that of the juveniles was slightly higher in winter (May–August) compared with that in summer (December–April), with mean values of 23.3 ± 0.2 g and 21.0 ± 0.8 g respectively. The main populations of the Southern Grey-headed Sparrow are rather similar in size to the *ugandae* race of the Grey-headed Sparrow, though where the two species overlap the species appears to be significantly smaller. This is certainly the case with the isolated *P. d. luangwae* population that is completely surrounded by *P. g. ugandae*, and from the few measurements available, it also appears to hold in the area of overlap in Angola. White and Moreau (1958) show that whereas the Grey-headed Sparrows in Angola and the Southern Grey-headed Sparrows in Namibia are virtually identical in wing length (males 77–87 mm and 79–86 mm, females 78–83 mm and 79–82 mm respectively), a sample of Angolan Southern Grey-headed Sparrows proved much smaller, males 78 mm (3), females 73–76 mm (4).

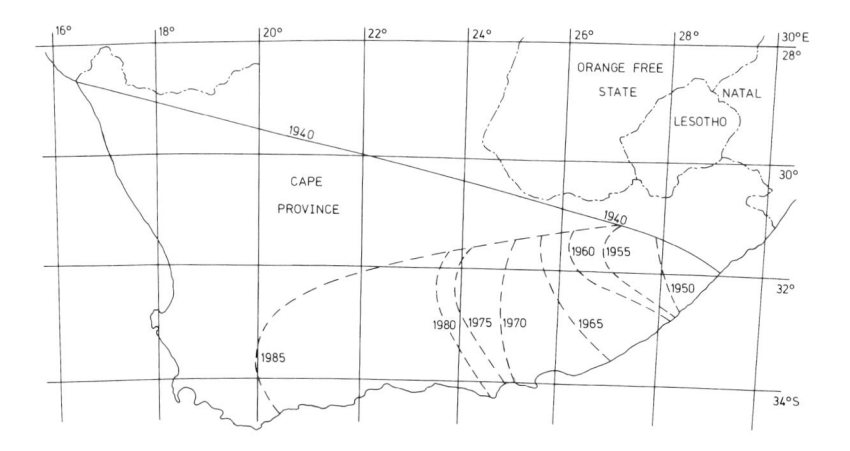

Fig. 9 Spread of Southern Grey-headed Sparrow in Cape Province, since 1940. Dashed lines show approximate position of the advancing front at 5-year intervals.

DISTRIBUTION

The nominate race of the Southern Grey-headed Sparrow occurs in the southern part of the continent. The northern limit runs from western Angola, on the coast north of Luanda, from there southeastwards to Kalabo in Zambia, and east to southern Mozambique. In the south it ranges through Namibia (though absent from the desert area in the south), Botswana and Zimbabwe; in South Africa it occurs just south of the Orange River in Cape Province in the west, through the Orange Free State, the Transvaal, Lesotho and Natal. This was the distribution about 1950, but in the last 35–40 years it has spread south from Natal, through the Transkei and westwards across eastern Cape Province, south of about 32°S, and by 1984 had reached about 23°E. A reconstruction of this spread is given in Fig. 9 (Craig *et al.* 1987).

The race *mosambicus* occurs from northern Mozambique, north to Tanzania, including Zanzibar and Pemba islands.

The third race, *P. d. luangwae*, is found in an isolated population in the upper

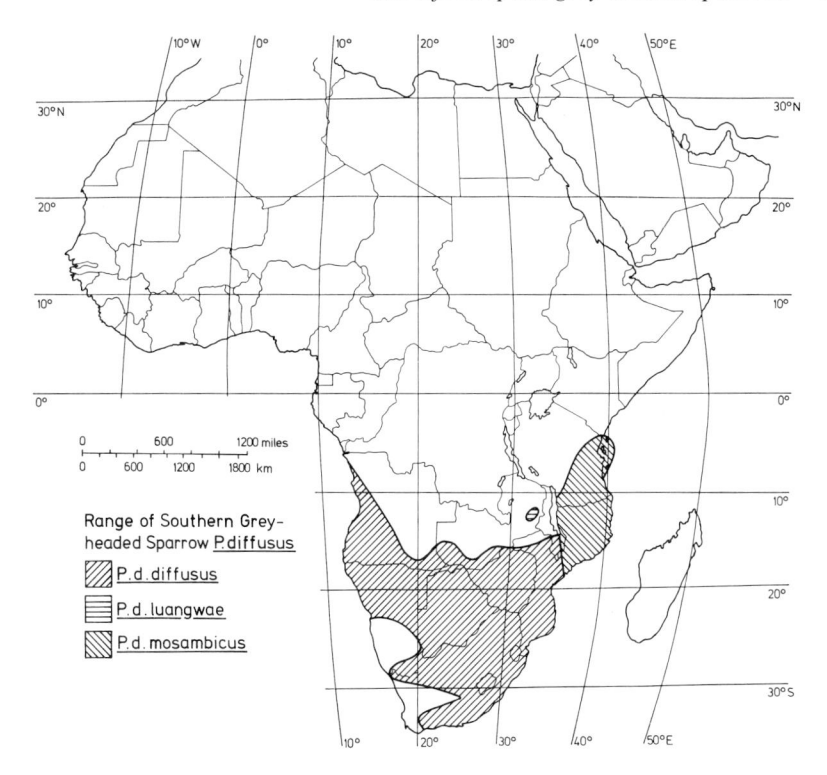

Fig. 10   Range of Southern Grey-headed Sparrow

Luangwa valley in eastern Zambia, extending for just over 100 km between 11°45'S and 12°37'S.

The complete distribution of the Southern Grey-headed Sparrow is given in Fig. 10.

HABITAT

The Southern Grey-headed Sparrow is primarily a bird of open acacia woodland, frequently near settlements. Although it penetrates into villages, it is much less of a 'house sparrow' than the Grey-headed Sparrow, possibly because over much of its range it is sympatric with the Cape Sparrow *Passer melanurus* that is very much more adapted to man's habitations and successfully competes for nesting sites against the Southern Grey-headed Sparrow.

It has been suggested that the recent spread of the House Sparrow over much of southern Africa (see Chapter 9) has had a deleterious effect on the numbers of the Southern Grey-headed Sparrow, through the displacement of the latter from villages by its more aggressive congener. The story is, however, not as straightforward as that; in the area of overlap the two species frequently breed side by side in isolated houses, and in disputes over nest sites the newcomer to South Africa is not always successful. Moreover, the recent spread of the Southern Grey-headed Sparrow in southern Cape Province has been into an area where the House Sparrow was already in occupation, though only a few years in advance of the Southern Grey-headed Sparrow. The spread of both species through the Transkei took place almost simultaneously, though the House Sparrow outstripped its rival in its spread to the west. Even more surprising, there is some suggestion (V. Pringle, *in litt.*) that as the Southern Grey-headed Sparrow has caught up, the House Sparrow has decreased in numbers.

In Malawi, south of Lake Malawi, the Southern Grey-headed Sparrow is almost completely associated with mopane woodland, as is the isolated population in the Luangwa valley. It prefers a drier habitat to the southern population of the Grey-headed Sparrow and occurs more frequently away from cultivation and habitations than that species.

BEHAVIOUR

This is rather a shy species, usually occurring in pairs, though outside the breeding season it may be found in small flocks, probably mainly family parties. It feeds on the ground, but is otherwise usually in trees. In the feeding areas it associates peacefully with House and Cape Sparrows, though there is considerable competition for nest sites.

An unusual habit of the Southern Grey-headed Sparrow is that it walks as well as hops. B. Every (*in litt.*) has drawn my attention to this feature, though it was noted earlier by Gill (1936), and it is described as 'walking on the ground with small, shuffling steps' in the 5th Edition of *Robert's Birds of Southern Africa* (McLean 1984). This behaviour is very abnormal in *Passer* species, though G. R. Cunningham-van Someren (*in litt.*) informs me that the Grey-headed Sparrow may take an occasional step; Drs G. and M.Y. Morel have confirmed this and have seen the same with the Golden Sparrow. It is quite evident, however, that this habit is much more pronounced in the Southern Grey-headed Sparrow. Dr A. J. F. K. Craig (also *in litt.*) notes that when House and Southern Grey-headed Sparrows are feeding together, the difference in the gait of the latter is quite obvious.

The bird appears to be fairly sedentary, five recoveries of ringed adults in South Africa all being under 10 km, though there appear to be some local movements judging from the report by Tree (1972) of temporary concentrations at a pan in Botswana in June 1969 and 1970.

BREEDING BIOLOGY

Nesting is mainly in holes, both in trees and in buildings. In the former site it uses the old nesting holes of other species, such as woodpeckers and barbets; in buildings it will adopt the old nests of swallows and swifts. Frequently very small cavities are used and it will occupy holes quite close to the ground, such as hollow metal gateposts.

The bird at the nest site calls repeatedly; when excited the calling rate increases, the rump and lower back feathers are ruffled and the tail is flicked from side to side in a rotary motion. It will defend a nest site against competing congeners as well as other members of its own species, threatening with the wings held away from the body, the tail raised and spread; at times fights even occur with the two birds, locked together, falling to the ground. At one nest under the roof of a barn that I watched in the Transvaal, where the dispute was with a pair of House Sparrows, threat and fighting appeared to be directed against the competitor of the same sex; at least it was the presumed male Southern Grey-headed Sparrow (*viz* the bird that was calling at the nest) that attacked the male House Sparrow, and the other, the presumed female, that attacked the female House Sparrow. In some of these disputes, at least, the Southern Grey-headed Sparrow is victorious. This was the case in the dispute

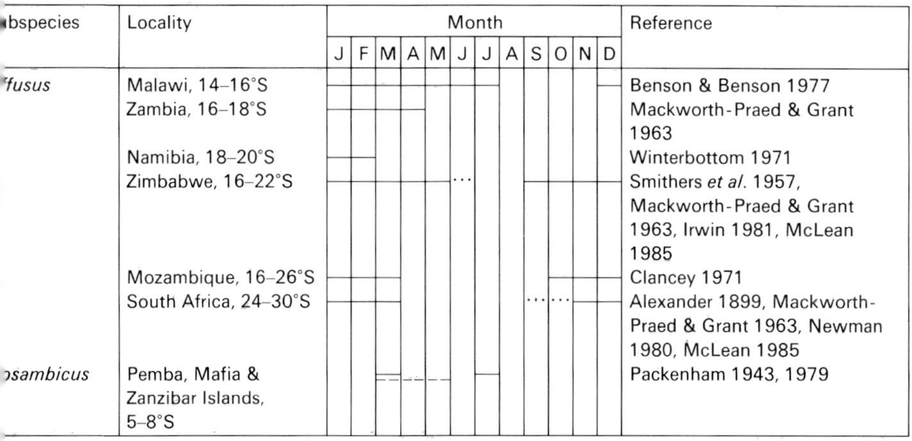

Subspecies	Locality	Month												Reference
		J	F	M	A	M	J	J	A	S	O	N	D	
diffusus	Malawi, 14–16°S													Benson & Benson 1977
	Zambia, 16–18°S													Mackworth-Praed & Grant 1963
	Namibia, 18–20°S													Winterbottom 1971
	Zimbabwe, 16–22°S					⋯								Smithers et al. 1957, Mackworth-Praed & Grant 1963, Irwin 1981, McLean 1985
	Mozambique, 16–26°S													Clancey 1971
	South Africa, 24–30°S							⋯⋯						Alexander 1899, Mackworth-Praed & Grant 1963, Newman 1980, McLean 1985
mosambicus	Pemba, Mafia & Zanzibar Islands, 5–8°S													Packenham 1943, 1979

Fig. 11   Breeding season data for Southern Grey-headed Sparrow (breeding: normal ——   exceptional ⋯   rains ----)

described above, the Southern Grey-headed Sparrow successfully rearing a brood from this nest.

In the pre-breeding period, pairs of Southern Grey-headed Sparrows investigating nest sites frequently resort to headlong display chases, similar to those described for other grey-headed sparrow species.

The nest is the usual untidy bundle of grass, lined with feathers, wool and hair. Both sexes co-operate in nest building. The species is double brooded, with the clutch normally 3–5 eggs. Both adults take part in feeding the young (Every 1976).

Breeding season data are summarised in Fig. 11. The breeding season, particularly in the sub-tropical populations, coincides with the austral spring and summer. In the tropical islands off the east coast, breeding follows the start of the rains.

MOULT

The birds moult following the breeding season. Moreau (1940) quotes July to November for Zanzibar. In common with other *Passer* species, the Southern Grey-headed Sparrow has a complete post-juvenile moult (Dean 1977).

VOICE

This is a quiet bird. I have transcribed the call as a repeated *chirp* or *chirrup*, rather softer and more musical than the chirp call of the House Sparrow. Like the other grey-headed sparrows, a series of call-note variants can be combined into a 'song'; McLean (1984) gives this as *chirrip cheeu chiriritit cheeu* and says it has a tinny quality, thinner and higher pitched than the similar notes of other sparrows.

FOOD

The food taken consists both of weed seeds and insects, but this has not been studied in detail. The birds have also been recorded feeding on nectar from two species of aloes (Oatley & Skead 1972).

The grey-headed sparrows: a summary

Although the grey-headed sparrows are a closely similar group, they can, however, be distinguished from each other in the field in good conditions, particularly in areas where there is overlap. It thus seems worth while completing this chapter with a summary of the main characteristics by which they can be distinguished and with a composite map, Fig. 12, showing areas of overlap. These regions of overlap will be discussed in more detail in Chapter 17.

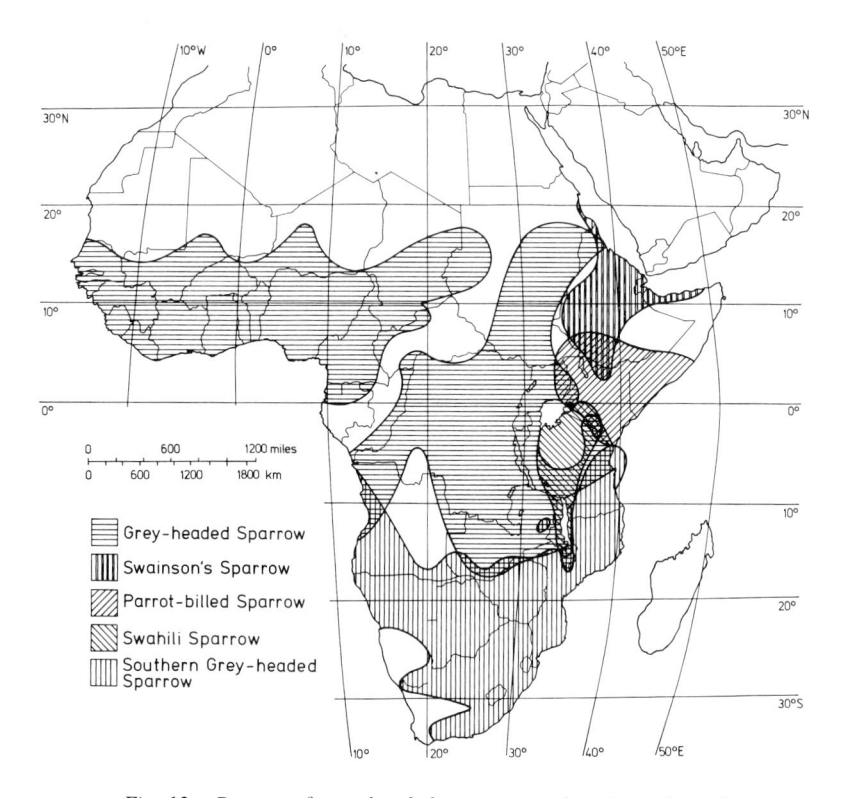

Fig. 12 Ranges of grey-headed sparrows and regions of overlap

Grey-headed Sparrow

The subspecies *griseus* and *ugandae* both show a distinct white throat patch contrasting with the grey breast and upper belly. A white bib is present on some specimens of the Swahili Sparrow, but it is poorly developed and is not detectable in the field. No white bib is distinguishable in the race *laeneni* as this has almost pure white underparts, but as this race is completely allopatric from all the other grey-headed sparrow species this does not present an identification problem.

Swainson's Sparrow

This species is readily distinguishable from the Grey-headed Sparrow by its darker

underparts, darker than the race *ugandae* of the Grey-headed Sparrow with which it is parapatric, and the complete absence of a white bib. It is much smaller, both in overall size and in bill dimensions, than the Parrot-billed Sparrow that it overlaps in southern Ethiopia and northern Kenya, though there is evidence of some interbreeding between the two species. Birds that I watched at Marsabit in northern Kenya were noticeably smaller than typical Parrot-billed Sparrows.

Parrot-billed Sparrow

This species is readily separable in the field from the other grey-headed sparrows with which it comes into contact. Not only is it much larger in overall size, with a noticeably heavier bill, but it is also quite the darkest of the group with completely dark grey underparts. The white wing bar stands out very prominently.

Swahili Sparrow

The Swahili Sparrow is almost as dark as the last species, but it is easily distinguished by its smaller size. It is darker below than the race *ugandae* of the Grey-headed Sparrow that it overlaps (even though *ugandae* is itself darker than the nominate race of the Grey-headed Sparrow) and, despite this, showing no sign in the field of a contrasting white throat patch, though this is detectable in museum skins. This last feature shows something of a northwest to southeast cline, being most distinct in *P. g. griseus*, less so but still obvious in *P. g. ugandae* and almost absent in *P. suahelicus*. A further feature of the Swahili Sparrow is that the underparts, unlike those of the Grey-headed Sparrow, are completely dark and do not become lighter on the belly.

Southern Grey-headed Sparrow

This species is rather similar to the race *ugandae* of the Grey-headed Sparrow that it overlaps, but the pale grey underparts show no sign of the white throat patch that is a distinguishing feature of the latter. I have not observed the two species in the area of overlap, but according to Dowsett and Dowsett-Lamaire (1980) the longer tail, paler coloration and heavier bill of the Grey-headed Sparrow allow them to be separated; there is insufficient published biometric data available to confirm the first of these differences, but this is shown for the bill by measurements that I have made on the British Museum specimens, particularly for the small race, *luangwae*, though there is some overlap *viz:*

	Males			Females		
	No.	Range	Mean	No.	Range	Mean
P. griseus ugandae	14	14.5–16.5	15.7	15	14.0–17.0	15.8
P. d. diffusus	10	13.0–15.8	14.2	10	13.9–15.0	14.3
P. d. luangwae	3	12.5–13.8	13.6	3	12.9–14.0	13.6

Dowsett and Dowsett-Lamaire also remark that the white bar on the median coverts of the Southern Grey-headed Sparrow is particularly striking, such a bar being absent, or very narrow in the Grey-headed Sparrow.

Again, I have not observed the Southern Grey-headed Sparrow in its area of overlap with the Swahili Sparrow, but museum specimens of the latter have much darker underparts (the Southern Grey-headed Sparrow appears pale underneath in the field) and a heavier bill.

2: Golden Sparrow *Passer luteus*

NOMENCLATURE
Fringilla lutea Lichtenstein, Verz. Doubl. 1823: 24.
    Dongola, northern Sudan.
Auripasser lutea Bonaparte 1851.
    Kunfuda [Al Qunfidhah], southwest Arabia.
Passer luteus (Lichtenstein) 1823.
  Subspecies: *Passer luteus luteus*.
    Synonym: *Auripasser luteus tilemsiensis* Bates 1932.
      Taberreshat [northeast of Bourem], French Sudan [Mali].
    Passer luteus euchlorus (Bonaparte) 1851.
      Kunfuda [Al Qunfidhah], southwest Arabia.

The Golden Sparrow differs sufficiently from the other sparrows for several authorities to have considered that it was not a member of the genus *Passer*. Bonaparte placed it in a new genus *Auripasser*, together with the Golden Sparrows from Arabia that he attributed to a separate species, *Auripasser euchlora*; Bates (1934) supported this view as far as African mainland birds were concerned. In contrast, Lynes (1924), who had extensive experience of *luteus* in the field, considered that it properly belonged to *Passer*, together with the closely related Chestnut Sparrow *Passer eminibey*. Meinertzhagen (1954) went even further and made *luteus, euchlorus* and *eminibey* conspecific, despite evidence from Lynes (1924) that *luteus* and *eminibey* have overlapping ranges in Darfur Province, Sudan, and hence must be specifically separate. Kunkel (1961), on the basis of a study of the behaviour of captive birds in an aviary, concluded that the Golden Sparrow was a *Passer*. In the most recent analysis of these

birds, Hall and Moreau (1970) treat *luteus, euchlorus* and *eminibey* as the three members of a superspecies, [*luteus*], on the grounds that they are alike in habits, structure and small details of plumage. However, since *luteus* is allopatric with *euchlorus* and differs only in the absence of chestnut on the back and wings, I prefer to treat these two birds as races of the same species, as Moreau himself did in an earlier paper (White & Moreau 1958).

DESCRIPTION

The Golden Sparrow is a small sparrow, length about 120 mm. Sexes are sesquimorphic.

Passer luteus luteus

Male. Head and underparts canary yellow; mantle and back chestnut; rump yellow; wings dark brown with scapulars chestnut and with two whitish wing bars; tail greybrown. Eyes dark brown. Legs dark flesh. Bill, upper mandible horn, lower pale horn, changing entirely to black in breeding season.

Female. Head and upperparts sandy brown; underparts cream, but in certain individuals so strongly tinted with yellow that they can be confused with first year males.

Immature. Similar to adult female, but lighter below and with grey flecks on back of head and nape. Plumage of first year male paler than that of older birds. Tarsus blue in immature, becoming flesh-coloured in adult.

Passer luteus euchlorus

Male. Differs from nominate race in that it is a darker yellow, back is yellow, not chestnut, and uniform with head and rump.
Female. As nominate race, but back ashy brown.
Immature. Similar to adult female, but greyer.

Morel and Morel (1978) suggest that the striking yellow plumage of the male Golden Sparrow, which is unique in the genus *Passer*, is an adaptation to colonial life in the open landscape of the bushy savanna, similar to that of many of the *Ploceus* weavers, though the Golden Sparrow differs from the latter in that the yellow plumage is retained throughout the year, whereas the ploceid weavers have two moults during the year and the yellow is only a nuptial plumage acquired for the breeding season.

BIOMETRICS

Only limited biometric data are available for this species. Table 11 gives typical figures.

Table 11: Biometric data for Golden Sparrow

Subspecies	Feature	Males		Females	
		Range	Mean	Range	Mean
luteus	weight		14.4		14.4
,,	wing	**60–70**	**63.5**	**58–67**	**60.1**
euchlorus	wing	57–64	(60.5)	55–64	(59.5)
luteus	tail	46–50	48	45–50	(47.5)
,,	tarsus	17–18	16.5		
,,	culmen	10.0–11.3	10.6	10.0–11.2	10.5

48 *Golden Sparrow*

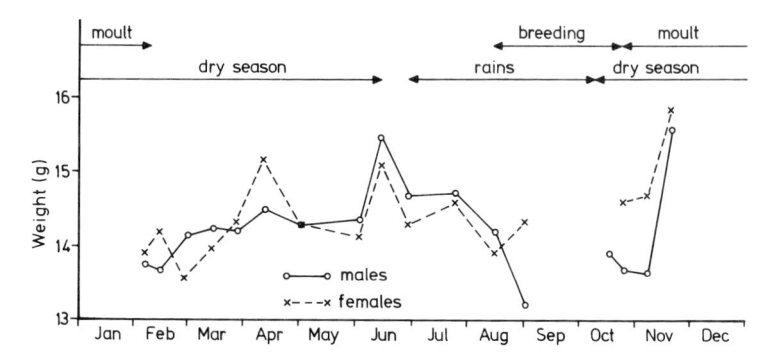

Fig. 13 Seasonal variation in mean weights of Golden Sparrow

This is one of the smallest sparrow species. The limited data for *euchlorus* suggest that it is slightly smaller than the nominate race. R. Klein (pers. comm.) has measured weights throughout the year for birds in Niger. According to his results there was an average increase in weight during the day of 1.1 g (*ca* 7.5%) through the presence of food in the crop and stomach. The yearly figures (Fig. 13) show no clear seasonal pattern, though it appears that there is a gradual increase in weight during the dry season, followed by a decrease at the onset of breeding and an increase at the beginning of moult. There is little difference between the sexes (Table 11), though Klein's measurements in Niger showed the females slightly heavier (*ca* 2%).

DISTRIBUTION

Passer luteus luteus

The nominate race of the Golden Sparrow is confined to a narrow zone of arid country lying to the south of the Sahara from the Sahel in the west to the Red Sea coast in the east. It is difficult to define the range precisely as the bird is largely nomadic, breeding opportunistically when conditions are suitable. With the changes in climate that this region has undergone since records are available*, it is perhaps not unexpected that significant changes have been recorded in the breeding distribution of the bird. This has been most closely studied in west Africa. Morel and Morel (1976) considered that the regular breeding range was limited to the south by the 500 mm isohyet and to the north by the 250 mm isohyet, that determines the limit of cultivation without irrigation. In favourable years, or cycles of climate, it can extend north of this limit and now does so increasingly where there is irrigation of cultivated land, in the extreme to the 100 mm isohyet. For example, in 1971 it bred as far north as Atar in Mauritania, 20°32′N, but the most northerly colony, found in 1975, was at 18°22′N. In the same period it was breeding in the Ferlo region of northern Senegal, *ca* 16°N, south to about Linguère (Morel & Morel 1976). Breeding continued intermittently in the Ferlo until 1981, and in 1979 was first recorded southwest of this, reaching Pékesse (15°07′N) in 1980, Baba Garage (14°57′N) in 1981 (Ruelle & Semaille 1982) and south of Diourbel at 14°30′N in 1984 (personal observation). The bird did not breed in the Ferlo in 1984.

* For the period 1964–1976 the rainfall in west Africa between 10°N and 20°N has been 60% of that recorded between 1931 and 1960, with the isohyets having moved as much as 400 km to the south (Jones 1976). In fact recent evidence suggests that this decline may have been going on for 200 years (D. Winstanley quoted by Gribben 1985).

birds, Hall and Moreau (1970) treat *luteus, euchlorus* and *eminibey* as the three members of a superspecies, [*luteus*], on the grounds that they are alike in habits, structure and small details of plumage. However, since *luteus* is allopatric with *euchlorus* and differs only in the absence of chestnut on the back and wings, I prefer to treat these two birds as races of the same species, as Moreau himself did in an earlier paper (White & Moreau 1958).

DESCRIPTION

The Golden Sparrow is a small sparrow, length about 120 mm. Sexes are sesquimorphic.

Passer luteus luteus
Male. Head and underparts canary yellow; mantle and back chestnut; rump yellow; wings dark brown with scapulars chestnut and with two whitish wing bars; tail greybrown. Eyes dark brown. Legs dark flesh. Bill, upper mandible horn, lower pale horn, changing entirely to black in breeding season.
Female. Head and upperparts sandy brown; underparts cream, but in certain individuals so strongly tinted with yellow that they can be confused with first year males.
Immature. Similar to adult female, but lighter below and with grey flecks on back of head and nape. Plumage of first year male paler than that of older birds. Tarsus blue in immature, becoming flesh-coloured in adult.

Passer luteus euchlorus
Male. Differs from nominate race in that it is a darker yellow, back is yellow, not chestnut, and uniform with head and rump.
Female. As nominate race, but back ashy brown.
Immature. Similar to adult female, but greyer.

Morel and Morel (1978) suggest that the striking yellow plumage of the male Golden Sparrow, which is unique in the genus *Passer*, is an adaptation to colonial life in the open landscape of the bushy savanna, similar to that of many of the *Ploceus* weavers, though the Golden Sparrow differs from the latter in that the yellow plumage is retained throughout the year, whereas the ploceid weavers have two moults during the year and the yellow is only a nuptial plumage acquired for the breeding season.

BIOMETRICS

Only limited biometric data are available for this species. Table 11 gives typical figures.

Table 11: Biometric data for Golden Sparrow

Subspecies	Feature	Males		Females	
		Range	Mean	Range	Mean
luteus	weight		14.4		14.4
,,	wing	60–70	63.5	58–67	60.1
euchlorus	wing	57–64	(60.5)	55–64	(59.5)
luteus	tail	46–50	48	45–50	(47.5)
,,	tarsus	17–18	16.5		
,,	culmen	10.0–11.3	10.6	10.0–11.2	10.5

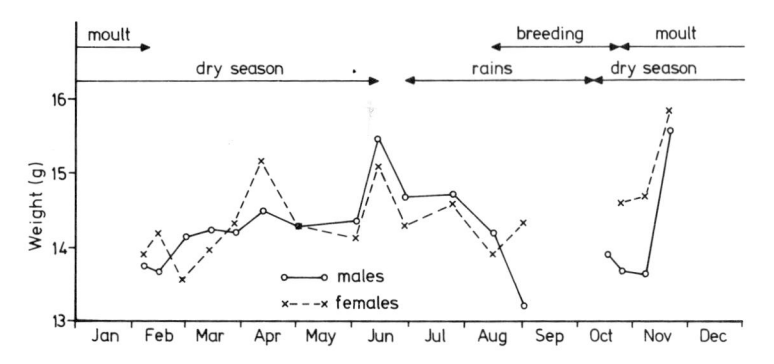

Fig. 13   Seasonal variation in mean weights of Golden Sparrow

This is one of the smallest sparrow species. The limited data for *euchlorus* suggest that it is slightly smaller than the nominate race. R. Klein (pers. comm.) has measured weights throughout the year for birds in Niger. According to his results there was an average increase in weight during the day of 1.1 g (*ca* 7.5%) through the presence of food in the crop and stomach. The yearly figures (Fig. 13) show no clear seasonal pattern, though it appears that there is a gradual increase in weight during the dry season, followed by a decrease at the onset of breeding and an increase at the beginning of moult. There is little difference between the sexes (Table 11), though Klein's measurements in Niger showed the females slightly heavier (*ca* 2%).

DISTRIBUTION

Passer luteus luteus
The nominate race of the Golden Sparrow is confined to a narrow zone of arid country lying to the south of the Sahara from the Sahel in the west to the Red Sea coast in the east. It is difficult to define the range precisely as the bird is largely nomadic, breeding opportunistically when conditions are suitable. With the changes in climate that this region has undergone since records are available*, it is perhaps not unexpected that significant changes have been recorded in the breeding distribution of the bird. This has been most closely studied in west Africa. Morel and Morel (1976) considered that the regular breeding range was limited to the south by the 500 mm isohyet and to the north by the 250 mm isohyet, that determines the limit of cultivation without irrigation. In favourable years, or cycles of climate, it can extend north of this limit and now does so increasingly where there is irrigation of cultivated land, in the extreme to the 100 mm isohyet. For example, in 1971 it bred as far north as Atar in Mauritania, 20°32′N, but the most northerly colony, found in 1975, was at 18°22′N. In the same period it was breeding in the Ferlo region of northern Senegal, *ca* 16°N, south to about Linguère (Morel & Morel 1976). Breeding continued intermittently in the Ferlo until 1981, and in 1979 was first recorded southwest of this, reaching Pékesse (15°07′N) in 1980, Baba Garage (14°57′N) in 1981 (Ruelle & Semaille 1982) and south of Diourbel at 14°30′N in 1984 (personal observation). The bird did not breed in the Ferlo in 1984.

* For the period 1964–1976 the rainfall in west Africa between 10°N and 20°N has been 60% of that recorded between 1931 and 1960, with the isohyets having moved as much as 400 km to the south (Jones 1976). In fact recent evidence suggests that this decline may have been going on for 200 years (D. Winstanley quoted by Gribben 1985).

Non-breeding vagrants have also been reported further south at Dakar, where it occurs erratically (Morel & Morel 1973a) and, in February 1976, in Banjul in The Gambia (Jensen & Kirkeby 1980).

To the east it breeds in Mali, Burkina Faso (Upper Volta), north of 14°N, and Niger. In 1985, R. Klein (pers. comm.) found three main centres of breeding in Niger: in the dry savanna south of the Aïr; in the west in the Niger valley near Tillabéri and in the east from Diffa to Lake Chad. Judging by the old nests, the first two of these had been used previously, the third area was a new one following favourable conditions created by extensive rains; thus suggesting a similar situation to that in Mauritania and Senegal. There are no breeding records for Nigeria, though it has become increasingly common there in recent years. Jones (1976) reported that, whereas in the 1960s it was only a rare visitor to Nigeria, by the mid 1970s Golden Sparrows could be seen regularly as far south as 12°N, and even at times at 11°30'N. According to Elgood (1982) they are now abundant in northern Nigeria during the dry season (October–June), with a few remaining at Malamfatori throughout the year, though as yet without any evidence of breeding. There is also a record of a single bird as far south as the Wazza National Park (11°25'N) across the border in northern Cameroun in April 1967 (de Greling 1972). Louette (1981), writing on the birds of Cameroun, states that Golden Sparrows remain about Lake Chad all the year round.

Oddly enough, coincident with this expansion south into Nigeria that has accompanied the drought in the Sahel, there has also been a northern extension of range from Niger into the extreme south of Algeria, where breeding was recorded in January 1984 (Jacob & de Schaetzen 1984). This possibly reflects a spread from the increased population in the Aïr, consequent on an increase in irrigated cultivation and the development of mining villages and their surrounding gardens in that area.

The bird has not been studied sufficiently further east to establish whether similar changes in breeding distribution have taken place at the same time in Chad, Sudan and Ethiopia.

With the proviso that periodic changes take place, *Passer luteus luteus* breeds in a narrow latitudinal zone from the Atlantic coast in Mauritania and northern Senegal, through central Mali and northern Burkina Faso, most of Niger north to the Aïr and into extreme southern Algeria. In Chad it occurs south to Lake Chad and north to Ennedi; the range continues east to northern Sudan, across to the Red Sea coast, and south down the Nile to Dongola and to Archico in the coastal plain of northern Ethiopia.

As already mentioned, there is some movement south in the non-breeding season in the west, but less is known about off-season wandering in the east, though Hall & Moreau (1970) give an extreme southerly record at 9°N in western Ethiopia, and Goodman and Watson (1983) report the first Egyptian record from Wadi Akwantra in February 1982; this latter occurrence is over 400 km north of the known breeding range in the Sudan.

Passer luteus euchlorus

This subspecies is confined to the Tihamah region of Saudi Arabia from Al Qunfidhah, through coastal Yemen and the Gulf of Aden coast of South Yemen as far east as Shuqra. According to Jennings (1981a) it is a common, but local, breeding resident in the Tihamah and southern Red Sea coastal region of Saudi Arabia. It also occurs in Somalia where Archer & Godman (1961) reported it along the coast from Zeila to Laba Geri; although Meinertzhagen (1954) failed to find any in this area in 1949. Ash and Miskell (1983) describe it as a common resident in the northwest coastal region of Somalia.

The distribution of the Golden Sparrow is given in Fig. 14, including the possible

50 *Golden Sparrow*

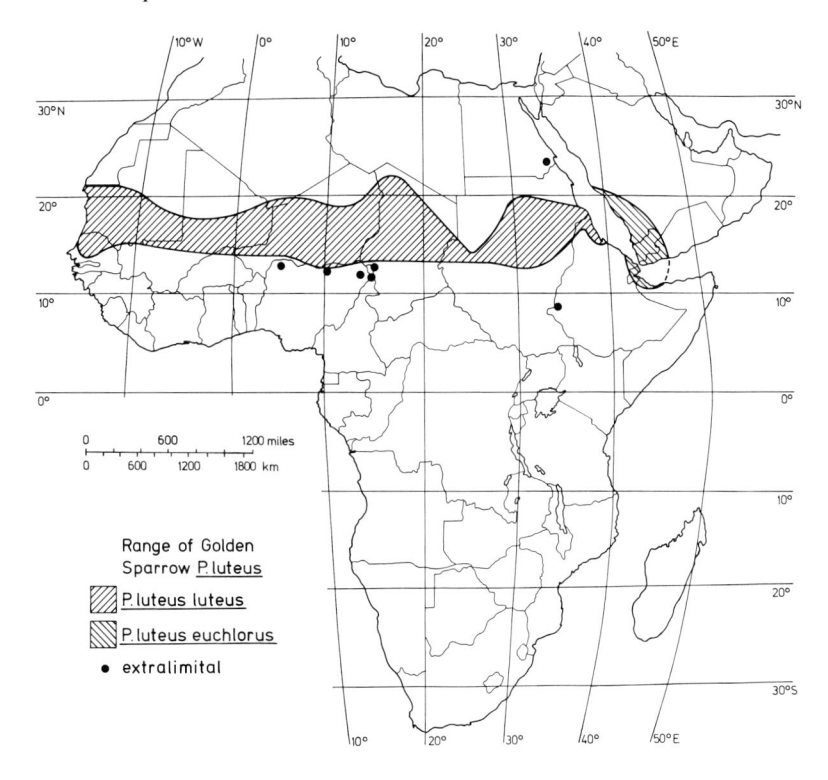

Fig. 14   Range of Golden Sparrow

zone of occurrence of *euchlorus* in Somalia.

The Golden Sparrow has been a subject of special study in Mauritania, Senegal, western Mali and Niger by Oubron (1967), Bortoli and Bruggers (Bortoli & Bruggers 1976, Bruggers & Bortoli 1976, Bruggers 1977), and, particularly, Morel and Morel (1973a, 1973b, 1976, 1978, 1980) and currently in Niger by R. Klein (pers. comm.), because of the possible economic importance of the species as a pest of agriculture. Much of the following sections is based on this work.

HABITAT

The Golden Sparrow occurs in arid, sandy savanna with a seasonal ground cover of annual plants and a low density of thorny bushes and trees (mainly *Balanites aegyptiaca, Acacia* sp. and *Ziziphus* sp.), with breeding occurring where there is permanent water or seasonal ponds. It is thus confined to the arid zone just south of the Sahara (the Sahel in west Africa) and the sub-desert, where mean annual temperature is about 30°C and the annual range 10–40°C. Comparison of the distribution of the species given in Fig. 14 with that of the main vegetational types in Africa (Brown *et al.* 1982); Fig. 15, shows how closely this conforms to the northern tropical zone of dry woodland and steppe, that stretches across Africa south of the Sahara Desert, with some extension into the southern edge of the desert itself. Unlike most of the sparrows, the Golden Sparrow is rarely associated with man. It does not frequent villages, but comes into cereal cultivation and this has allowed some penetration into desert areas, particularly where there is irrigation. One obvious divergence between the two figures is the absence from the dry savanna country of southern Sudan and western Ethiopia north of Lake Turkana, and the one specimen from this area makes one wonder if its distribution should extend south of the area given in Fig. 14 into southern Sudan and southwestern Ethiopia.

BEHAVIOUR

The Golden Sparrow is highly gregarious; it breeds in large colonies and wanders widely during the non-breeding season, turning up rather erratically depending on the availability of food. Oubron found a distinct pattern of movement in northern Senegal and southern Mauritania, Klein a very similar pattern in Niger, with a clear movement towards the breeding areas at the onset of the rains. Breeding occurred during the rains in areas where rain had fallen and produced a flush of green vegetation. After the end of breeding at the beginning of the dry season, the birds dispersed over the cultivated area and then, following the harvest, into places where there was some water – lakes, marshy areas and pools. The degree of dispersion depended on the availability of food; the birds formed large concentrations in cultivated crops when these were available, but after the harvest they scattered more widely in search of the seeds of wild grasses, though even when widely dispersed through the day, when searching for food, the social nature of the species is shown

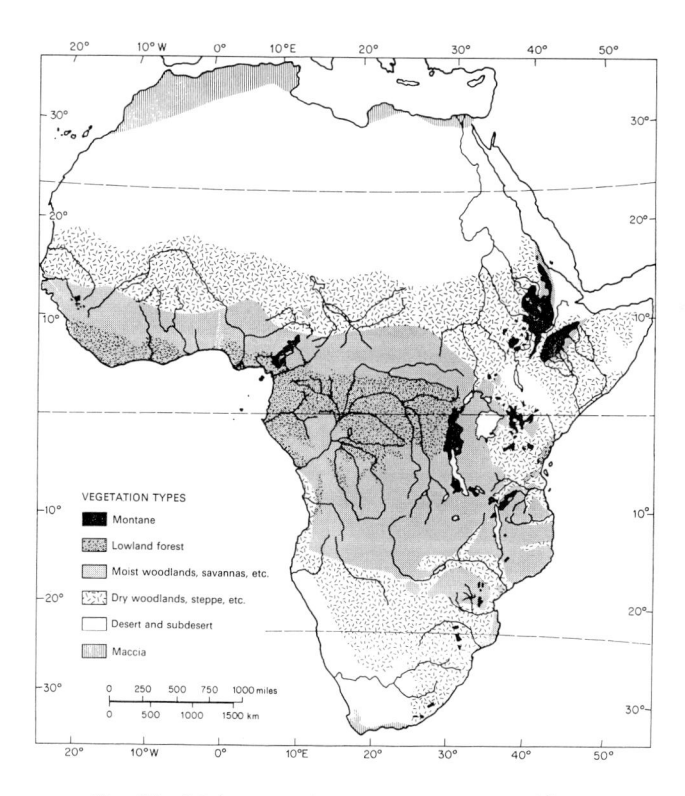

Fig. 15   Main vegetation types occurring in Africa

by the way the birds collected to form large roosts at night. Bannerman (1951) reported how the birds came into Khartoum at night from the surrounding countryside to roost in large numbers in the trees of the river-front gardens and neighbouring woods. Oubron gave 400,000 as a maximum for the roosts he encountered, Klein a maximum of 350,000, but Bruggers and Bortoli estimated a million birds in a roost in sugar cane near Richard Toll, Senegal, in April. The roost may be widely dispersed: Oubron gives an example of a roost covering 20 ha, with the birds choosing the densest trees and bushes into which they could retreat for maximum security. Klein found the same roosts were occupied for up to five months during the non-breeding season.

The daily behaviour is much influenced by the weather conditions. In hot, dry weather Oubron found the birds left the roost at dawn, moved quickly to the feeding areas and spent the morning searching for food; by midday, however, when the temperature became very high, they moved into the trees for shade, forming a daytime roost. When food was available in the neighbourhood of the night-time roost this was used as the daytime roost, otherwise a separate day roost near the feeding place was occupied. When the temperature abated, about 4 pm, the birds resorted to places where there was water and indulged in drinking and bathing before going to roosting places and settling for the night. In rainy weather, and during the breeding season, the birds remained active throughout the day and did not form a daytime roost. Klein did not find daytime roosts in Niger, the birds departing from the roost in the half hour before sunrise and returning to it $1-1\frac{1}{2}$ hours before darkness fell.

During the day the birds occur in loose flocks of up to about 100 individuals which feed on the ground by roller feeding (*ie* the flock faces in one direction, with birds at the rear continuously flying over those in front to take the lead). Feeding bouts are interspersed with pauses when the birds rest in the cover of trees or bushes.

Like all members of the genus, Golden Sparrows frequently indulge in sand bathing as well as water bathing. The normal gait is by hopping, but for short distances the bird can run for a few steps.

BREEDING BIOLOGY

Breeding takes place in large, dispersed colonies that form when conditions are suitable. This is a time of great excitement among the birds. The behaviour postures have been described by Kunkel (1961) from his studies of captive birds and, although this is a somewhat abnormal situation, with the birds in close confinement and therefore unable to distance themselves from one another, his descriptions are largely confirmed by observations on free-living birds.

The breeding colony builds up in a number of stages. First, the males collect in a suitable area of thorn trees, individuals adopting possible nest sites in the trees. The birds utter a series of calls from the chosen sites; these are the familiar sparrow-type chirps, though fuller and more musical than those of the House Sparrow, that can be strung together into a song – *chitta chitta churr chitta chit* – the birds at the same time showing their excitement and nervousness by flicking the tail down, then back up again, possibly a flight intention movement. At this stage there is much disputing over the sites and the males chase each other through the branches of the trees and from one tree to another. In threat, the bird takes up a horizontal or slightly downward sloping posture, with the wings held out from the body and drooped, the opened bill directed at the opponent. At higher intensity the wings are stretched momentarily upwards and fights may occur, with the two birds falling to the ground locked together. In these threat postures the body feathers, except for those on the crown, are sleeked in readiness for flight, showing the ambivalence of threat and fear of the rival. Once a nest site has been established, nest building begins, and at this time the savanna resounds with the calling of the males at the partially completed nests. The noise is great enough to make it easy to locate a breeding colony from the road over the noise of a car when driving along.

In contrast to the sleeked posture used in threat, the male, trying to attract a female to the nest in the pair formation display, fluffs up his feathers: the head is held up to expose the ruffled, bright yellow breast feathers, the tail is raised and the wings held out to expose the ruffled yellow rump and lower back. Throughout, the bird calls incessantly. In the presence of a female the calling becomes even more intense and the male flicks his wings (this is a sharp incisive movement, in contrast to the wing shivering that occurs with most other *Passer* species). When the female approaches, the displaying male hops around, bowing in this posture and flies up repeatedly to the nest entrance (nest showing).

Once the pair has been formed, the female becomes the dominant partner (throughout the remainder of the year the normally larger males are dominant), and to solicit copulation the male takes up a submissive crouched posture, with the wings raised and shivered, that is similar to the food-begging behaviour of the young birds. If possible, he takes up a position below the female so that his bright plumage is well displayed. If the female is not ready for mating she pecks at the male; this attracts nearby males and the hen flies off pursued by the males in a group display. When the female is ready for mating she takes up a somewhat similar invitation posture to that of the male, with drooped and shivering wings, calling softly. The male approaches the dominant female, cautiously, with feathers sleeked; he then stretches up and mounts the female. This may be repeated 15–20 times, with the male pecking at the female's nape.

The Golden Sparrow is dependent on a supply of insects for rearing its young. In the arid area that it occupies, it has to wander extensively to find food, settling down to breed only when conditions become suitable. This normally follows the start of the seasonal rains (Fig. 16) and, as this can vary from year to year, the bird is an opportunistic breeder both in time and in space. For example, in 1984 no breeding occurred in northern Senegal where the rains failed and there was a complete absence of ground cover beneath the thorn trees. Whereas in central Senegal, where some rain had fallen, there was a flush of green vegetation beneath the thorn trees and the birds formed breeding colonies in the trees. The changing pattern of rainfall has

Subspecies	Locality	Month												Reference
		J	F	M	A	M	J	J	A	S	O	N	D	
luteus	Senegal													Morel & Morel 1973a, 1976, 1982, Bruggers & Bortoli 1976, Ruelle & Semaille 1982
	Mauritania													Morel & Morel 1973a, 1976, 1982, Ruelle & Semaille 1982
	Mali													Bannerman 1948, Bortoli & Bruggers 1976
	Burkina Faso													L. Bortoli (pers. comm.)
	Niger													R. Klein (pers. comm.)
	Algeria													Jacob & de Schaetzen 1984
	Chad													Bannerman 1953, Guichard 1955, Heim de Balsac & Mayaud 1962
	Sudan													Bannerman 1948, 1951, 1953, Mackworth-Praed & Grant 1960, Allen & Jackson 1973, Bortoli & Bruggers 1976
	Eritrea													Smith 1955a, b, Mackworth-Praed & Grant 1960, Urban & Boswall 1969, Urban & Brown 1971
euchlorus	Somalia													Smith 1955a
	Saudi Arabia													Meinertzhagen 1954

Fig. 16 Summary of breeding season data for Golden Sparrow (breeding (main) ——— rains – – – – exceptional ···)

apparently resulted in a southward extension of the breeding range, though still within the area of the thorn tree savanna. The gregarious behaviour results in the bird becoming a colonial nester, forming extensive colonies that can occupy an area from a few to several hundred hectares.

The rains, in addition to stimulating a flush of vegetation on which the insect life depends, create seasonal ponds that provide water for drinking. Less frequently, breeding has been recorded in the dry season, but only in the vicinity of permanent water. This has been reported at both limits of the breeding range, namely in Senegal and Mauritania along the Senegal river in the west (Bruggers & Bortoli 1976) and Ethiopia in the east (Smith 1955a).

The opportunistic nature of breeding is shown by an observation by K. D. Smith in Eritrea in 1953. Here the Golden Sparrow normally breeds during the rains in December–January; in the winter of 1952–53 the rains failed and the birds did not breed, but a flourishing breeding colony was found on the plateau, in late September 1953, after some light rain that gave rise to a flush of green vegetation (Smith 1955b).

The nests are built predominantly in thorny trees, mainly *Balanites aegyptiaca* and *Acacia* sp. Large trees are preferred, though in the absence of trees scrubby bushes only a metre or two high are used. Tree densities of 5–200 trees/ha (trees 5–50 m apart) were found in Senegal, most frequently in the range 20–100 trees/ha, but Klein reported breeding colonies in southern Niger where there were 100–700 trees/ha. The males occupy the breeding trees first and build the external part of the nest. This takes about 10 days and involves about 50 hours of building work. Stiff dead twigs are used to build the nests rather than the grass used by most sparrows. This is probably an adaptation to the climate – twig nests withstand heavy downpours of rain more effectively than those of grass and also allow air currents through the open structure that may help in preventing the eggs from overheating in nests directly exposed to the tropical sun; these twig nests are also an effective defence against predators. The use of twig nests enables colonial breeding that would not be possible if nesting were restricted to holes.

The male starts by building a platform in a fork; the walls are raised from this and finally it is roofed over. Not only is the nest firmly supported on a bough, but it is securely worked into the surrounding branches. The entrance is in the upper part of the side, with a tunnel leading into the nest cup. The whole is a bulky, untidy, oval structure. With the thorny branches of the tree and the strength of the nest, it is nearly impossible to examine the contents without removing the nest from the tree and destroying it. Nests measured in Senegal had the following external dimensions: height 370–430 mm, diameter 230–310 mm, volume 17–33 dm^2, with a weight of 220–470 g (*ie* 15–32 times the weight of the bird) and consisted of 700–1,000 separate pieces of twig up to 500 mm long, average 150 mm. Further south the nests may be even larger; Klein describes a medium-sized nest for Niger as weighing 470 g and containing 1,209 twigs. One nest from Mauritania at 20°N, where because of the aridity less material was available, weighed only 129 g and had only 341 pieces of twig.

A problem of colonial nesting is that a considerable amount of stealing of nest material by neighbouring males takes place. However, once the male has attracted a mate she helps in defending the nest. When the external structure has been completed the female co-operates in building the nest cup, the male bringing the material and the female arranging it. The cup consists of pieces of green or dried grass, lined with hair, wool, feathers and even strips of cloth and bits of string.

In west Africa the trees in a colony hold from 1–8 nests, with an increasing number of nests per tree further south as the trees become larger and more widely separated (Ruelle & Semaille 1982). Klein found up to 15 nests/tree in Niger and in other places up to 20–30 nests in one tree having been reported. 65,000 nests were estimated in one colony in Senegal that extended over 630 ha (100 nests per ha), giving a breeding population of 130,000 birds. Ruelle and Semaille (1982) report that the nest density in Senegal can range from 10–200 per hectare, most frequently in the range 20–100 nests per hectare; Klein counted up to 570 nests per hectare near Lake Chad in Niger.

The birds are territorial only as far as the nest and its immediate surroundings are concerned. Morel and Morel (1978) suggested that the fairly low density of nests, compared with some of the social weavers, allows the birds to collect the insect food for their young in the close vicinity of the nest, whereas the social weavers breed much more closely together and have to fly outside the colony to collect food, the size of a social weaver colony being then determined by the distance that the adults have to fly to obtain food. Golden Sparrows are quite tolerant of each other in the food collecting areas. However, although not densely packed, the nests are well within sight of each other and, as with many colonial species, the act of building by one male tends to stimulate its neighbours to the same activity.

Data on the breeding season are summarised in Fig. 16. Although the breeding season normally extends over three months, at any one location conditions for breeding are unlikely to be suitable for as long as this and any colony will probably only be occupied for about eight weeks. Egg-laying in any one colony is not closely synchronised and extends over about four weeks. As the flush of insects on which the birds depend for breeding is not necessarily associated with the onset of rain, the colonies may be abandoned even after egg-laying has started if the insects do not appear. At the colonies studied in Senegal, the period when conditions were favourable for breeding was too short to allow two broods to be raised, though a second clutch was attempted by birds which had started early; and pairs that had lost or abandoned an early clutch could be successful with a second one. In exceptional circumstances, when conditions remain favourable, a second brood may be raised. For example, in Burkino Fasa, where the normal breeding season is from August to September, L. Bortoli (pers. comm.) found the birds breeding in October 1985, and examination of the nests that had a new lining on the top of the old dirty one indicated that nests were being used for a second time. This exceptional breeding was considered to have occurred because of the abnormal abundance of grasshoppers present that year after the rainy season. More normally, however, favourable conditions are of short duration, and when conditions for breeding deteriorate so the nesting colony is abandoned, even by those pairs still having eggs or nestlings, and there are no successful second broods.

There is, however, some evidence from the west and centre of the range of the birds moving to a new area to rear a second brood. In 1979 to 1981, Ruelle and Semaille (1982) found that new colonies were formed in west central Senegal 2–7 weeks after breeding had occurred in the Ferlo. Moreover, in 1981 they observed a strong southwestern passage of the birds at Keur Momar Sarr, south of the Lac de Guier, after breeding had been completed in the Ferlo colonies, amounting to an estimated 1.5–2 million birds between 27th September and 30th October. Klein found that the conditions at the most northerly breeding colony that he studied in Niger, 80 km west of Agadez, where the rains were of short duration, allowed time only for one brood to be raised, after which the birds moved off; breeding then occurred in two more southerly locations, 20 km NW of Tillabéri in the Niger valley and between Diffa and Lake Chad in the southeast of the country, where the vegetative growth was later and the rains more prolonged. He presumed that these involved birds from the north that had moved south for second and even third broods when the conditions there became favourable for breeding.

In 1984 the birds apparently failed to breed in the Ferlo, but I located seven breeding colonies in September in an area of approximately 8,000 km^2 lying between 14°30'N and 15°30'N, 16°30'W and 15°45'W in the Diourbel Region.

The breeding areas in Senegal in 1975, 1979–80, 1981 and 1984 are given in Fig. 17. This shows how the Golden Sparrow is an irregular breeder, wandering during the breeding season and settling down opportunistically where conditions are suitable. In the Ferlo, where records are available for over 30 years, breeding has been recorded only in seven years: 1948, 1961, 1967, 1975, 1979, 1980 and 1981.

In the west of its range the clutch size is 3–4 eggs for the normal rainy season breeding, but only 2–3 in abnormal dry season nesting. Klein found a mean clutch size of 4.2 eggs at the beginning of the breeding season and 4 or less for the late, presumed second and third, clutches. The mean is larger in the east for the nominate race, with clutches up to 5 in the Sudan, and for the subspecies *euchlorus*, with clutches of 4–6 in Somalia and 3–5 in Saudi Arabia. Kunkel (1961) had clutches of 3–4 with his captive birds. Laying starts as soon as the nest is completed. Both sexes attend the nest, the female more than the male, though because of the high daytime temperatures it is probably not necessary for the eggs to be covered; the female incubates the eggs overnight. In captivity, in Germany, Kunkel found that the female alone had a brood patch and she alone incubated the eggs, though spending less time on them than a House Sparrow would do at the same latitude. Attendance at the nest

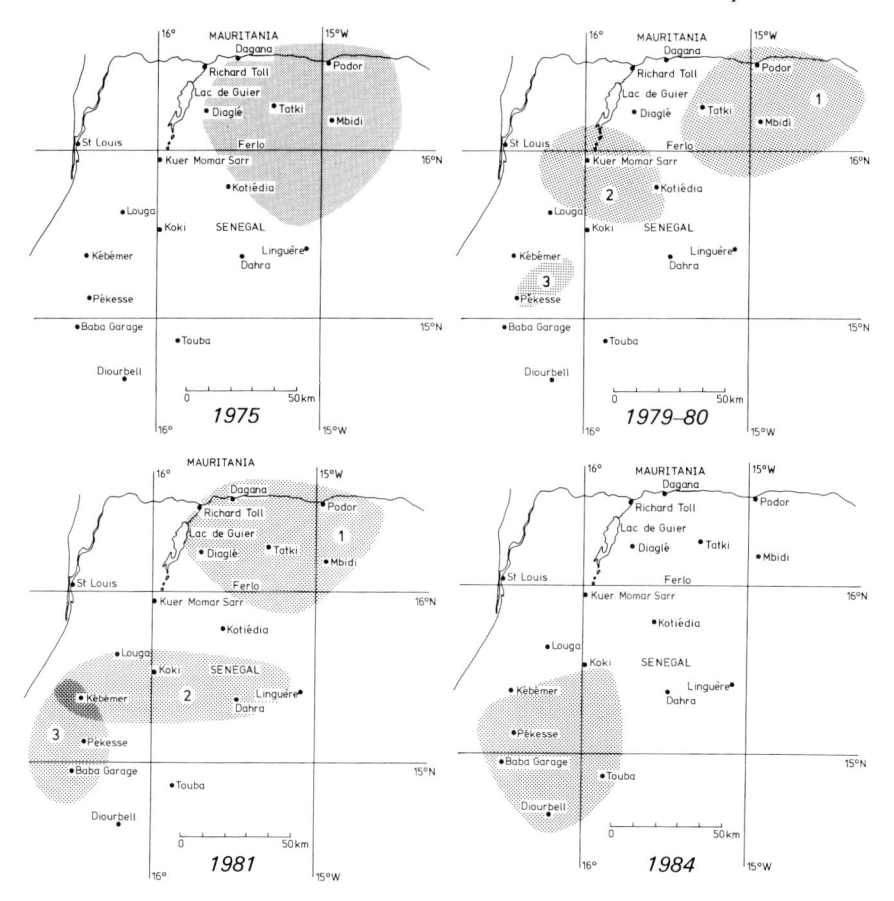

Fig. 17 Zones of breeding of Golden Sparrow in Senegal, 1975–84. Zone 1: 1979 5–10 September; 1980 15–25 August; 1981 25–30 August. Zone 2: 1979 1–5 October; 1980 15–30 September; 1981 5–15 September. Zone 3: 1979 20 October; 1980 1–5 October; 1981 20–25 October

is important to prevent the stealing of material by males that have not yet completed their nests. The mean incubation period for broods in the wild is 11 days (range 10–12); Kunkel found the same in captivity. The addition of nest material, and copulation, continue during the incubation period.

Both sexes feed the young in the nest, though the female does so on average twice as frequently as the male, the pair making a total of about 11–12 visits per hour with food. In the absence of a male the female is capable of rearing the young by stepping up her feeding rate to that achieved by the pair. Removal of faecal sacs is solely by the female. The fledging period is 13–14 days (14–15 days in captivity); just prior to fledging, adults attempt to entice the young from the nest by hopping and wing-shivering in front of the nest with food. Golden Sparrows leave the nest with a weight of 14–15 g (Klein pers. comm.). The combined period with eggs and nestlings in the nest is 1–6 days shorter than that for any other of the sparrows for which data are available. This could possibly be an adaptation to the short period for which conditions are favourable in any one area, leaving the nest early giving the birds the opportunity to move away to another area where conditions are better.

The few results available on breeding statistics are summarised in Table 12. The low success rate in the colony studied by Morel and Morel (1973a) was mainly caused by abandonment of the nests (30% of the nests in the colonies were abandoned) as the conditions for breeding deteriorated. Because the high ambient temperatures leads to the development of the egg as soon as it is laid, there is also asynchronous hatching and this may be a contributory factor; certainly Morel and Morel noted that clutches of 4 eggs normally resulted in the last-hatched young dying early. Clutch size was lower in dry-season nesting. Bruggers and Bortoli (1976) suggested that this may have been associated with the age of the birds, many of which were less than a year old. In the particular colony that was studied it is of interest that non-breeding females, or females that had lost their clutches, assisted in feeding the young, with as many as four females bringing food to one nest.

Too much should not be read into these few results. Without more definite information on second, and even third, broods it is not possible to speculate on the average number of young produced per pair per year, though it must be greater than the figure given in column 6 of the table. Considerable variations in breeding success must be expected in a species living in such a capricious environment.

SURVIVAL

In the absence of ringing data there is no information on the survival of Golden Sparrows in the wild. There are records, however, of individuals surviving in captivity for 9–14 years (Flower 1925).

Table 12: Golden Sparrow breeding statistics

Area	Comment	Average nest contents		Hatching success %	Av. No. of young fledged	Fledging success %	Breeding success %	References
		eggs laid	young hatched					
Senegal	normal season	3.2	3.0 (2.8*)	86	2.3 (1.6*)	61	52	Morel & Morel 1973a
Senegal	presumed 2nd clutches				2.5–2.8			Ruelle & Semaille 1982
Mauritania	dry season nesting	2.5						Morel & Morel 1973a
Burkina Faso	presumed 2nd clutches	3.0†	2.5†					L. Bortoli (pers. comm.)
Mali	normal season	2.3						Bortoli & Bruggers 1976
Niger	presumed 1st clutches	4.2	3.9	93	2.5	64	60	R. Klein (pers. comm.)
	presumed 2nd clutches	3.7	3.17	86	1.8	47	41	R. Klein (pers. comm.)

* including clutches or broods totally lost
† estimated values

Table 13: Moult of primaries by Golden Sparrow. (*Percentage of birds in moult in Senegal December–March*)

Primary No.	Adults		Juveniles (post-juvenile moult)			
	Dec	Jan	Dec	Jan	Feb	Mar
1	0.0	0.0	9.5	0.0	0.0	0.0
2	0.6	0.0	42.8	0.0	0.0	0.0
3	1.2	0.0	23.8	0.0	0.0	0.0
4	2.3	0.0	0.0	6.5	0.0	0.0
5	20.3	0.0	0.0	18.5	3.0	0.0
6	40.0	0.0	0.0	26.5	11.5	0.0
7	28.4	3.0	0.0	17.5	26.0	0.0
8	6.9	6.0	0.0	24.0	25.5	22.0
9	0.6	20.0	0.0	0.0	24.0	48.0
complete	0.0	71.0	0.0	0.0	0.0	30.0
No. of birds	486	335	21	45	807	36

MOULT

The moult begins as soon as breeding is over. In adults it is completed quite rapidly, but, as the post-juvenile moult of the young starts later and is not synchronised throughout the population, there is a period of several months when individuals will be at some stage of moult. This is illustrated in Table 13, which is due to Drs G. & M.-Y. Morel (*in litt.*), where the percentage of birds moulting the different primaries is shown for December to March (data from Senegal). Similarly in Niger, Klein found the moult extended over a period of 3–4 months (October–February) following the breeding season. In the Sudan the birds were still moulting in February, five months after the end of breeding, but moult was complete by the end of May (Lynes 1924). The plumage of young birds after the post-juvenile moult is somewhat duller than that of full adults; this is not acquired until the second year.

VOICE

The calls are typical of sparrows. Mackworth-Praed and Grant (1960) mention a sparrow-like chirp; no doubt this is the *schilp* of Kunkel (1961), the disyllabic *tchirrup* of the Morels, and the call at the nest that I have transcribed as *chitta*. These calls can be strung together to form a song with the phrases, according to Klein, containing up to 20 notes; this song is given by the male at the nest as part of the display used to attract a female. Kunkel refers to a long drawn-out rattle used in threat and I heard a typical sparrow alarm *churr* when I approached males at their nests. Mackworth-Praed and Grant also mention a twittering flight call.

FOOD

The Golden Sparrow is primarily a seed-eater, taking the medium-sized seeds of grasses (*Panicum* sp.) and cereals – the small millet (*Pennisetum* sp.) and sorghum (*Sorghum* sp.). These medium-sized seeds are dehusked in the bill before swallowing. The bird does not seem able to dehusk rice grains, and rice is only taken after it has been dehusked and becomes available at the threshing grounds; feeding on rice does not occur in the rice fields. According to Klein, millet and dehusked rice are the preferred grains and sorghum is only rarely taken.

Klein obtained data on the principal types of food consumed by adults throughout the year in Niger by analysis of crop contents; these were divided approximately as follows into different categories: weed seeds 55%, grain 30%, leaves, blossoms and fruits 8%, animal matter 7%. Breeding begins shortly after the start of the rains and, as the seeds begin to germinate and become unavailable to the birds, the diet of the adult becomes mixed with an increasing amount of invertebrate food. Fruits are also taken at this time. Both Klein and Guichard (1955) noted the berries of *Salvadora persica* and Bortoli and Bruggers (L. Bortoli *in litt.*) found the birds eating tomatoes in the Senegal river valley. Klein's analysis of the stomach contents of nestlings showed a very different picture, with animal matter accounting for 88%, grain and weed seeds 8% and 4% respectively.

Invertebrates from no less than 10 orders have been recorded, with Hemiptera (bugs of the family Coreidae), Lepidoptera (mainly caterpillars), Hymenoptera (mainly ants), Coleoptera (mainly weevils of the family Curculionidae) and Orthoptera (mainly locust larvae) being the most significant (data from Oubron, the Morels and Klein). The proportion of the different invertebrates varies, no doubt depending on availability. In Klein's study the amount of animal food fed to the nestlings at one colony amounted to 89% for the first four days, and then decreased to less than 50% for the remainder of the nestling period. At two other colonies the animal food remained by far the most important part of the diet throughout the whole nestling period. Ruelle and Semaille (1982) found grain of the small millet in the gizzards of nestlings and, in the case of breeding in the dry season, Bruggers and Bortoli (1976) found that a mixed diet of ants and grass seeds (*Panicum* sp.) was fed to the young for the first eight days, and from then to fledging they were fed exclusively on seeds. Similarly for a dry-season nesting in Ethiopia, Smith (1955a) considered that the young must have been fed exclusively on seeds as insects were almost non-existent at that time.

The bird can be a pest of cereal cultivation, particularly when the ripening of the grain coincides with the fledging of the young at the end of the rainy season. Morel and Morel (1976) give an example from two fields of small millet: in one 4 ha field, 45–50% of the ears were damaged; in a larger one of 42 ha the losses were from 15–20%. This is not, however, invariably the case. In his study in Niger, Klein found that sorghum was little taken, rice only at the threshing grounds after it had been dehusked, and the birds were never seen in the rice fields. Only in one area, where breeding coincided with ripening, was millet taken, but the damage done was not excessive as the concentration of birds was small.

3: Chestnut Sparrow *Passer eminibey*

NOMENCLATURE
Sorella eminibey Hartlaub, J. Ornith. 1880 26: 211, 325.
    Lado, northern Uganda.
Passer eminibey (Hartlaub) 1880.
  Subspecies: none.
    Synonyms: *Sorella emini* Hartert 1881.
      Sorella eminibey guasso Van Someren 1922
        Archer's Post, northern Kenya.

The Chestnut or Emin Bey's Sparrow was considered conspecific with the Golden Sparrow by Meinertzhagen (1954), but there is a zone of sympatry in western Sudan (Lynes 1924) and hence it must be recognised as a good species (White & Moreau 1958). Hall and Moreau (1970) place the Chestnut Sparrow in a superspecies with the Golden Sparrow, *Passer luteus* and the Arabian Golden Sparrow, *Passer euchlorus*, which they recognise as a separate species. G. R. Cunningham-van Someren (*in litt.*), on the other hand, considers the Chestnut Sparrow sufficiently distinct from other *Passer* species to justify replacing it in the monotypic genus *Sorella*. I retain it in a superspecies with *Passer luteus*.

DESCRIPTION
A small sparrow (length *ca* 115 mm) that shows affinities with Golden Sparrow,

yellow in male of latter being replaced by chestnut; females and immatures rather similar. Sexes sesquimorphic.

Male. Almost completely uniform dark chestnut except for wings and tail blackish brown. Outside breeding season male shows flecks of white, particularly on underparts. Bill black in breeding season, otherwise horn.

Female. Head grey-brown, mantle brown with dark streaking, rump chestnut-brown. Wings brown with chestnut scapular patches. Breast chestnut brown, becoming buff to white on belly. Bill horn.

Immature. Similar to adult female, but lacking warmer brown colour and chestnut throat patch. Bill horn.

BIOMETRICS

The limited data on measurements available for the Chestnut Sparrow are summarised in Table 14. These suggest it is even smaller than the Golden Sparrow.

Table 14: Biometric data for Chestnut Sparrow

Feature	Males		Females	
	Range	Median	Range	Median
wing	60–65	62.5	57–60	58.5
tail		*ca* 40		*ca* 40
tarsus		*ca* 15		*ca* 14
bill		*ca* 10		*ca* 10

DISTRIBUTION

The Chestnut Sparrow is confined to east Africa. Its breeding range extends in the north from Darfur Province in western Sudan, south to the extreme east of Uganda, most of Kenya and northern central Tanzania to the east of Lake Victoria and into extreme southeast Somalia. In the northeast it extends to the lower lying country (below 600 m) in the southwest and Rift valley of Ethiopia. The breeding range is shown in Fig. 18.

It wanders to some extent outside the breeding season and there are extralimital records from Kabalega Falls National Park at 31°50'E in northern Uganda, and Dar-es-Salaam on the coast of Tanzania (Britton 1980).

HABITAT

This species is to be found in dry savanna, though, in general, not in such arid parts as the Golden Sparrow. It also, however, occurs near water in papyrus swamps and, much more than the Golden Sparrow, regularly comes into gardens and human settlements.

BEHAVIOUR

Most authorities describe the species as gregarious, often associating with Queleas and other weavers. While usually in flocks, this is, however, not invariably the case; I have seen small groups of a few birds, and a single male was watched feeding on crumbs on a bird table at Lake Baringo, Kenya, in February, with a mixed collection of weavers.

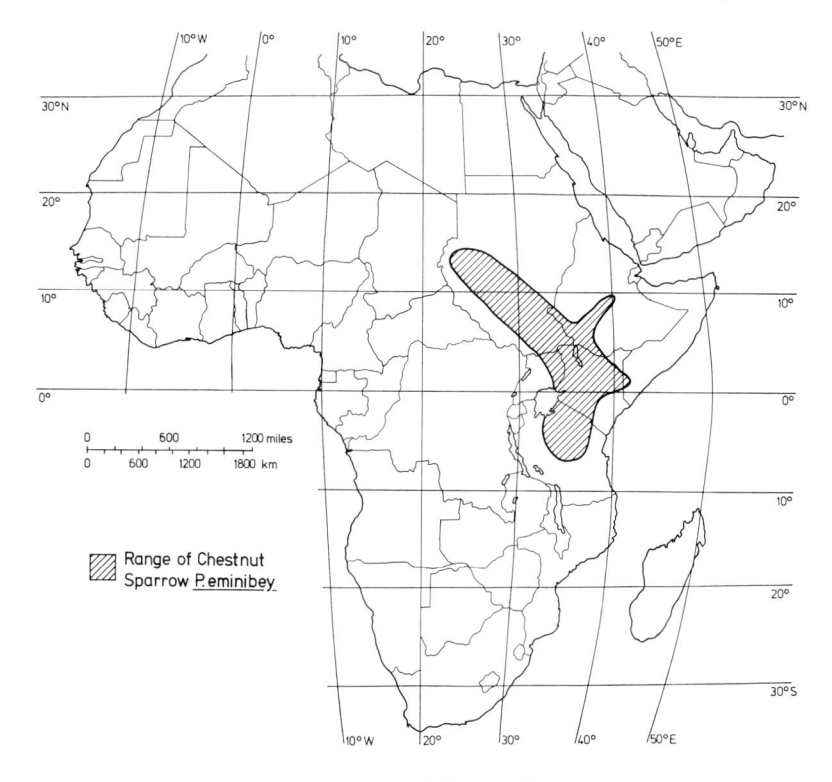

Fig. 18 Range of Chestnut Sparrow

BREEDING BIOLOGY

The exact details of the nesting behaviour of the Chestnut Sparrow have been a matter of some confusion. Lynes (1924) described the nest as 'sparrow-like' – an untidy grass oven, lined with feathers, and with a side entrance; whereas according to Mackworth-Praed and Grant (1960) it has a weaver-like nest with a side entrance, usually suspended from the branches of trees some distance above the ground. This apparent contradiction can be explained by the fact that the Chestnut Sparrow builds its own nest but also takes over those of weavers.

Betts (1966) described a breeding colony of at least 100 pairs near Narok, in the Kenyan Rift valley, in June, that was in a clump of trees filled with the nests of Speke's Weaver *Ploceus spekei* and the Masked Weaver *P. intermedius*. In some cases the Chestnut Sparrows had taken over the old weavers' nests, but in others they had built typical, globular, sparrow-type nests with a side entrance.

Payne (1969) studied the breeding of Chestnut Sparrows near Lake Magadi, Kenya, in 1967. Here, in contrast, they were breeding entirely in nests of the Grey-capped Social Weaver *Pseudonigrita arnaudi,* in a colony of some 6,000 pairs of the latter species. Payne commented that the range of the Chestnut Sparrow coincides with that of the Grey-capped Social Weaver and its congener, the Black-capped Social Weaver *P. cabanis,* implying that it was an obligate nest parasite on these two species. (The term nest parasite is used when a bird actually usurps a nest of another species while it is still in occupation, not merely the use of an old nest of another species.)

This is not the whole story, however, as not only are the nests of the two species of Social Weavers used, and those of a number of Ploceid Weavers, *Ploceus rubiginosus* and *P. velatus* (G. R. Cunningham-van Someren, *in litt.*) in addition to those of *P. intermedia* and *P. spekei* already mentioned, but, also as described above and confirmed to me by Cunningham-van Someren, the Chestnut Sparrow will breed in colonies that consist entirely of self-built, typically untidy, sparrow-type nests.

Even with the nests of other species there is some doubt as to how far these are always usurped from the rightful owner (nest parasitism) or merely taken over after they have been abandoned.

Leaving aside the actual nest, the most complete published account of the breeding behaviour is that given by Payne (1969) and, even if this is not the full story, it is worth recounting in some detail, particularly as he describes the courtship display of the male that leads to pair formation. In the colony studied by Payne the nests of the Social Weavers were in acacias, with up to seven active nests in one tree. Chestnut Sparrows were seen to display at, and presumably appropriate, one in ten of the weavers' nests. The Chestnut Sparrows took over the nest at the end of building or when it contained eggs. It seems most likely that this was initiated by the male sparrow, who flew up to the nest and displayed there, and, although chased away by the owners, repeatedly came back again to display at the nest. The displaying male was soon joined by a female, which in turn was chased away by the weavers, until in time the latter gave up and abandoned their nest. Payne remarked particularly that the sparrows only appropriated active nests; old abandoned nests were completely ignored.

As a preliminary to breeding, the male postured in front of the appropriated nest by crouching, shivering the wings in a shallow V above the body, and uttering a high twittering trill. According to Payne this is used to advertise the nest and attract a mate. When a female approached, the intensity of the display was heightened, the rate of wing-shivering increased and the bird bowed deeply, at the same time spreading and depressing the tail; at highest intensity the wings were held almost upright over the back. When the female was ready for mating, she flew up to the male at the nest, crouched with head drawn in and pointed upwards; the wings were held out, spread slightly and shivered. The male mounted, still shivering his wings, and pecked at the nape feathers of the female. The male could also, apparently, stimulate the female by going into the nest, and a female that had not solicited prior to this frequently did so immediately the male went into the nest and came out again.

Payne noted that the birds never added material to the nests, nor were they seen carrying nest material, but other observers have found appropriated nests to be lined by the sparrows with fine grasses and feathers.

Published data on the breeding season of the Chestnut Sparrow are given in Fig. 19.

In general, breeding coincides with the rains. In those colonies in which it is a nest parasite of one of the weavers, clearly the breeding season must be determined by that of the host species. Payne (1969) describes how, in the colony he studied, breeding lagged behind that of the weavers and he even went as far as to suggest that the Chestnut Sparrows were stimulated to reproductive condition by the nest-building activity of the host species.

It may be that the Chestnut Sparrow is in the course of evolving from normal breeding to nest parasitism and possibly at an intermediate stage on its way to becoming a brood parasite. Clearly the nesting behaviour of this species warrants further study.

The clutch size is typically 3–4 eggs (Lynes 1924, Betts 1966). Nothing appears to have been published on the incubation and fledging periods nor on other aspects of the nesting cycle.

Locality	Month												Reference
	J	F	M	A	M	J	J	A	S	O	N	D	
Sudan													Lynes 1924, Mackworth-Praed & Grant 1960
Ethiopia													Mackworth-Praed & Grant 1960
Kenya													Mackworth-Praed & Grant 1960, Betts 1966, Payne 1969, Brown & Britton 1980, T. Stevenson (pers. comm.)
Uganda													Mackworth-Praed & Grant 1960
Tanzania													Mackworth-Praed & Grant 1960, Brown & Britton 1980

Fig. 19 Summary of breeding season data for Chestnut Sparrow (breeding ——— rains - - - -)

VOICE

Little has been published on the calls. Williams and Arlott (1980) mention subdued chirping; the high, twittering trill of the male at the nest has already been referred to.

FOOD

The adults appear to feed mostly on grass seeds, though they will also take household scraps as I have mentioned above. Payne noted that the nestlings were fed both on insects (small beetles) and on soft grass seeds.

4: Cape Sparrow *Passer melanurus*

NOMENCLATURE
Loxia melanura Müller, Syst. Nat. Suppl. 1776: 153.
  Cape of Good Hope; restricted to Cape Town by Macdonald, Contr. Orn. W. Africa, South Africa 1957: 157.
Fringilla arcuata Gmelin 1788.
  Cape of Good Hope.
Passer arcuatus (Gmelin) 1788.
Passer melanurus (Müller) 1776.
  Subspecies: *Passer melanurus melanurus* (Müller) 1776.
    Synonyms: *Loxia melanura* Müller 1776.
      Fringilla arcuata Gmelin 1788.
  Passer melanurus damarensis (Reichenow), Orn. Monatsb. 1902 10: 77.
      Brakwater, 12 miles north of Windhuk, Damaraland, [Windhoek, Namibia].
    Synonym: *Passer arcuatus damarensis* Reichenow.
  Passer melanurus vicinis Clancey, Bull. Br. Orn. Cl. 1958 78: 59.
      Bethlehem, eastern Orange Free State.
Moreau (1962) accepts this race, although Mackworth-Praed and Grant (1963) treat it as synonymous with *Passer melanurus melanurus*.

DESCRIPTION
The Cape Sparrow or Mossie very distinctive; medium to large, length 140–160 mm; sexes sesquimorphic.

Cape Sparrow

Passer melanurus melanurus
Male. Head mainly black except for a broad white band starting from behind eye, curling round back of cheek to front, where it almost meets on throat, leaving narrow black band connecting black of head to black bib, the latter spreading over breast; nape and upper mantle dark grey; rump, back and lower mantle bright chestnut, conspicuous in flight. Scapulars chestnut, bordered by white wing-bar; flight and tail feathers streaked black and grey. Underparts whitish except for grey on flanks. Bill black in breeding season, otherwise horn.
Female. Similar in plumage to male except that black replaced by grey. Head and breast patch donkey grey; lower back and rump dull chestnut. Shoulders chestnut, bordered below by narrow black line and broad white bar. Underparts pale grey. Bill horn.
Immature. Very similar to female, but male shows a darker throat.

Passer melanurus damarensis
Male. Paler than nominate *melanurus* on nape and mantle; black on head and nape tinged with brown.
Female. Much lighter than nominate race.

Passer melanurus vicinis
Slightly larger than nominate *melanurus*.
Male. More contrastingly black and white than nominate *melanurus*.
Female. Darker than nominate *melanurus*.

BIOMETRICS

Typical values of body measurements are given in Table 15. There is a clear increase in linear dimensions from west (*damarensis*) to east (*vicinis*).

Table 15: Biometric data for Cape Sparrow

Feature	Subspecies	Males		Females	
		Range	Mean	Range	Mean
weight	damarensis		25.5		25.1
	melanurus	17.4–24.6	23.5	17.3–21.0	20.3
	vicinis	25–34	29.6	22–38	29.4
wing	damarensis	74–79.5	76.4	73–79	75.7
	melanurus	73.5–82.0	78.9	70–80	75.1
	vicinis	80–86	83.2	77–82	78.7
tail	damarensis	51–62	57.5	53–59	56.7
	melanurus	55–64	59.1	52–60	56.5
	vicinis	61–65	62.5	59–63.5	61.1
tarsus	—	17.5–21*			
culmen	damarensis	13.5–15	14.2	13.5–15	14.4
	melanurus	13.0–15.5	14.5	13–16	14.4
	vicinis	14.5–16	15.2	14.5–15.5	15.1

*not sexed

DISTRIBUTION

Passer m. melanurus
The nominate race occurs in southern Cape Province, except for the extreme east, north to about 30°S and east to southwestern Orange Free State.

Passer m. damarensis
This race occurs in Cape Province and northwestern Orange Free State, north of the nominate race, with which there is a narrow transition zone, and through most of Namibia, but not north of about 30°S except for the coastal region, continuing to the north in the coastal plain of Angola as far as Benguela. It is rather patchily distributed in the west and east of Botswana south of about 21°S, though apparently lacking in the arid central Kalahari desert. Further east it occurs over northern and central Transvaal (apart from the extreme east), and penetrates into the extreme southwest of Zimbabwe. There is an extralimital record from Harare (Salisbury), well outside the normal range.

Passer m. vicinis
The third race is found in southern Transvaal, the eastern Orange Free State, Lesotho, eastern Cape Province, the extreme west of Swaziland and western Natal, where it is locally common above 1,000 m, rather uncommon below 500 m, and absent from the lower lying country to the west.

The complete distribution is shown in Fig. 20, with an indication of the boundaries between the subspecies.

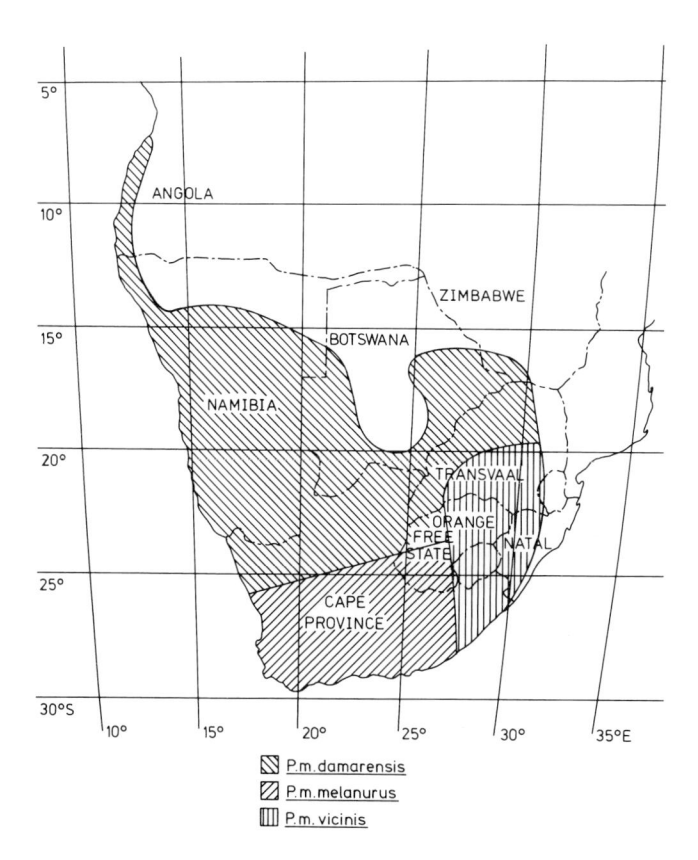

Fig. 20   Range of Cape Sparrow

HABITAT

The Cape Sparrow is basically a bird of semi-arid regions with annual rainfall of less than 750 mm, being absent from the low lying area of eastern Natal, where it is in excess of 800 mm. It occupies open grass savanna with trees, thornveld thickets or light woodland; in the drier areas it occurs particularly along both wet and dry watercourses where there are lines of trees, though it also penetrates the desert margins, again as long as there are a few trees or bushes, particularly in the neighbourhood of water holes. It is now to be found in agricultural land and is frequently in the vicinity of settlements, particularly in the more southerly parts of its range. This must have been a comparatively recent development as settled agriculture only came to the southern parts of Africa, where the Cape Sparrow occurs, in the last thousand years. It has also moved into built-up areas, where it is found particularly in parks, orchards and gardens in the drier suburban areas, though also to a lesser extent in the completely built-up urban areas. Earlé (in press), however, has shown that the latter is a suboptimum habitat for the species with breeding significantly later and clutch size significantly smaller in urban Bloemfontein, Orange Free State, than in the neighbouring suburban and rural areas.

About 1956, it moved into the vineyards in southwestern Cape Province. This followed a change in the cultivation technique when grass and weeds were allowed to grow between the rows of vines to conserve moisture. The abundant weed seeds attracted the Cape Sparrow into the vineyards and the birds turned to the grapes when the seeds had become exhausted; it has now become a significant pest in such areas. This may, however, be a suboptimum habitat, as Siegfried's (1973) study suggests that the breeding success is insufficient to maintain the population, far from allowing for expansion, so it presumably depends on continued immigration from more favourable areas.

As indicated above, the Cape Sparrow, while not an obligate commensal of man, is now, particularly in the southern parts of its range, closely associated with man-altered environments – cultivation, settlements, gardens, commonly living in urban areas as well as villages and isolated habitations – and is commoner in this habitat than it is in its original habitat of grassland savanna.

BEHAVIOUR

The Cape Sparrow is normally gregarious, living in flocks and for the most part breeding in loose colonies; this is, however, not invariably the case and isolated breeding pairs are not uncommon. In uncultivated, semi-arid country, the original habitat of the species, the flocks can number up to 200 individuals that wander widely, settling in an area where food is available and moving on again when it becomes exhausted. In cultivated areas the birds collect in farmyards where grain is put out for stock, together with other sparrows and seed-eaters, such as the Cape Weaver *Ploceus capensis* and bishops, *Euplectes* sp. These flocks, and the ones that form in built-up areas, tend to be somewhat smaller than those in the more open country and are more sedentary, living out their lives within a range of a few kilometres, no doubt an adaptation to a more assured food supply. Like the House Sparrow in Europe, the urban birds quickly recognise gardens where food is regularly put out for birds and gather expectantly in the trees at the appropriate time of day.

The sedentary birds of cultivated land and built-up areas build special nests for roosting that are used throughout the year. The birds of the wandering flocks also prefer nests for roosting and where they are available will use old nests for this purpose; failing a suitable supply of old nests they roost in thick bushes in the area in which they have temporarily taken up residence. Like the House Sparrow in Europe, birds of the built-up areas close to cultivation move out to the ripening grain fields to feed during the day and return at night to their roost. Although by no means exclusively associated with man, Cape Sparrows have adopted many of the habits of the House Sparrow. Some competition between these two species must therefore have occurred when the House Sparrow, introduced to Durban and East London at the beginning of this century (see Chapter 9), began to spread into the range of the Cape Sparrow in the late 1940s. The House Sparrow has spread right through the range of the Cape Sparrow and is now well established in Zambia and Zimbabwe, but there is some suggestion that its penetration of urban areas has been impeded by the Cape Sparrow. Johannesburg, where the Cape Sparrow is abundant, is a good example of this; House Sparrows reached the outskirts of Johannesburg in the early 1950s, but according to Harwin and Irwin (1966) had only got to within 8 km of the centre by 1961 and were still greatly outnumbered by their congener. Despite the fact that the Cape Sparrow is less dependent on man than the House Sparrow, it does not appear to have been displaced from the towns. For example, Mrs M. K. Rowan informed me (*in litt.*) that the two species exist side by side in many little Karroo towns and villages, both behaving as 'house sparrows', with no evidence that the House Sparrow is ousting its indigenous relative. No doubt the fact that the Cape Sparrow is predominantly a tree nester reduces the risk of competition over nest sites and I have seen both species with nests in trees within 5 m of each other; on the other hand it seems inevitable that in the urban surroundings these birds must be in competition for food. Harwin and Irwin consider that the Cape Sparrow may now have reached a state of equilibrium with the House Sparrow and may be holding its own with the latter in built-up areas. This suggests that there are subtle differences in the ecological requirements of these two species in the built-up environment, though these have not yet been elucidated. A similar situation exists between the House and the Tree Sparrow in northern continental Europe and between the House and the Cinnamon Sparrow in the Indian Himalayan hill stations. In contrast, the Cape Sparrow appears to be largely dominant over the Southern Grey-headed Sparrow in competition over nest sites in holes in buildings.

J. Davies of Johannesburg described to me (*in litt.*) the following social behaviour, which he regularly observed in his garden in the summer months. A group of 20–50 Cape Sparrows, separates itself from the other birds and collects on the ground, standing close together with heads held high and tails touching the ground. Sometimes the group moves slowly along by irregular (*ie* not concerted) single hops and quite often one bird or several will fly and hover over the group for a few seconds at a height of 30–60 cm. The gatherings last from one to three or four minutes and break up with the birds gradually dispersing. Both sexes take part. Throughout, the birds are quite silent and there is no trace of any agonistic behaviour. I have no idea of the significance of these 'gatherings', nor have they been described for any other sparrow species.

BREEDING BIOLOGY

The Cape Sparrow is mainly a social breeding species, forming loose colonies of 50–100 pairs, though a number of pairs (possibly amounting to 10% of the birds) also nest solitarily. The reason for this difference is not clear.

Immelmann (1970) made a study of Cape Sparrows in the Gemsbok-Kalahari National Park in South Africa for a period covering the two months September–October, during which some of the birds were breeding, while others were still in non-breeding flocks. This study gives us an insight into the formation of a breeding colony by the members of a wandering flock. According to Immelmann's observation the birds had already paired up in the non-breeding flock, but from the short period of observation he could not determine how pairs were formed, whether just prior to leaving the non-breeding flock to join the breeding colony or whether, once formed, pairs remain together for life. Some authors suggest that pairs remain together throughout the year, though this has not been confirmed by studies on marked birds. My own observations in the Transvaal showed that just prior to breeding the pair stays very close, flying up to the nest together and staying side by side during the feeding forays.

According to Immelmann a pair that is ready for breeding leaves the flock to seek a suitable nesting place, in his study area a high acacia (*Acacia giraffae*). At first this was only in the morning hours, the birds returning to the flock later in the day. However, once a nest site has been selected (frequently the remains of an old nest that serves as the base for the new one) and the birds have started to build, they leave the flock and spend the whole day in the vicinity of the nesting tree. They are joined there by other pairs and a breeding colony builds up quickly in this way.

The bird is very catholic in its choice of places to nest. The preferred site is in trees or bushes, where the nest is placed openly in the branches, though similar free-standing nests are also built on telegraph poles and electricity pylons. In addition, however, it will also nest in a variety of covered sites, such as creepers on house walls, under house eaves, holes in haystacks and in earth banks, and the nests of other species, such as swallows and weavers. Solitary nests are usually in low bushes or on telegraph poles.

The bird probably has a preference for thorny trees and bushes (*Acacia* sp.), though these are not exclusively selected and almost any type of tree, native or exotic, can be used. The nests can be at any height from a metre upwards and are often quite close together; I have counted 14 nests in a low, isolated acacia only about 2 m high, with some of the nests less than a metre apart. Like the other colonial nesting sparrow species, the defended territory is only the nest itself and its immediate vicinity.

In the Barberspan Nature Reserve in southwest Transvaal, the birds were using cans that were hung on power cables as a visual warning to wildfowl flying into the pan. The cans were open ended and attached to the cables in pairs with the open ends facing each other, a few centimetres apart. There were nests in 112 of the 200 cans. This meant that in some cases, where both pairs of cans were in use, the nest entrances were very close together. When I made these observations the birds were only at the nest-building stage; there was much threatening between neighbouring males and even fights wherein the two birds fell to the ground, locked together bill-to-bill. The cables ran above bare ground with only low vegetation and no bushes or trees.

The nest built openly in trees is a large untidy, domed structure, constructed mostly of rather coarse dry grass stems and small twigs, though often containing other material, such as string and pieces of rag, and warmly lined with feathers and plant down. In thorny trees the nests are built round twigs and thorns; whereas in gum trees the leaves of the tree are worked into the nest, which is thus clearly distinguishable from House Sparrow nests, which were frequently in the same tree. The structure, however, has little coherence and falls to pieces if removed from the tree. The entrance is on the side, sometimes being extended to form a short funnel. When holes are used they are filled with a shapeless mass of grass with a central cup for the eggs. Where the nest of a weaver is taken over, this is merely lined with feathers without the addition of further building material. There may be some competition for hole sites near buildings with the indigenous Southern Grey-headed Sparrow and the introduced House Sparrow. Building is carried out by both sexes and during spells of building the pair keeps close, flying up together to the nest with material and away again to search for more.

In its original arid or semi-arid biotope, the Cape Sparrow is an opportunistic breeder dependent on suitable climatic conditions. Breeding is thus sporadic, presumably triggered off by the availability of the insects required for the formation of the eggs by the female and for the rearing of the young; it can, moreover, be patchily distributed as suitable conditions may occur only over an area of no more than a few square kilometres (Siegfried 1973). Unlike some opportunistic breeding species, there is no general movement into the favourable zone and, although there may be some concentration, it appears that it is only the local birds that breed.

In cultivated land, where there is a more regular pattern of vegetational growth through irrigation and a consequential flush of insects, and similarly in gardens in urban areas, the bird has been able to adopt a more regular seasonal pattern of breeding.

Little has been published on courtship display. In a ceremony that I observed, the male hopped round a female in the branches of a tree with the wings held out and the chestnut feathers on the back fluffed up. There is another pre-breeding display in which two or more males can be seen in close pursuit of a female, similar to the group display in some of the other sparrow species (*eg* House Sparrow, Golden Sparrow, Somali sparrow). In these other species, when the female is not ready for mating and flies off pursued by the importuning male, the pursuit flight attracts other nest-owning males in the neighbourhood; I have not, however, been able to make sufficient observations on such chases in the Cape Sparrow to determine if they originate in the same way. The female, if ready for mating, crouches and is then mounted by the male.

Published breeding season records are summarised in Fig. 21. These come from the extremes of the Cape Sparrow range – from the Cape Province in the south to the Transvaal in the northeast. It can be seen that, although breeding has been

Subspecies	Locality	Month												Reference
		J	A	S	O	N	D	J	F	M	A	M	J	
damarensis	Namibia		—	—	—	—	—	—	—	—				Mackworth-Praed & Grant 1960, Smithers 1964, Winterbottom 1971
	Botswana						—							Beesley & Irving 1976
	Transvaal		—	—	—	—	—	—	—	—	—			Rowan 1966, Tarboton 1968, Skead & Dean 1977
	Cape			—										Rowan 1966
melanurus	Cape		—	—	—	—	—							Rowan 1966, Winterbottom 1971, Siegfried 1973
vicinis	Natal			—										Clancey 1964c

Fig. 21 Breeding season for Cape Sparrow (main ——— secondary _ _ _ _)

recorded in all months of the year, it is mainly concentrated in the austral spring and summer. There is, however, a difference between the Cape Province and the Transvaal. In the Transvaal the main breeding season is quite protracted, with about 95% of nests in the eight-month period August to March and the main breeding activity spread from October to February. In the Cape, in contrast, over 90% of nests are in August to December, with a clearly defined peak in October, Fig. 22. More detailed analysis by Rowan (1966) indicates that there is some seasonal variation within the Cape region, the maximum breeding activity occurring on the west coast in September and becoming progressively later towards the east, though she was unable to find any climatic factor that could be correlated with this. Earlé (in press) in a study of the species in Bloemfontein found that breeding started some four weeks later in the urban area than in the surrounding suburban and rural areas.

Clutch size ranges from 2–5, rarely 6, eggs, with an average of 3–4 eggs. Rowan (1966) found an average of 3.55 eggs based on 104 clutches and Siegfried 3.45 eggs based on 228 clutches. More detailed analysis suggests both a latitudinal (Table 16) and seasonal (Table 17) variation in clutch size that is presumably correlated with the availability of food for the young.

Incubation of the eggs during the day is shared almost equally between the sexes with the birds alternating in 10–15 minute spells; at night the female alone covers the eggs, though the male may also roost in the nest. Immelmann (1970) noticed a difference in nest relief between colonial and solitary nesters. In the latter the incu-

Table 16: Variation in average clutch size of Cape Sparrow with latitude (Rowan 1966)

Latitude range	25–27°S	27–31°S	33–34°30S
No. of clutches	27	18	59
Average size	3.11	3.38	3.81

Table 17: Variation in average clutch size of Cape Sparrow with season (Siegfried 1973)

Month	Aug	Sep	Oct	Nov	Dec	Jan	Feb
Average size	3.08	3.81	3.92	3.42	3.13	3.30	3.00

bating bird flew off when it heard the call of its partner in the distance; whereas in the former it waited in the nest until the partner went in, possibly to prevent other members of the colony from interfering with the nest. The incubation period lasts 12–14 days, with the young hatching over 2–3 days; this suggests that incubation starts before the clutch is complete.

Once the young have hatched the parents continue to share equally the brooding and feeding. According to my observations this was also the case with nest sanitation, though at nests watched by Immelmann it was almost exclusively carried out by the male. The young are brooded for the first few days with the incubation rhythm continued, each bird doing spells of 10–15 minutes in the nest. This is probably related to the development of feathers, which only begin to appear on the fifth day; the eye slits begin to open about the same time.

At nests observed by Immelmann in the Kalahari and by me in the Transvaal, the young were fed exclusively on insects, with caterpillars of lepidoptera forming the major part of their diet. These were collected in the nest tree and at up to 500 m away.

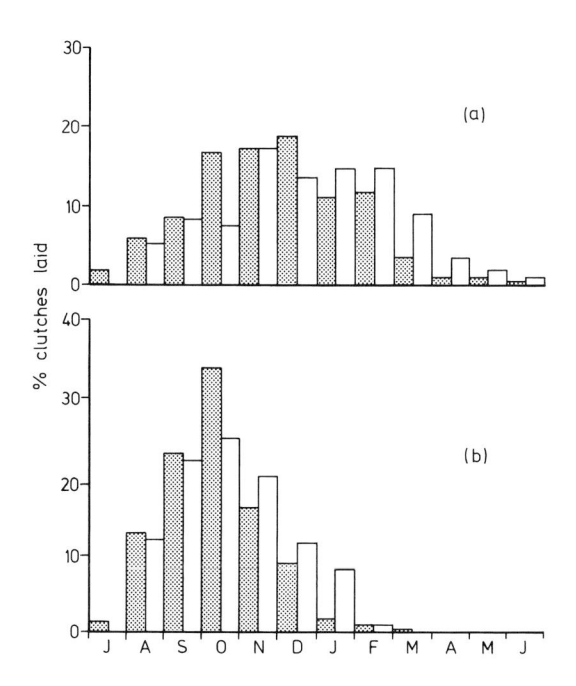

Fig. 22 Seasonal distribution of nests of Cape Sparrow in (a) Transvaal, (b) Cape Province

The female is clearly dominant at the nest, threatening the male when both arrive at the same time with food. At one nest that I watched, the pair made 10–20 visits per hour with food, equally split between the sexes. The nestling period is typically 17 days (range 16–25 days). After fledging, the young are fed by the adults for a further 1–2 weeks. There is no positive information on the average number of broods raised in a season by each pair of birds. Rowan (quoted by Siegfried 1973) found that in urban areas the same nest was normally used more than twice in the one season; earlier, Rowan (1966) suggested that some urban pairs might raise as many as five broods in one season. In contrast, Siegfried (1973) in his rural study found that only a small number of nests (6) had successive broods. By implication each pair in the arid acacia country had only one brood. In the Kalahari, Immelmann stated that the birds always built a new nest for each clutch of eggs. As in neither Rowan's nor Siegfried's studies were the birds marked, there was no way of knowing whether pairs continued to use the same nest for subsequent broods or whether they were appropriated by other pairs, nor even if individual pairs used more than one nest. The general conclusion is that most pairs of birds living in cultivated and built-up areas raise at least two broods and a minority may raise three; in the dry areas where there is opportunistic breeding it is unlikely that conditions remain suitable for long enough for more than one brood to be attempted.

Siegfried's study gave an overall breeding success of 25% (638 eggs hatched 410 chicks, of which 160 survived to fledging). This is much lower than the 50% found by Rowan (1966) for mainly urban nesting birds and 52–53% for urban and rural birds, and 69% for suburban birds by Earlé (in press). Siegfried's low figure, mainly the result of the death of nestlings by starvation, suggests that the birds are not yet well adapted to the agricultural area that presumably has been exploited only comparatively recently by birds from urban areas. Probably the opportunistic breeding in the bird's original habitat is more successful, though there are no reported studies from such localities. In the urban areas, where the birds had a similar seasonal breeding cycle to Siegfried's rural population, it can only be assumed that the food supply was more dependable. Dean (1978), in a semi-arid cultivated area in southwest Transvaal, estimated an average annual production of 3.5 young per pair.

In the Transvaal the Didric Cuckoo *Chrysoccyx caprius* is a frequent brood parasite of the Cape Sparrow, laying in possibly 10–20% of nests.

SURVIVAL

Dean (1978) has summarised a number of studies on the life expectancy and mortality of the Cape Sparrow based on recapture of ringed individuals. These gave life expectancies ranging from 1.09–2.20 years for adult birds. Life expectancy was about 30% greater in males than in females. The longest lived birds at Barberspan in southwest Transvaal were a male recaptured after 6 years and a female after 4 years.

MOULT

According to Dean (1977) the moult at Barberspan starts in October and continues through into August of the following year, with the period for any one individual lasting an average of 150 days (range 90–225 days for males, 134–216 days for females). Thus the moult coincides with the main breeding season and, in some individual females, moult was found to have begun during incubation. Dean suggests that the protracted moult may have evolved in the Cape Sparrow to compensate for the extended breeding season at Barberspan, thus spreading the energy demand. On the other hand, there was no significant loss of weight during moult, suggesting that it did not place any great physiological stress on the birds.

Juveniles, in common with those of other *Passer* species, have a complete post-juvenile moult; this extends over a similar period to that of the adults. Juvenile males adopt partial adult plumage in a period of 16–167 days after leaving the nest, that is within the breeding season of hatching (Rowan 1964).

VOICE

Prozesky (1970) gives the normal call as *chissip* or *chirrup*, McLean (1984) as *chreep*, *chirreep* or *chirrichreep*. I have transcribed the nest ownership or advertisement call of the male as a distinctive, loud and far-carrying, *tweeng* or *twilleeng*, that at times goes over to a jerky, repetitive song, with variants of the above call notes; McLean represents the song as *chip cheerup, chip cheerup* or *chreep chroop, cheep chroop* or *chip chollop tlip tlop*. The timbre is much more mellow and musical than the familiar chirping of the male House Sparrow.

FOOD

The Cape Sparrow is mainly a seed eater, primarily specialising on the large seeds of grasses, other small plants (the introduced 'Khaki weed' seemed a favourite in the Transvaal), wheat and similar cultivated grains. It also takes the soft shoots of plants, including the buds of fruit trees and thus can cause damage in orchards and gardens. Some soft fruit is also taken and, as already described, in recent years (*ca* 1956) the bird has moved into the wine-growing areas of the Cape and is attacking the grapes and causing considerable damage (estimated to amount to 1 million Rand in 1963). Grapes seem, however, to be a second preference to seeds, as it is only the varieties ripening in December, when weed seeds are not available, that are attacked. Considerable economic damage is also caused in orchards, pears being particularly mentioned, by their depredations on the buds.

The adults will also take insects, mainly small caterpillars, and these appear to be the principal food brought to the nestlings, as already described in the section on breeding biology, perhaps offsetting to some extent the damage caused to fruit trees in gardens and orchards, and the grapes in the vineyards.

Cape Sparrows have also been recorded probing in aloes for nectar (Oatley & Skead 1972), but this seems to be a casual activity rather than an important source of food.

5: Rufous Sparrow *Passer motitensis*

NOMENCLATURE
Pyrgita motitensis A. Smith, Rep. Exped. Explor. Cent. Afr. 1836: 50.
    Old Lakatoo, 60 miles south of Orange River.
    [*error:* Motita, near Old Lakatoo, 135 miles north of Orange river according to Winterbottom, Ostrich 1966 37:138].
Passer motitensis (A. Smith) 1836.
  Subspecies: *Passer motitensis motitensis* (A. Smith).
    Passer motitensis benguellensis (Lynes) 1926.
      Synonym: *Passer iagoensis benguellensis* Lynes 1926.
        Huxe, Benguella [southern Angola].
    Passer motitensis cordofanicus (Heuglin) 1871.
      Melspez, Kordofan, Sudan.
    Synonym: *Passer cordofanicus* Heuglin 1871.
    Passer motitensis insularis (Sclater & Hartlaub) 1881.
      Socotra Island.
    Synonyms: *Passer insularis* Sclater & Hartlaub 1881.
      Passer hemileucus Ogilvie Grant & Forbes 1900.
      Abd el Kuri Island.
    Passer motitensis rufocinctus (Finsch & Reichenow) 1884.
      Lake Naivasha, Kenya.
    Synonym: *Passer rufocinctus* Finsch & Reichenow 1884.
    Passer motitensis shelleyi (Sharpe) 1891.
      Lado, southern Sudan.
    Synonym: *Passer shelleyi* Sharpe 1891.
    Passer motitensis subsolanus Clancey 1964
        Ingwezi Ranch, Syringa, Matabeleland, Southern Rhodesia [Zimbabwe].

According to Clancey (1964a) this last subspecies was first recognised by Dr G. Rudebeck in 1956, but he did not publish his findings. I have not been able to examine specimens and have thus had to rely entirely on the information given by Clancey.

The relationships between these sparrows and the closely similar rufous sparrows of the Cape Verde Islands, together with their nomenclature, is rather confusing. They form six distinct allopatric populations, as shown in Fig. 23, that were originally considered to be six distinct species. More recent opinion, however, tends to treat them as one species, including the bird of the Cape Verde Islands. The latter was collected by Darwin in 1832 on the voyage of the *Beagle* (Darwin 1841) and named by Gould (Gould 1837) as *Pyrgita iagoensis*. As Dr Andrew Smith's original 1836 description of the rufous sparrows from South Africa was overlooked, and attributed to a later publication in 1848, the name *Passer* (= *Pyrgita*) *iagoensis* took priority for this species. This error was corrected by Clancey (1964b, 1965) and *motitensis* restored as the specific name. Meantime Macdonald (1957) had identified the type locality, which is not completely clear from Smith's diaries, as Hopetown, northern Cape Province. However, as there are no records of this species from south of the Orange River, Clancey considered the original locality given by Smith an error and put forward Kuruman, north Cape Province for the type locality. Winterbottom (1966) suggested it should be Motita, near Old Lakatoo; this seems consistent with the specific name given to the bird by Smith.

I consider the Cape Verde Islands birds to be specifically distinct from the other rufous sparrows (Summers-Smith 1984a). This species, for which the name *iagoensis* is available, will be dealt with in the next chapter, where I shall enlarge on the reasons for treating it as a full species.

The birds to be covered in this chapter fall into five discrete populations; these are widely scattered over Africa and, as can be seen from Fig. 23, are allopatric*,

---

*The populations of Socotra and Abd-al-Kuri Islands are allopatric, *sensu stricto*, but are considered to belong to the same race.

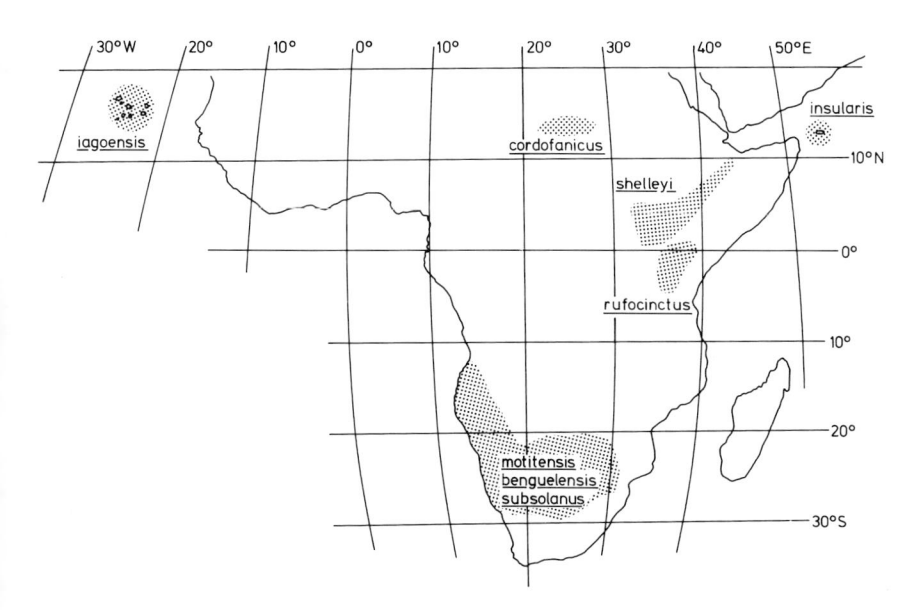

Fig. 23 Distribution of African rufous sparrows

although two (*shelleyi* and *rufocinctus*) are probably better described as parapatric (*ie* contiguous without intergradation), with the southern African one split into three subspecies, *motitensis*, *benguellensis* and *subsolanus* that intergrade with one another. The birds of all these populations are closely similar in appearance. I shall give first a complete description for the nominate race and then show how the others differ from it.

DESCRIPTION

Passer motitensis motitensis
Birds of the southern African population have the name Great Sparrow. With an overall length of 150–160 mm, they are similar in size to the sympatric Southern Grey-headed Sparrow, although much heavier and justify their appellation on this account. They are also the largest of the five populations of Rufous Sparrows now being considered. The birds belong to 'black-bibbed' sparrows, with sexes sesquimorphic.

Male. Crown grey, with grey extending to nape and fanning out to front; grey is bordered by chestnut, which begins behind eye and circles cheek, almost reaching black bib on throat; sides of forehead white. Black from bill to eye and extending backwards a short distance as narrow black line; cheeks white. Back and rump chestnut, upper back boldly streaked with black. Small black bib, fanning out on breast. Shoulders chestnut, bordered at scapulars with row of black spots and narrow white wing-bar; flight feathers streaked black and grey. Tail dark brown. Underparts white. Bill thick, slightly arched, black in breeding season, otherwise horn.

Female. Similar basic pattern to male, but lacks chestnut on head, this replaced by pale, buffy cream, starting as broad supercilium and curving down behind cheek, bordered below by darker grey line through eye and outlining cheek. Grey on crown paler than male, fans out over nape in same pattern as male. Bib is grey and indistinct. Bill dark horn to almost black.

Table 18: Comparison of key features of the subspecies of Rufous Sparrows

Subspecies	*motitensis*	*benguellensis*	*subsolanus*	*rufocinctus*	*shelleyi*	*cordofanicus*	*insularis*
Males							
ear coverts	pale ashy grey	white	darker than *motitensis*	grey	white	white	white
black from eye to ear coverts	trace			absent	present	present	present
rump	chestnut		chestnut darker than *motitensis*	chestnut	chestnut	chestnut	grey
back	no contrast with rump		no contrast with rump	no contrast with rump	darker than rump	no contrast with rump	no contrast with rump
black streaks on mantle			heavy and dark		heavy, broad	well defined	
underparts	white, darker on flanks		grey, slightly streaked	pale grey, darker on flanks	whitish, grey on flanks	white	ashy grey whitish on breast
Females							
bib	pale grey, indistinct			dark grey, distinct	dark grey, distinct	dusky grey, distinct	absent

Table 19: Biometric data for Rufous Sparrow

Feature	Subspecies	Locality	Males Range	Males Mean (Median)	Females Range	Females Mean (Median)
weight	motitensis	—	34.0–35.8	34.7	30.6–32.0	31.5
wing	benguellensis	—	77–87	(82)	81–82[a]	81.5[a]
	motitensis	—	82–88	84	79–85	81
	subsolanus	—		89.5[b]		
	rufocinctus	—	70–85	77.5	73–81	76
	shelleyi	—	70–78	(74)	68–73	70
	cordofanicus	—	74–81	(77.5)	69–76	72.5
	insularis	Socotra	72–80	(78)		
		Abd-al-Kuri	69–73	(71)		
tail	benguellensis	—	54–58	(56)	55–56[a]	(55.5)[a]
	motitensis	—	56–67[c]	61.5[c]		
	subsolanus	—		59[b]		
	rufocinctus	—		ca 51		
	insularis	—		ca 58		
tarsus	motitensis	—	18–21	(19.5)[b]		
	subsolanus	—		22[b]		
	rufocinctus	—		ca 19		
	insularis	—	19–22	(20.5)		
culmen	benguellensis	—		ca 16[c]		
	motitensis	—	13–15[c]	(14)[c]		
	subsolanus	—		17[b]		
	insularis	—		(15)[c]		

[a] very small sample   [b] type specimen   [c] not sexed

Immature. Young birds resemble adults, but paler with more washed out appearance.

The key differences between the seven subspecies are highlighted in Table 18. For more detailed discussion it is convenient to treat the five allopatric populations separately, considering the nominate race and its contiguous subspecies together. There is more information about the birds of this population and it provides a suitable starting point. It is probable that these three subspecies are more closely related to each other than to any of the others; by treating the populations in this way it makes it easier to examine their relationships.

BIOMETRICS

Typical biometric data for the seven races of Rufous Sparrow are summarised in Table 19 so that they are readily available for comparison with the other sparrow species. There appears to be a systematic trend in size (wing length) with the smallest birds close to the equator and the largest ones in the southern African population, comprising nominate *motitensis* and the parapatric races *benguellensis* and *subsolanus*.

Great Sparrow *Passer motitensis motitensis/benguellensis/subsolanus*

The Great Sparrow with an average weight of 35 g is, with the possible exception of the Parrot-billed Sparrow for which no weight data are available, the heaviest of the sparrows. As with the Cape Sparrow there is a cline in wing length from west to east, with the largest birds in Natal.

DISTRIBUTION

The Great Sparrow, which as we have seen is separated into three races, occurs in coastal Angola as far north as Benguela, through most of Namibia, Botswana except for the extreme north, the southwest corner of Zimbabwe, through the Transvaal to Swaziland (but not in the Kruger National Park in the east). In the south it occurs in Cape Province, north of the Orange River, and the north of the Orange Free State.

It is nowhere a common bird and may be decreasing. For example, Farkas (1966) noted a slight, but progressive, decrease near Barberspan, Transvaal, and nine years later Milstein (1975) reported that it had declined drastically in that area. In contrast, Traylor (1963) described it as abundant in the coastal plain of Angola, and Macdonald (1957) found it fairly common in parts of Namibia.

The distribution of the Great Sparrow is shown in Fig. 24. It has not been possible to indicate the ranges of the three races of the southern African Great Sparrow as

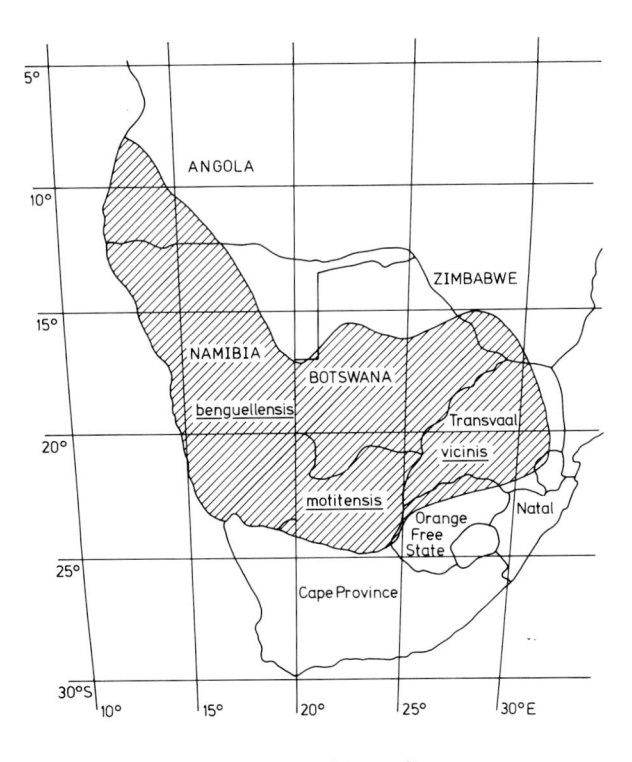

Fig. 24   Range of Great Sparrow

these have not been clearly defined. In general terms, *benguellensis* occurs in the west in Angola and Namibia, east to about latitude 20°E; *motitensis* in Botswana and the northern Cape Province; *subsolanus* to the east in southwest Zimbabwe (Matabeleland), the Transvaal, northern Orange Free State and Swaziland (Clancey 1964a).

HABITAT

The Great Sparrow is a bird of dry acacia savanna, frequenting trees; it is usually to be found in the wilder parts and, although it may occur near habitations, it is never to be found in built-up areas. It occupies arid country where the average rainfall is about 550 mm (range 360–780 mm), though in its range, where rainfall is capricious, averages based on records over a number of years tend to be misleading and in some years the rainfall can be negligible.

BEHAVIOUR

The Great Sparrow is a solitary species, thinly distributed over what appears to be suitable country. For example, in two areas where I have watched them breeding – a dry acacia valley some 50 km north of Molepolole in Botswana, and the Nyl flood plain near Naboomspruit in central Transvaal – pairs were no closer than about a kilometre and had a density of no more than one pair to 100–500 ha in apparently suitable habitat. Most authorities describe it as a shy species, though Ogilvie-Grant (1912) remarks that it is a rather conspicuous bird, frequently to be seen perching on top of a bush uttering a loud call. This certainly agrees with my experience of it in the breeding season. At the nest I found it even to be confiding, allowing a close enough approach for photographs to be taken, and when I drove a car up to within a few metres of one nest, in order to make close observations, the male worked his way down the tree from branch to branch to look into the car and then the pair carried on apparently unconcerned by my presence.

It is clearly a bird that has been little watched as nothing has been recorded about its displays. At the nests I watched, it did not react to other species, such as Longtailed Shrikes *Corvinella melanoleuca* perching in the nest tree, or even aggressively towards White-browed Sparrow Weavers *Plocepasser mahali* and Quail Finches *Ortygospiza fuscocrissa* stealing feathers from the nest entrance, though the male chased them when they flew off.

Again, nothing has been recorded about what happens to the young after fledging; whether they remain in family parties or whether small flocks form outside the breeding season is not clear. Macdonald (1957) mentions small flocks in Namibia in April and Tree (1972) found 'immensely large concentrations' there in May and June 1970. These are the only records of flocks that I can find and from the information available it would appear to be one of the least gregarious of all the sparrow species.

In Botswana, according to Beesley and Irving (1976), the bird disappears at the end of the breeding season (December) and is rarely seen from then until May. Seasonal wandering in Botswana, for the eastern part of the country at least, is confirmed by the more recent records of the Botswana Bird Club (1975–82, per Mrs J. Barnes): ten of the 13 records fall in the period October–January, and the other three in June–July. The paucity of records again emphasises the relative scarcity of the species in that area.

BREEDING BIOLOGY

The Great Sparrow builds a free-standing nest at heights of 1.5–4 m in the branches of a bush or thorn tree. The nest is an untidy, elongated bundle, about rugby football size, domed and with the entrance near the top at one end. It is built mainly of dried grass, but other materials, such as green leaves and pieces of string, may be worked into the structure. The nest cup is lined with feathers and plant down. Building is by both sexes.

At nests in Botswana and the Transvaal that I watched for about ten hours, both sexes took part in covering the eggs, though the female spent about twice as much time as the male in this activity. The birds did remarkably long spells on the eggs for small song birds, with maxima of 32 minutes for the male and 44 minutes for the female.

The clutch size ranges from 3–6 eggs, with 5–6 typical in Namibia and 4 in other parts of the range. Both sexes feed the young in the nest. At one nest that I watched, feeding was shared almost equally; the low feeding rate of only seven visits per hour suggested that the parents might have been having difficulty in finding food.

Information on the breeding season is summarised in Fig. 25; breeding takes place

Locality	Month												Reference
	J	A	S	O	N	D	J	F	M	A	M	J	
Angola													Traylor 1963
Namibia													Mackworth-Praed & Grant 1963
Botswana & Zimbabwe													Smithers et al. 1957, Mackworth-Praed & Grant 1963, Beesley & Irving 1976, Irwin 1981, McLean 1984
South Africa													Tarboton 1968, McLachlan & Liversidge 1978

Fig. 25 Breeding season data for southern African populations of Rufous Sparrow (breeding ——— rains - - - -)

in the austral summer and coincides with the rains, such as these are for the arid region in which the bird occurs.

MOULT

Little has been published on the moult of this bird. Macdonald (1957) stated that the first sign of post-breeding moult in the birds he collected in Namibia was in a male obtained in late April.

VOICE

The main call is a twangy chirrup, *churr-chirrup*, rather deeper in tone than the call of the House Sparrow; Mackworth-Praed & Grant (1963) describe this call as a rather deep *chissick*. The pair at the nest greet each other with a variety of soft conversational notes that I have variously transcribed as *chee-ti-cheet, ti-cheet-it, ti-chee-tit-tit* and *chee-wee*.

FOOD

No proper study has been made of food taken by the Great Sparrow. Most authors give it as grain and weed seeds. At the nests I watched, the young were being fed on small caterpillars.

Kenya Rufous Sparrow *Passer motitensis rufocinctus*

DISTRIBUTION

The Kenya Rufous Sparrow has a restricted range in the Rift valley highlands at altitudes of 1,000–3,000 m (mainly 1,400–2,200 m) from central Kenya to the extreme north of Tanzania, as shown in Fig. 26. It is a local resident, moderately common in certain localities, but by no means plentiful.

HABITAT

The Kenyan race of the Rufous Sparrow is found mainly in acacia savanna, but also occurs in more open country than the Great Sparrow, such as grazed land and cultivation. Unlike the latter, it also occurs around human habitations and even comes into towns. Sharpe (1891) described it as plentiful in the vicinity of kraals and I have seen it in Nakuru town, though I find it difficult to agree with Archer and Godman (1961) who state that 'in Naivasha ... it is entirely domesticated and closely resembles the European House Sparrow in its habits'.

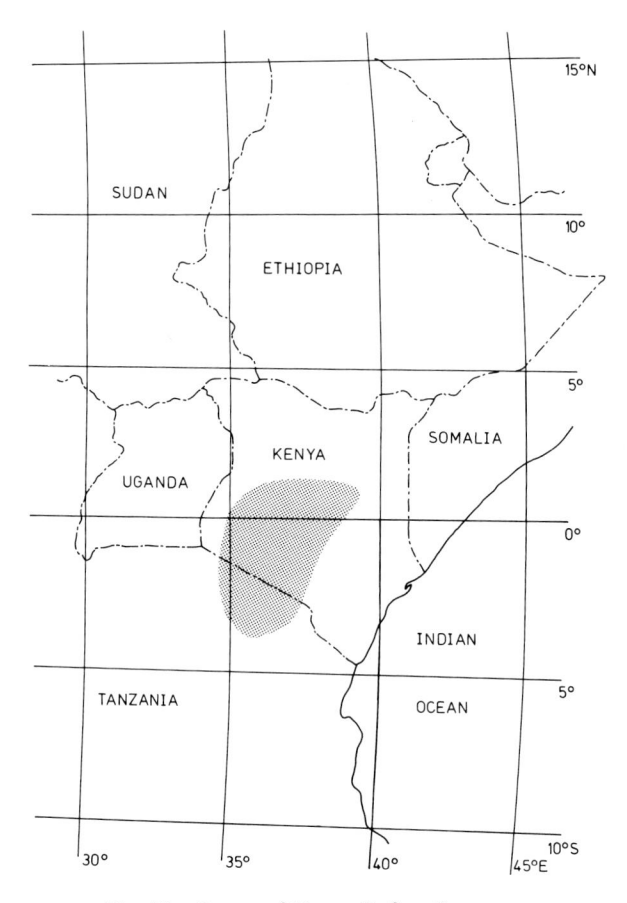

Fig. 26 Range of Kenya Rufous Sparrow

BEHAVIOUR

This subspecies of the Rufous Sparrow is mainly a solitary bird, never more than one pair breeding in the one place, and is usually seen in pairs throughout the year, though outside the breeding season it can be found in small flocks numbering up to about ten individuals. The birds that occur in built-up areas are very confiding and allow close approach without taking flight. It feeds on the ground more frequently than the Great Sparrow.

BREEDING BIOLOGY

The nest is usually built in the top branches of small thorn trees from about a metre upwards; less commonly it is placed in covered sites, such as under the eaves of a house, in holes in trees or in deserted nests of weavers. I saw one nest about three metres from the ground in a cavity in a concrete electricity pylon in a busy street in Nakuru; this cavity was covered by a creeper.

The free-standing nest is loosely built; it is wedged among the branches of the tree and is not particularly hidden. It is an ovoid in shape, according to Collias and Collias (1964) about 200 mm high and 250 mm long, domed over with the entrance on the side or near the top. It is composed mainly of dry grass and plant stems, the nest cup lined with feathers and animal hair.

The most complete data on the breeding season are given by Brown and Britton (1980). Fig. 27 shows that breeding has been recorded in Kenya in every month except for August; it can be seen, however, from the numbers of nests recorded in each month which are included in the figure, that there are clear peaks in April–July and November–December, coinciding with the 'long' and 'short' rains respectively. The clutch size is 3–5 or 6 eggs; no information has been published on the part played by the sexes in the breeding activities, nor on the lengths of the incubation and nestling periods. Betts (1966) describes a nest that suggests that two clutches may be laid; this nest contained half-grown young on 12th January; on 25th January it had been relined and contained fresh eggs.

VOICE

The call is described as a typical sparrow chirp. In my opinion it is deeper in tone than that of the House Sparrow and I have transcribed it as *chwupp* or *chwuppup*. The male is said to have a short song, presumably formed by stringing together various chirping notes.

FOOD

Nothing beyond the bald statement 'seeds and insects' has been published about the food taken by this bird. I have watched it come down to feed on crumbs at one of the Kenyan Game Lodges.

Month												
J	F	M	A	M	J	J	A	S	O	N	D
1	2	3	11	13	6	4	0	1	1	3	3

Fig. 27 Breeding season data for Kenya Rufous Sparrow (after Brown & Britton 1980). (*The figures show number of nests found.*) (breeding ——— rains ----)

White Nile Rufous Sparrow *Passer motitensis shelleyi*

The White Nile Rufous Sparrow occurs immediately to the north of the Kenya Rufous Sparrow, though the two populations appear to be distinct, without any intergradation between them. It was first collected at Lado in southern Sudan by Emin Pasha in 1884.

DISTRIBUTION

The White Nile Rufous Sparrow occurs from the Karamoja district (Moroto) of eastern Uganda and Equatoria Province in the extreme south of the Sudan (south of about 6°N), through northwestern Kenya and in a strip through Ethiopia, north of the highland plateau, in arid plains at 1,000–1,500 m, just penetrating across the border into northern Somaliland. It is local in Uganda, common in Ethiopia, but again rather uncommon in Somalia. The distribution is shown in Fig. 28.

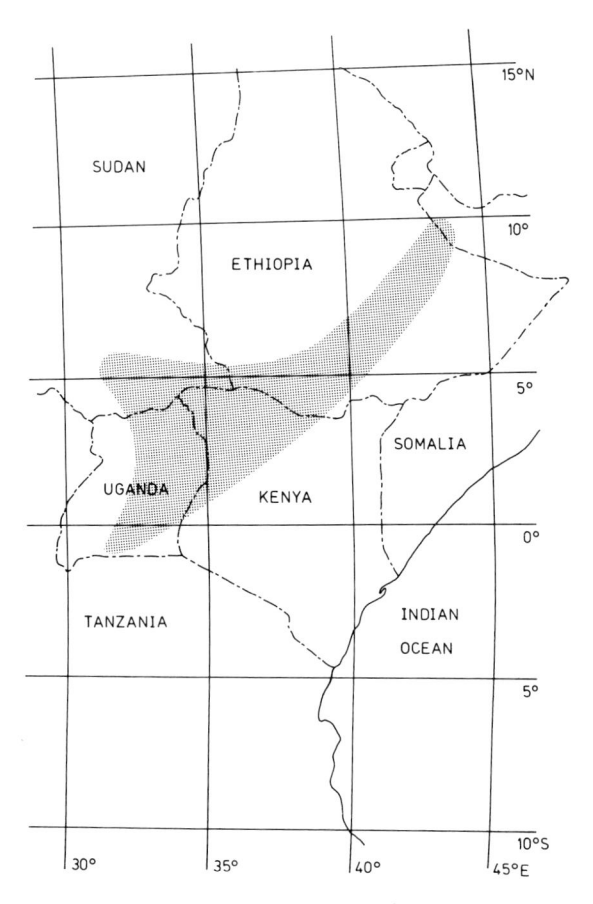

Fig. 28   Range of White Nile Rufous Sparrow

HABITAT
This race of the Rufous Sparrow is mainly found in open, grassy savanna with scattered trees, but extends into semi-desert on the one hand, and light woodland on the other; it also occurs in cultivated land and near human habitations. Unlike the Kenya Rufous Sparrow it does not occur in built-up areas and does not associate with man.

BEHAVIOUR
Practically nothing has been reported about its behaviour except that, like the Kenyan race, it regularly feeds on the ground.

BREEDING BIOLOGY
The nest is placed openly, at 1–2 m, in the branches of small, thorny acacias. It is similar to that of the Kenya Rufous Sparrow, loosely built of grasses, ovoid in shape (200 × 250 mm), domed over with an entrance in the side. The nest chamber is lined with feathers.

The few published data on the breeding season are published in Fig. 29. The clutch size is reported as 3–5 eggs.

Locality	J	F	M	A	M	J	J	A	S	O	N	D	Reference
Sudan						—							Mackworth-Praed & Grant 1960
Ethiopia									—				Benson 1947, Mackworth-Praed & Grant 1960, Urban & Brown 1971
Somalia			—	—									Mackworth-Praed & Grant 1960, Archer & Godman 1961

Fig. 29 Breeding season data for White Nile Rufous Sparrow

Kordofan Rufous Sparrow *Passer motitensis cordofanicus*

There is a gap of about 1,000 km between the Kordofan Rufous Sparrow and the next nearest population of Rufous Sparrows – the White Nile Rufous Sparrows.

DISTRIBUTION

The Kordofan Rufous Sparrow is a common resident in central Darfur Province of the Sudan, between 12°45′N and 14°45′N, extending westwards into the extreme east of Chad, and eastwards into the Kordofan Province of the Sudan, where it is said to be rather uncommon. The distribution is shown in Fig. 30.

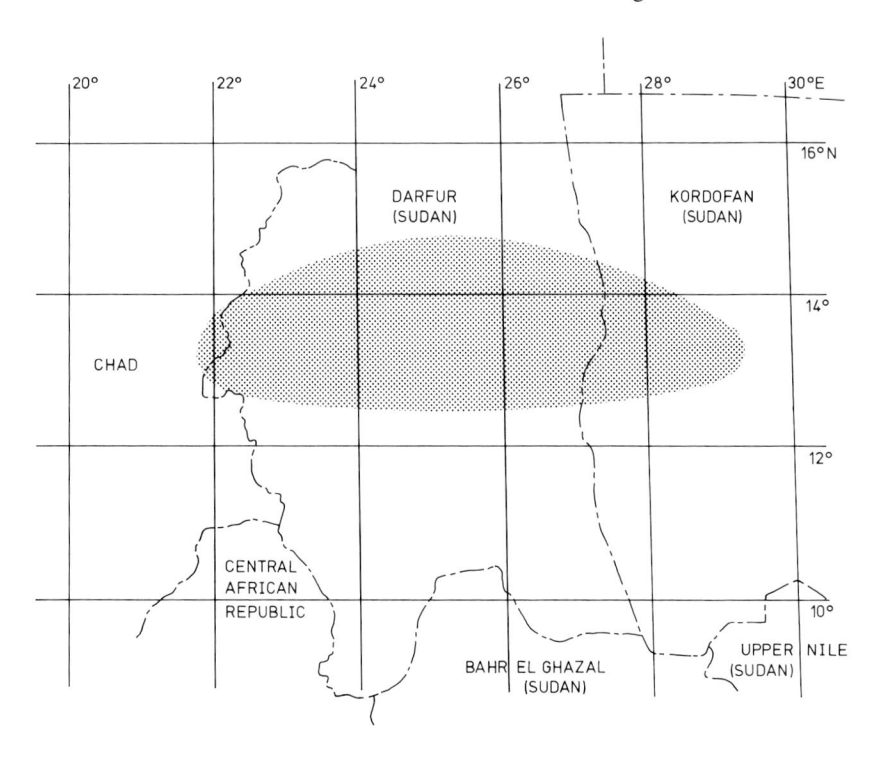

Fig. 30 Range of Kordofan Rufous Sparrow

HABITAT

The Kordofan Rufous Sparrow lives in a very arid area where the average rainfall is only about 300–400 mm. It stays around villages and the surrounding cultivated land. There is some wandering in the non-breeding season, though it seldom penetrates far into the open bush country (Lynes 1924).

BEHAVIOUR

Little is known about its habits, though it is said to behave as a typical 'house sparrow', hanging around villages. This is somewhat surprising as its range overlaps with that of the Grey-headed Sparrow, which is very much at home in built-up areas, whereas the other races of the Rufous Sparrow largely avoid man's habitations.

BREEDING BIOLOGY

The nest is built in the branches of a thorny bush or tree, at heights of 2–3 m. According to most authorities it closely resembles that of the other Rufous Sparrows, made of hay, domed and with an entrance in the side. Lynes (1924) alone described a foot-long entrance-tunnel extension. The nest cup is lined with soft grass and feathers. Published data on the breeding season are summarised in Fig. 31. Lynes (1924) found many nests were built in the Sudan in June, but these were abandoned and breeding did not begin until September.

Locality	Month												Reference
	J	F	M	A	M	J	J	A	S	O	N	D	
Chad						⊢−−⊣							Mackworth-Praed & Grant 1960
Sudan									⊢−⊣				Lynes 1924, Mackworth-Praed & Grant 1960

Fig. 31 Breeding season data for Kordofan Rufous Sparrow

Socotra Sparrow *Passer motitensis insularis*

DISTRIBUTION

A separate race of Rufous Sparrows occurs on Socotra Island, lying about 240 km off the Horn of Africa. The birds from 'Adb-al-Kūrī Island, lying about half way between Socotra and the African coast, were first described as a separate species from those on Socotra, but although paler and slightly smaller the differences are not great enough to warrant their separation from the birds on Socotra, even as a different race.

GENERAL

Socotra is an arid island with an average rainfall of only 150 mm. There are a number of grazing animals, but no cereal crops. Apart from the original description of these birds, little is known about their habitat preference, behaviour or breeding. The latter is said to occur in December.

GENERAL COMMENTS

The Rufous Sparrows stand out from all other members of the genus by the absence of social behaviour. Not only are they solitary breeders, they seem to form little more than family flocks outside the breeding season. It is an arid-loving species and its distribution, as shown in Fig. 23, follows closely the African dry woodland, steppe zone (see Fig. 15, Chapter 2), the one obvious exception being the absence of the bird in the western Sahel.

There are some striking differences between the different populations: birds of the southern *motitensis*-group (*motitensis*, *benguellensis* and *subsolanus*) are significantly larger than those of the other populations (in line with Bergmann's ecogeographical rule); the island population, *insularis*, is distinctive in that it is the only one that lacks the chestnut rump and any sign of a dark bib in the female; *rufocinctus* is distinct in being the only race that lacks the black line from the eye to the ear-coverts. There are, however, no obvious geographical clines in plumage. On the other hand, it is perhaps no coincidence that the greatest difference between any two of the races is that between *rufocinctus* and *shelleyi*, the two parapatric populations.

There would appear to be some ecological differences between the populations, though insufficient information is available as to whether any of these populations could be recognised as specifically distinct. The one obvious difference is that the most typical sparrow characteristic, the tendency to associate with man, appears to be completely absent in the southern population, thought it is present to a minor extent in the equatorial ones.

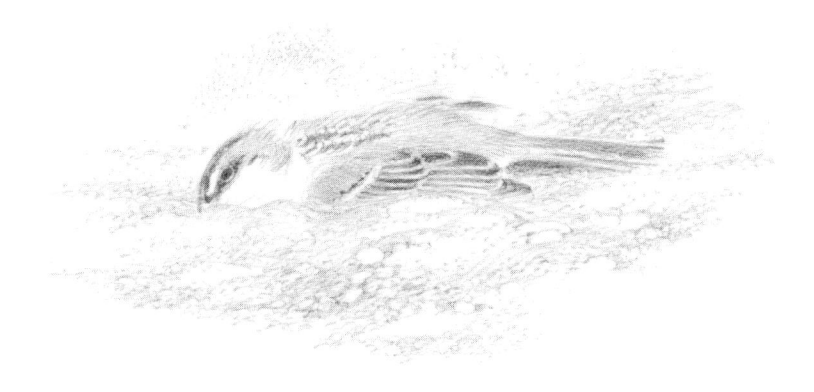

6: Iago Sparrow *Passer iagoensis*

NOMENCLATURE
Pyrgita iagoensis Gould, Proc. Zool. Soc. 1837: 77.
    São Tiago, Cape Verde Islands.
Passer iagoensis (Gould) 1837.
    Synonyms: *Passer brancoensis* Oustalet 1883.
        Branco, Cape Verde Islands.
    Passer erthrophrys Temminck.
        Senegal [*error* = Cape Verde Islands].

The Iago Sparrow was first collected by Darwin on São Tiago Island in the Cape Verdes in 1832 during the voyage of the *Beagle* (Darwin 1841). Sharpe (1888), van Someren (1922), Lynes (1924), Bannerman (1948), Macdonald (1957) and Mackworth-Praed and Grant (1963) all considered *iagoensis* to be specifically distinct from the Rufous Sparrow *Passer motitensis,* though it was merged with the latter by Moreau (1962) and again in *An Atlas of Speciation in African Passerine Birds* (Hall & Moreau 1970).

The Iago Sparrow is very similar in appearance to the African Rufous Sparrows and is obviously closely related. It is separated physically by about 5,000 km from both of the nearest populations of the Rufous Sparrow, those of *Passer motitensis cordofanicus* in Chad and *Passer motitensis benguellensis* in Angola; it differs in a number of characteristics that I consider are sufficient for specific separation (Summers-Smith 1984a). For non-migratory species, wing length can be taken as an indication of size. The male wing length data plotted in Fig. 32 show that there is no overlap with any of the races of *motitensis*. Size within a species is influenced by climate. For example, Bergmann's rule states that body size tends to be larger in the cooler parts of the range. The male wing length ranges for *iagoensis* and *motitensis* are plotted in Fig. 33 against the range of mean monthly temperature for the areas in which the different populations occur. It will be seen that, again, *iagoensis* does not fit the relationship for the four races of *motitensis* for which the necessary data are available.

Although the Iago Sparrow closely resembles the Rufous Sparrow, it has a number of features that distinguish it. Head of male is blackish grey, merging into dark grey on nape; whereas in all races of Rufous Sparrow head is mid grey with no contrast with nape. Black bib of male in *iagoensis* very small and no variation in breadth, whereas in the others it is much more conspicuous and fans out over breast. *Iagoensis* shows very clear sexual dimorphism, with female lacking any hint of chestnut and any sign of dark bib; in *motitensis,* on the other hand, the sexual difference in plumage is much less obvious, with female merely duller version of male, showing grey bib and dull chestnut on back, rump and scapulars. In general, *iagoensis* is a much brighter looking bird, contrary to the usual situation that island representatives tend to be duller than their mainland relatives, as is the case with the island race of Rufous Sparrow *P. m. insularis.*

Rufous and the Iago Sparrows both occupy an arid habitat, but whereas the Rufous Sparrow is dependent on trees in which to build its nest in the branches*, the Iago Sparrow is quite at home in completely treeless areas and predominantly nests in holes.

Finally, and most importantly, there is a great difference in social behaviour. Rufous Sparrows are the least social of all the *Passer* sparrows, nesting solitarily and outside the breeding season forming at most very small flocks, possibly little more than family parties; the Iago Sparrow, on the other hand, usually nests in small colonies of up to about ten pairs and forms large flocks outside the breeding season.

These differences suggest that the Iago Sparrow has been separated from the African mainland Rufous Sparrows longer than the isolated populations of the latter from each other. The specific separation of allopatric populations must inevitably be a matter for personal opinion, but while none of the above differences by itself would be enough to warrant separation, it is considered that, taken together, this is justified and that evolution has proceeded far enough in the Cape Verde Island birds to give reproductive isolation, should they ever come into contact again with Rufous Sparrows.

There seems to be no justification for the separation of the birds from Branco Island by Oustalet (1883) on the basis of one specimen that was not even sexed with certainty.

I use the vernacular name, Iago Sparrow, for this species in preference to the alternatives: Cape Verde Island or Rufous-backed Sparrow.

DESCRIPTION

Iago or Cape Verde Rufous Sparrow closely similar to Rufous Sparrows found on mainland Africa, but significantly smaller (length 125–130 mm). One of the 'black-bibbed' sparrows; sexes dimorphic.

Male. Crown dark grey, almost black, fading to paler grey on nape and upper back; cinnamon supercilium, broadening round back of cheek and spreading on sides of neck. Small white patch between eye and bill. Cheeks pale grey, separated from cinnamon at back by narrow black line. Back and rump chestnut, with bold black streaks on upper back. Chestnut shoulder patch ending in broad white wing bar, remainder of wing streaked brown and buff. Small, narrow, straight-sided bib; underparts silvery grey. Tail brown. Bill thinner and more pointed than in any of race of *Passer motitensis,* black in breeding season, otherwise horn. (Bills of birds in the British Museum are horn for birds collected March–June and black for October–December; no specimens available for other months.)

Female. Crown grey brown, becoming brown on nape. Creamy supercilium extending back from eye, corresponding to cinnamon of male, very conspicuous. Wings and tail light brown, the former with white bar. Underparts pale silvery grey with no sign of dark bib. Bill horn.

* *P. motitensis rufocinctus* does make use of tree holes and buildings for nesting, but, as we saw in Chapter 5, only to a very limited extent.

Juvenile/Immature. Some young birds that I saw being fed by adults had cinnamon supercilia, whereas in others it was pale cream. Female-plumaged birds in the British Museum collection with cinnamon supercilium also showed dark bib (all collected December–June). It seems likely that cinnamon supercilium is a plumage character of young male.

It will be seen that the male differs from the Rufous Sparrows principally in having a much darker crown and a smaller black bib that does not fan out on the breast. The female differs by the absence of any sign of a bib. The bill of both sexes is finer than that of the Rufous Sparrow.

Table 20: Biometric data for Iago Sparrow

Feature	Males		Females	
	Range	Mean (Median)	Range	Mean (Median)
wing	**57–69**	64.6	55–61	(58)
tail	48–58	(53)	43–52	(47.5)
tarsus	17–21	(18.7)	18–19	(18.5)
culmen	**12.5–16**	**14.7**	12–14	(13)

BIOMETRICS

Typical data on body measurements are given in Table 20. These show the Iago Sparrow to be significantly smaller than any of the races of Rufous Sparrow. This has already been demonstrated in Fig. 32 for the wing lengths of the males, using only Lynes' (1926) data to minimise the bias arising from comparison of measurements made by different workers; it will be seen that there is no overlap with any of the Rufous Sparrows, which overlap among themselves.

Bourne (1957b) suggested that there were two clinal trends running in different directions across the Cape Verde Archipelago, the birds getting darker from north to south, and smaller outwards from the African coast, but this was not confirmed by Vaurie (1958) from examination of 34 specimens in the American Museum of Natural History.

DISTRIBUTION

The Iago Sparrow is confined to the Cape Verde Island Archipelago; it occurs on

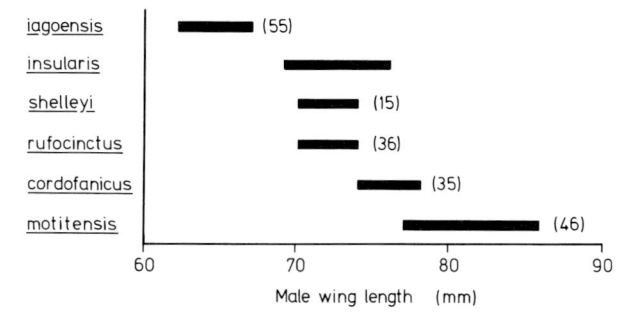

Fig. 32   Wing lengths of male rufous sparrows. Numbers of specimens measured in parenthesis; no date for insularis *is given by Lynes.*

96 *Iago Sparrow*

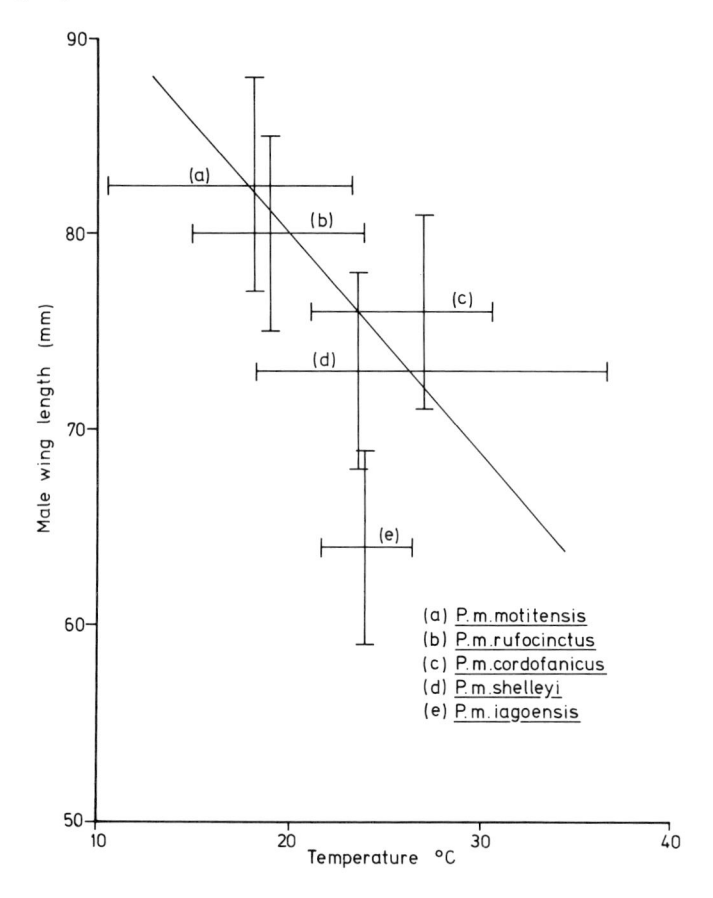

Fig. 33 Male wing lengths for Iago Sparrow and four races of Rufous Sparrow as a function of mean monthly temperature

(a) P.m.motitensis
(b) P.m.rufocinctus
(c) P.m.cordofanicus
(d) P.m.shelleyi
(e) P.m.iagoensis

all nine inhabited islands, with the exception of Fogo, and, in addition, on the uninhabited Desertas (Santa Luzia, Branco and Raso) and Rhombos Islets. Fig. 34 shows the first recorded visits by ornithologists and the first records for the Iago Sparrow. In most cases, as would be expected, they coincide, but those for Sal and Maio do not; Alexander (1898a), who recorded the species from the other islands he visited, made no mention of Iago Sparrows on these islands. There is contradictory evidence concerning the occurrence of the Iago Sparrow on Fogo. It was collected neither by Fea in 1898 (Salvatori 1899), who spent two months there, nor by the *Blossom* expedition, members of which apparently visited the island on two occasions in 1924 (Bannerman & Bannerman 1968); the latter seems to have been particularly assiduous and collected large numbers of Iago and Willow Sparrows on all the islands visited. Bourne (1955) did not visit Fogo, but listed Iago Sparrow as 'reported by a reliable observer.' I searched for it on my visit in 1983, but found only Willow Sparrows present. Thus it was something of a surprise to find subsequently that Nørrevang and den Hartog (1984), who had been there some 18 months previously, reported it as the commonest sparrow in São Filipe, the principal town. I spent five days on Fogo and, while Willow Sparrows were numerous, found no Iago Sparrows;

this was also the experience of M. A. S. Beaman (pers. comm.) on his visits in March 1985 and 1986, and Hazevoet (1986) in March 1986. It seems inconceivable that the Iago Sparrow could have arrived and subsequently disappeared in such a short time and one cannot help but wonder if this has been a misidentification. In the absence of confirmatory evidence, I am doubtful if the Iago Sparrow has occurred on Fogo since the first ornithological records were made on the island in 1898.

HABITAT

On the Cape Verde Islands the Iago Sparrow occurs in all suitable habitats, from arid lava plains and valleys, coastal cliffs and cultivated land to villages and even urban areas. On São Vicente, where the House Sparrow has been present since 1923 and is still to be found in Mindelo, the only town on the island, and in the surrounding impoverished cultivated land, the Iago Sparrow still occurs in both these habitats, despite the presence of its larger congener. Although there was some overlap in the town, Iago Sparrows seemed dominant in trees in the squares, and House Sparrows in the more closely built-up areas. On São Tiago, where the Willow Sparrow is present, I found the two species largely separated, with the latter alone in Praia, the major town on the island, the larger villages and in the areas of larger trees in the richer cultivated land, such as occur in the San Domingos valley. The Iago Sparrow occurs in the poorer cultivated land with smaller trees and in all the more arid parts.

Although found in all available habitats on the islands, the Iago Sparrow is most

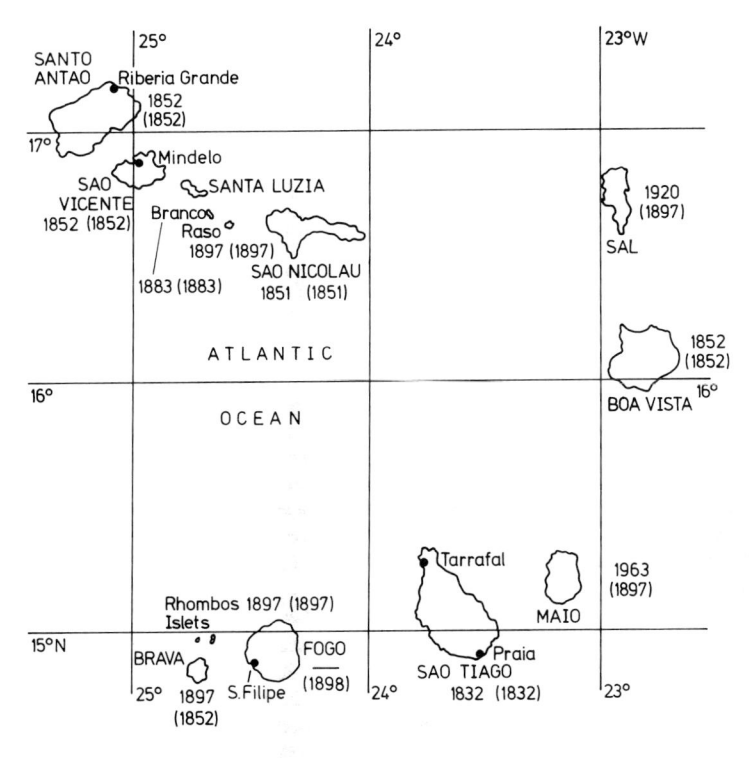

Fig. 34 *First records for Iago Sparrow on different islands of the Cape Verde Archipelago. Date of first visit by an ornithologist in parenthesis*

BEHAVIOUR

The Iago Sparrow is a gregarious species. I found it breeding in small loose colonies, though also occasionally in isolated pairs. Both Bourne (1955) and Bannerman and Bannerman (1968) found the birds in flocks outside the breeding season; Alexander (1898a) even described the flocks as immense. The birds appear equally at home on the ground, hopping about among the stones in the lava flows, on cliff faces and the roofs of buildings, as well as in the branches of trees. On the ground or a cliff face when searching for food, they move around restlessly, looking very much like small mammals. They are not shy and allow very close approach, both out in the open and when they are visiting their nests.

Although adapted to a very arid environment with little or no open water, the birds are very attracted to water, taking it from dripping pipes in irrigated land and they are sufficiently tame to approach closely human beings providing it; Bannerman remarks how, when he approached an unused well, an Iago Sparrow hopped up and eagerly drank the water when it was hauled up in a tin can; de Naurois records how, when workmen building a house threw buckets of water on the site, immediately hundreds of sparrows came from all sides to drink. In the absence of open water the birds can be seen frequently dust bathing in small parties in the dry powdery soil. They have the typical sparrow tail flick, though this is much less exaggerated than that of the House Sparrow.

BREEDING BIOLOGY

Most Iago Sparrows breed in small loose colonies of up to 10 or so pairs, with nests at times only a metre or so apart, though the occasional isolated breeding pair can also be found. The first choice for a nest site is a hole; this can be either in the ground under a boulder, a chink in a stone wall or a crevice in a cliff – coastal and inland lava cliffs on the islands are honeycombed with holes that provide suitable sites – or under the eaves or roof of a man-made building. The sites in the towns and villages are typical of those used by House Sparrows: at least eight pairs were nesting under the tiles on the roof of the hospital in Mindelo when I was there in October 1983. Pairs in Mindelo, on São Vicente, and Porto Novo, on Santo Antão, were using street lights, another very typical House Sparrow site.

Alexander (1898a) reported that on São Tiago 'these sparrows breed in large numbers, many of the acacia trees in the plains being covered with their untidy nest structures of the previous season'. On his second visit in 1897 (Alexander 1898b) he stated that when placed in a tree the nest is domed, but when in a hollow in the ground it is an open compact structure, and often lined with feathers. Bourne (1955) wrote that the Iago Sparrow prefers to nest in crevices or, failing crevices, in trees and, failing trees, on the ground under stones in the open desert. Despite extensive searching on São Tiago and São Vicente during a period when the birds were breeding, I did not find any tree nests; all the nests that I found were in holes. The untidy nests that Alexander reported as covering the trees on São Tiago sound more like those of the Willow Sparrow, a bird that has been present on São Tiago since Darwin's visit in 1832 at least, and one that is well known for its bulky tree nests. Bourne did not actually see Iago Sparrows using tree nests, but he has informed me that he saw a number of scattered open nests in thorns in the bare parts of São Tiago, which were similar to those used by the allied forms (*viz P. motitensis*) living on the African mainland.

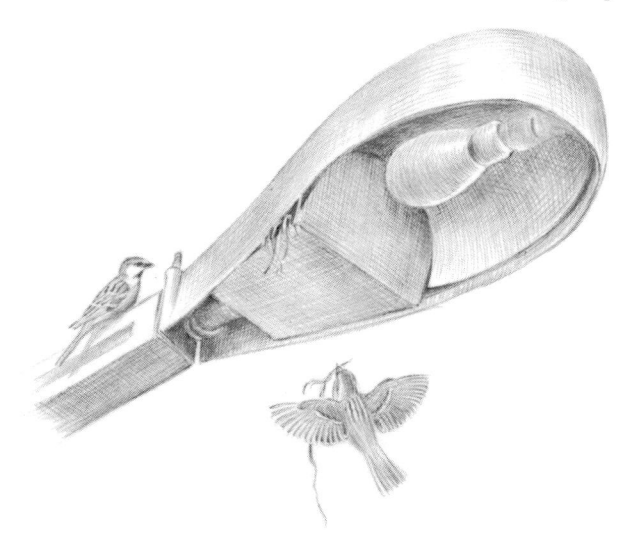

During my visit, Willow Sparrows were seen at nests in trees, but sites in the crowns of palms and in buildings were much more common. The fact that the Willow Sparrow has decreased markedly on São Tiago and the Iago Sparrow has increased since the time of Alexander's visit, provides further circumstantial evidence that the tree nests he described belonged to the former species. The smaller numbers of Willow Sparrows now present appear to be satisfied by the preferred sites in the crowns of palms and holes in buildings and do not nest to any great extent openly in trees. It is my view that the Iago Sparrow breeds almost exclusively in holes.

Bourne (1955) found on Cima, Rhombos Islets, that males had taken up territories in the rocks at the end of August, while females were still in flocks. During my visit in October breeding was well under way, though by no means synchronised, with some males attempting to attract a mate, while others were feeding newly fledged young. The unmated male calls regularly from the nest; this increases in intensity if a female approaches and the male hops round her in a crouched posture, with the wings drooped to expose the fluffed up chestnut rump feathers and the chestnut scapular patch directed to the front. Once the pair has formed, the birds spend much of the time close together, at the nest and when feeding on the ground. Building is carried out by both sexes. The cavity used for nesting is filled with a loose accumulation of dry grass and, in the inhabited areas where chickens are kept, copiously lined with feathers.

Data on the breeding season are summarised in Fig. 35. The peak of breeding takes place in October to November, coinciding with and immediately following the rains, though these are by no means regular. In 1983 rains fell on São Tiago on 24/25th

Month												Reference
J	F	M	A	M	J	J	A	S	O	N	D	
												Keulemans 1866, Dohrn 1871, Alexander 1898a, Bourne 1955b, Bannerman & Bannerman 1968

Fig. 35 Breeding season data for Iago Sparrow (breeding ——— rains ----)

August, 13th and 30th September (with the totals varying from 200 to 500 mm, depending on the part of the island); in October many pairs were feeding young in nests, but there were few fledged young to be seen, suggesting that breeding had been triggered by the flush of ground vegetation following the first rain. On São Vicente, where no rains fell, the birds were feeding young in nests in Mindelo, but there was no sign of breeding activity in the surrounding countryside.

Both sexes cover the eggs and feed the young, though the female takes the major share. Incubation periods are short, averaging 9 minutes (maximum 16 minutes) for the female and 6 minutes (maximum 13 minutes) for the male. Again, the female makes twice as many feeding visits as the male, who often accompanies her to and from the nest without taking in food. Feeding rates at the nests I watched, ranged from 20–40 visits per hour, but only 10 visits per hour at a nest where no male was seen. By the time the pairs nesting in holes in lava cliffs have reared their young, the adults' feathers are in a very tattered condition with rubbing against the rough lava on their many visits to the nest, to such an extent that it seems almost sufficient to impair their performance in flight until after the next moult. Out of the nest the male takes a more active role in feeding the fledglings.

MOULT

According to Alexander (1898a), moult starts in February, at the end of the breeding season, and lasts at least until the end of May.

VOICE

I transcribed the call of the male as a twangy *cheesp, chew-weep* and *chew-leep*, with that of the female a similar, but a more sibilant, *chisk*. A male calling at the nest site sometimes strings the notes together into a song, *cheep chirri chip cheep chirri chip cheep*. Bourne (1955) describes a slurred *chirrp* given by a courting male. These calls are similar in timbre to those of the Great Sparrow, but rather higher pitched.

The birds also used a typical sparrow churring alarm call, *chur-chur-chur*, and a speeded up version *chur-it-it-it-it*, when I approached them at a nest site they were reluctant to leave.

FOOD

The adults feed on grass seeds and grain, principally maize, the main cereal crop of the islands. In addition, according to Bannerman (1948), they do much damage to crops by eating the young leaves. In urban areas they feed on scraps round the houses, just like House Sparrows. Bourne (1955) saw the young being fed on flying insects; my observations showed caterpillars and orthoptera as the principal food for the nestlings.

7: Somali Sparrow *Passer castanopterus*

NOMENCLATURE
Passer castanopterus Blyth, J. Asiat. Soc. Bengal 1865 24: 302.
  Somaliland.
Subspecies: *Passer castanopterus castanopterus* Blyth 1865.
  Passer castanopterus fulgens Friedmann 1921, Occ. Papers Boston Soc. Nat.
    Hist. 5: 428.
      Indunumara Mountains, Kenya.

Meinertzhagen (1951) considered *P. castanopterus* to be a race of the Asian Cinnamon Sparrow, *Passer rutilans,* which it superficially resembles. This suggestion has not been accepted; not only is this inherently unlikely in view of the wide separation – the nearest Cinnamon Sparrows occur in the western Himalayas – but also the two species occupy different biotopes. The Somali Sparrow is a bird mainly of semi-arid, lowland country, whereas the Cinnamon Sparrow is a montane species.

DESCRIPTION
This is one of the smaller sparrows with an overall length of 130–140 mm. A 'black-bibbed' sparrow, with the sexes dimorphic.

Passer castanopterus castanopterus
Male. Head chestnut from bill to nape. Lores black, with black extending slightly to back of eye; cheeks pale grey washed with yellow. Upperparts grey with black longitudinal streaks on upper back, lower back and rump pale grey. Wings chestnut

with black streaking. Tail brown. Small black bib, spreading to sides of breast; remainder of underparts pale grey washed with yellow. Bill black in breeding season, otherwise horn. In fresh plumage after moult bib is less distinct on breast and is flecked with white.

Female. Very similar to female House Sparrow with broad pale supercilium.
Immature. Similar to female, but paler.

Passer castanopterus fulgens
Male. The chestnut of head is brighter and yellow on cheeks and underparts more distinct.

BIOMETRICS
The few published data on body measurements are summarised in Table 21.

Table 21: Biometric data for Somali Sparrow

Feature	Subspecies	Males		Females	
		Range	Mean (Median)	Range	Mean (Median)
weight	*castanopterus*		ca 18		
wing		70–75	(72.5)	67–?	
	fulgens	66–68	(67)	62.5–67	(65)
tail	*castanopterus*		ca 50		
tarsus			ca 18		

DISTRIBUTION

The Somali Sparrow occurs in two allopatric populations that lie over 300 km apart. The nominate race, *castanopterus*, is found in the Horn of Africa, south to about Mogadiscio in Somalia, and extending westwards to about 41°E in the Ogaden region of Ethiopia. It is common in the coastal plains and is also found at altitudes up to about 1,500 m.

The race *fulgens* was considered to be restricted to the neighbourhood of Lake Turkana in northern Kenya and extreme southern Ethiopia, west to Lodwar, north to Lake Chew Bahir, east to the edge of the Dida Galgalla desert and south to South Hor. In 1982, however, T. Stevenson (pers. comm.) found 30 birds at Kapedo (Karpedo), about 200 km to the south of Lake Turkana and since then has found other colonies in the same general area. Whether this indicates a recent extension of range or merely reflects the previous lack of observations in the area is not known.

Fulgens is a somewhat elusive bird, said to be most frequent to the west of Lake Turkana, though the first record of breeding, established as recently as 1981, was in the Kaisut desert to the southeast of Lake Turkana (Lewis 1981).

The distribution of the Somali Sparrow is given in Fig. 36.

HABITAT

In Somalia the species is primarily coastal, frequenting rock cliffs, but is also found in dry, open inhabited country and commonly in villages and towns, where it becomes a complete 'house sparrow'. In Kenya, it is found in open, arid country with acacias, often in the vicinity of villages, even coming into the villages themselves, where it lives around the houses, again behaving like a House Sparrow.

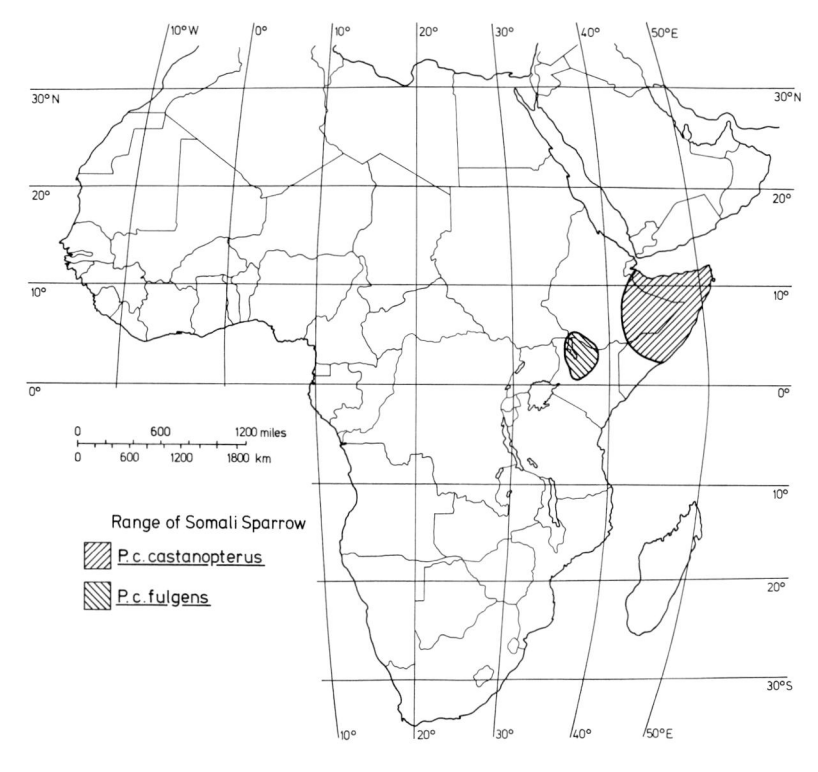

Fig. 36   Range of Somali Sparrow

BEHAVIOUR

The Somali Sparrow is gregarious, forming large flocks outside the breeding season and normally nesting in small colonies. It appears to be somewhat nomadic in the non-breeding season as it turns up in numbers in places where it is not normally seen or has not previously been recorded. This is no doubt a consequence of variation in the availability of suitable food and its opportunism is shown by the way it became extremely abundant along the lines of communication of the army during the campaign in British Somaliland in 1903–04, feeding on the camp litter and the droppings of the baggage animals (Witherby 1905).

In towns and villages it takes over the role of the House Sparrow and behaves in a similar way. In Kapedo in northern Kenya, where I watched a flock of about 30 birds, the similarity to the House Sparrow was very striking, the bird showing the same mixture of wariness and bravado as the more familiar species. The flock feeding close together on the ground would fly up to the safety of trees at any real or imagined danger, to trickle back down again in twos and threes after the threat had passed. I also watched them slipping into some open army tents and foraging for scraps, just as a House Sparrow would do in a similar situation.

BREEDING BIOLOGY

The Somali Sparrow builds its nest in a variety of situations from holes in buildings and caves to free-standing nests in the branches of thorn trees and bushes, the latter sometimes in colonies of Golden Sparrows. The nest is loosely built of grass and copiously lined with feathers; those in open sites are flask-shaped, placed horizontally (J. S. Ash, *in litt.*).

The resemblance to the House Sparrow is also shown by the way they have an identical group display in which a number of males, with wings held out, head raised, chest thrust forward and tail elevated, hop round a single female. I saw this in February and, although there was no evidence of nesting, about a quarter of the males had completely black bills and the remainder showed some darkening. This suggests that they were coming into breeding condition, so the display probably arises as it does in the House Sparrow, with a male courting a female that is not yet ready for mating.

Data on the breeding season are summarised in Fig. 37. The clutch size for Somalia, given by Archer and Godman (1961) is 4–7 eggs. No information is available on the roles played by the sexes in nest building, incubation or feeding young.

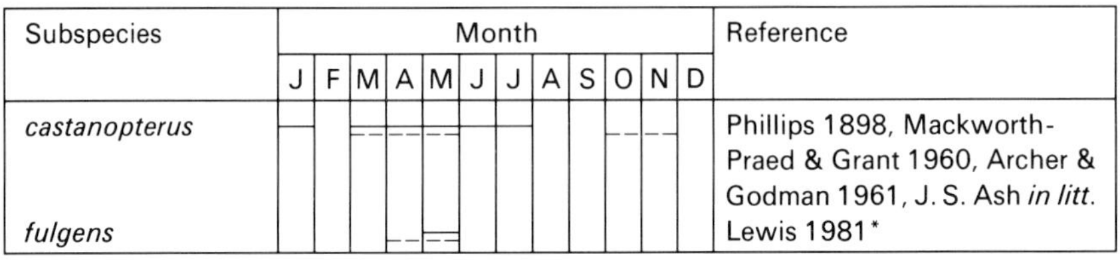

Subspecies	J	F	M	A	M	J	J	A	S	O	N	D	Reference
castanopterus		—	—	—	—			—	—	—			Phillips 1898, Mackworth-Praed & Grant 1960, Archer & Godman 1961, J. S. Ash *in litt.*
fulgens					= =								Lewis 1981*

* Only one record.

Fig. 37 Breeding season data for Somali Sparrow (breeding ——— rains - - - -)

VOICE

Most authorities state that the call is almost identical with that of the House Sparrow. I transcribed it as a soft, disyllabic *chirrip*.

FOOD

The birds associating with the army in British Somaliland fed on spilled grain from the fodder bags and on undigested grain in the droppings from animals (Witherby 1905). The birds I watched in Kapedo were feeding on scattered grain and household scraps. At the one nest recorded for Kenya, the young were being fed on caterpillars (Lewis 1981).

GENERAL COMMENTS

As we shall see later, the male closely resembles some of the Asian sparrows. In my brief acquaintance with it, I found it in its general behaviour to be strikingly similar to the House Sparrow, in my experience much more so than any other member of the genus. The close relationship between these two species is shown by the collection by Dr J. S. Ash, on 6th February 1980, of a hybrid male House × Somali Sparrow at a coastal cave near Hal Hambo, almost at the southern limit of the breeding distribution of the Somali Sparrow in Somalia (Ash & Colston 1981). This is completely outside the normal breeding range of the House Sparrow and Ash presumed that it was the offspring of a House Sparrow that had arrived on a ship-assisted passage and a local Somali Sparrow. That this is eminently possible is shown by the arrival of three House Sparrows in Mogadiscio in November 1981, 30 km northeast of the cave at Hal Hambo where the hybrid was collected.

8: Desert Sparrow *Passer simplex*

NOMENCLATURE
Fringilla simplex Lichtenstein, Verz. Doubl. Zool. Mus. Berlin 1823: 24.
    Ambukol [Ambikol], Dongola, Sudan.
Passer simplex (Lichtenstein) 1823.
   Subspecies: *Passer simplex simplex* (Lichtenstein).
     Synonym: *Passer simplex saharae* Erlanger, J. Ornith. 1899 47: 472.
         Tunisian Sahara [Jebel Dekanis according to Hilgert, Kat. Coll. Erlanger 1908: 80].
   Passer simplex zarudnyi Pleske, Ann. Mus. Zool. Acad. Sci.
     St Petersbourg 1896: 32.
        Transcaspia.

Hartert (1921) compared the Nubian skins from the type locality of *simplex* with skins from Tunisia that Erlanger had separated as *saharae,* and though he agreed that those of Nubian males were darker than Tunisian-Algerian ones, and females more rufescent, he pointed out that the Nubian skins were almost a century old and thus of doubtful value for comparison. A more recent specimen from Tekro in Chad, lying within the range of *simplex,* is identical with skins from Tunisia-Algeria, the type locality for *saharae* (Bannerman 1948). As there are no more recent skins from the Sudan and the bird appears now to be extinct there, Bannerman casts doubts on the validity of the separation of *saharae* from *simplex*. Both Niethammer (1955) and Vaurie (1959) have examined skins from the Aïr and Ennedi (within the traditional range of *simplex*): Niethammer could find no difference in plumage from the Algerian skins, though the latter tended to be larger, but Vaurie thought them intermediate between specimens from southern Sahara (*saharae*) and nominate *simplex*, though he had only one old pair of the latter available for comparison. In view of the scarcity and poor condition of existing authenticated *simplex* material available for study, it seems best to treat all the African birds as belonging to one race and to suppress *saharae* until such time as fresh material from Ennedi can be collected, a political difficulty at the present time, even though the African population seems to divide almost naturally into two (see Fig. 38).

Hall and Moreau (1970) place *simplex* in a superspecies with *Passer domesticus* and *P. iagoensis* (more correctly *motitensis*, see Chapter 5). It is even suggested that the African and Asian population of *simplex* might have evolved separately from *domesticus*. I find this difficult to accept. This would surely have resulted in different species, convergent through their adoption of the same habitat. It seems more probable that *simplex* once had a more or less continuous distribution from North Africa through the desert regions of the Middle East to Russian Turkestan, and that the population of the central part of this range has become extinct. The African and Asian subspecies are well differentiated (Vaurie 1959), suggesting that the populations have been separated for some time.

DESCRIPTION

A 'black-bibbed' sparrow; sexes dimorphic; medium-sized, length about 140 mm.

Passer simplex simplex
Male. Upper parts pale, yellowish grey, becoming cream on rump and upper tail coverts; lores black, black extending a short distance behind eye. Cheeks pale buffish white. Black bib fanning out sideways on upper breast, remainder of underparts pale buffish white. Wings black and grey with black on secondaries and tips of primaries, together with black band in centre of wing, standing out clearly in flight; two white wing bars. Tail grey. Bill black in breeding season, otherwise horn.
Female. Upperparts very pale, sandy buff. Underparts pale whitish buff. Wings as upperparts, but with dusky tips to primaries. Bill very pale brown.
Immature. Very similar to female.

Passer simplex zarudnyi
Smaller and generally paler than the nominate race, with upper parts in the male a paler, purer grey and underparts white. Bill smaller and more globular in shape.

BIOMETRICS

Typical body measurements are given in Table 22. The wing measurements suggest that the subspecies *zarudnyi* is smaller than nominate race.

DISTRIBUTION

The Desert Sparrow has a disjunct distribution with populations in North Africa, Iran and Russian Turkestan.

The type locality is in the Sudan, where, according to Vaurie (1956), it was reported by Heuglin in 1868 from Baiyuda, in the north, to Sennar on the Blue Nile, but there are no records from that area since the last century and it now appears to be extinct

108 Desert Sparrow

Table 22: Biometric data for Desert Sparrow

Feature	Subspecies	Males Range	Mean (Median)	Females Range	Mean (Median)
weight	*simplex*	19–20	(19.5)	18–21	(19.5)
wing	,,	72–81	(76.5)	69–77	(73)
	zarudnyi	69–74	72	68–72	69.9
tail	*simplex*	54–66	(60)	50–67	(58.5)
tarsus	,,	19–20	(19.5)	19–20	(19.5)
bill	,,	10–11.5	(10.8)	9–10.5	(9.8)
	zarudnyi		ca 10*		

* not sexed

in the Sudan. In Africa today it is found in northern Chad, west in a narrow latitudinal band through central Niger, Mali and Mauritania to within about 15 km of the Atlantic coast (Browne 1981, Gee 1984). From Chad it extends to the north, through western Libya and central eastern Algeria, probably also to southern Tunisia, though there are no recent records from there (Thomsen & Jacobsen 1979). The distribution in North Africa is given in Fig. 38, with hatched areas indicating where there have been recent records of breeding. In this range it is a bird of somewhat scattered occurrence. It is completely absent from the vegetationless ergs. But even in apparently

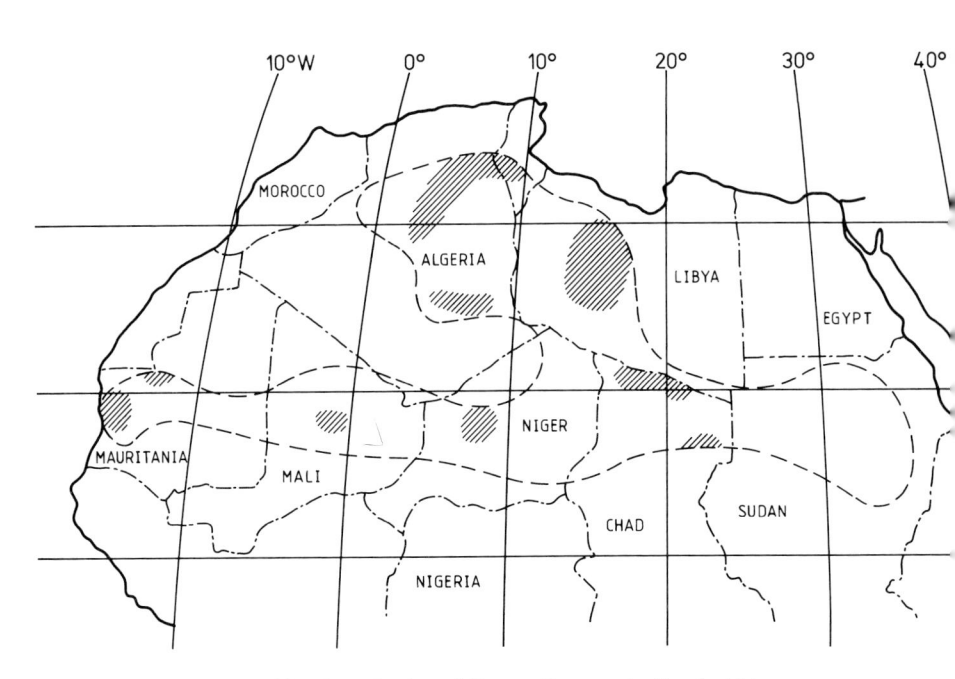

Fig. 38   Distribution of Desert Sparrow in North Africa

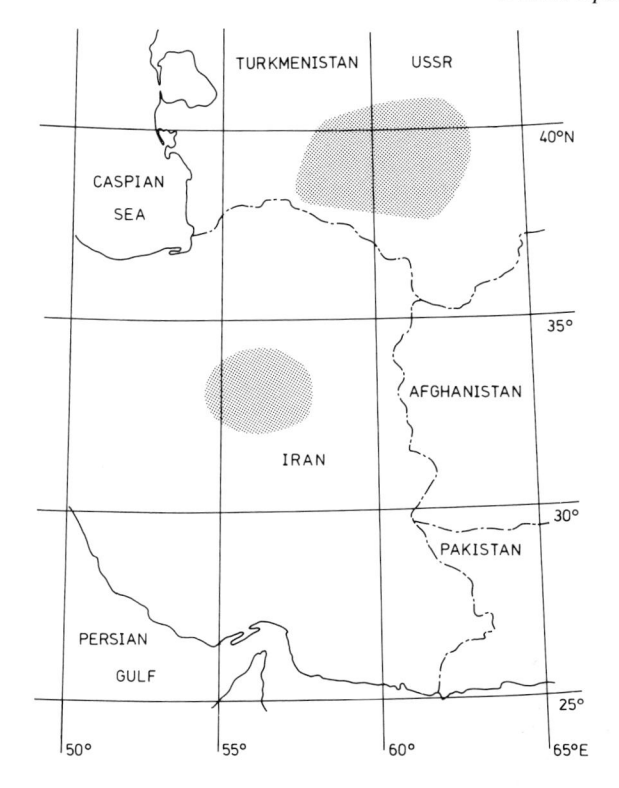

Fig. 39   Distribution of Desert Sparrow in Asia

favourable places, it is by no means regular and observers have failed to find it in some areas where it had previously been reported as common. For example, Meinertzhagen (1934) spent a month in 1931 in the Ahaggar and did not see any; Bannerman (1948) travelled from Touggourt to Souf El Oued in 1938 with the same result, although both Rothschild and Hartert (1911) and Heim de Balsac (1929) had earlier found it nesting in that area; and I travelled 5,000 km in March–April 1971 in a circuit from Ghardaia to Tamanrasset across the Ahaggar to Djanet and north to Ouargla, covering ground where it is reported to occur, and found them only at the last-named place. This could be an indication that the species is decreasing, as Niethammer (1955) considered was the case in Algeria, but it could also be that the bird is somewhat nomadic in view of the capricious availability of food in the area of its range.

There is, or was, a population of the Desert Sparrow on the western edge of the Great Sand desert or Dash-e-Lut in the Kuhistan (Kermān) region of southeastern Iran. Very little is known about this population, though according to Hüe and Etchecopar (1970), it ranged from about 32–34°N and 54–58°E. Recent searching in this area, however, has failed to find any trace of the bird (L. Cornwallis, pers. comm.) and the Iranian population may well now be extinct.

The Desert Sparrow turns up again in Russian Turkestan, mainly in Turkmenistan in the Karakum (Black Sands Desert) in an area lying between Yaradzha, Darvaza, Chardzhou and Mary, but also, according to recent reports (Sopyev 1965), in Uzbekistan in the extreme west of the Kyzylkum (Red Sands Desert). It is described by

Dementiev et al. (1970) as rare and not numerous in Turkmenistan, though more recently Sopyev (1965) found it to be a common bird in eastern Karakum. He described it as extremely sedentary. The distribution of the Asian race of the Desert Sparrow is shown in Fig. 39.

HABITAT

The habitat of the Desert Sparrow in the Sahara has been described in some detail by Heim de Balsac (1929), He states that it is mainly a bird of trees and bushes – *Tamarix articulata* (Tamarisk), *Phoenix dactylifera* (Date Palm), *Colligonium comosum, Raetama raetam* (Retama), *Aristida pungens* (Drinn) – though it freely comes down to the ground to search for food. Thus it is not a bird of open desert, but of wadis with bushes and trees, and the palmeries of the oases, the larger the trees the greater the density of the bird. It also occurs in rocky banks of wadis, walls of wells and other man-made structures; though where trees are absent, the Desert Sparrow, if it occurs at all, is only present in isolated pairs. Throughout, it is associated with sand, though it avoids the vast, treeless sand dune ergs of the Sahara.

In Turkmenistan, Sopyev (1965) described the habitat as bare, empty hilly sand plains with isolated trees and bushes – *Ammodendron conellyi* (Sand Acacia), *Holoxylon ammodendron* (White Saxaul).

BEHAVIOUR

Not much has been written about the habits of the Desert Sparrow. It is not one of the more gregarious sparrows, probably because it is nowhere common enough to form the large flocks that are found with some other species in the genus. Outside the breeding season it usually occurs in pairs or small groups of up to 10 individuals, though in the more favoured places flocks may amount to as many as 50 birds (Guichard 1955). It spends most of the time in trees, in which Tristram (1859) described it as shy and silent, more like a *Sylvia* warbler, as it hops from leaf to leaf and skulks in the crowns of palms, than a sparrow. It comes to the ground only for feeding sorties and can be found about buildings at such times. The small flocks roost together in the foliage of dense bushes (Heim de Balsac 1929).

BREEDING BIOLOGY

The Desert Sparrow nests in a great variety of situations, with holes the preferred site in the Sahara according to Heim de Balsac (1929), and branches of trees the commonest site in Turkestan according to Sopyev (1965). Of the hole sites, those in trees are most frequently used, but those in stone walls, deserted buildings, stone pyramids acting as desert landmarks, the walls of wells, etc, are not uncommon. In trees it will use the crowns of palms as well as building openly among the branches and in the understorey of old nests of corvids and birds of prey. Sopyev (1965) reported on 23 nests in the Karakum: 18 of these were in Sand Acacias, 9 under crows' nests, 7 under eagles' nests and 2 under old sparrows' nests, 1 in the wall of a hut and 4 in isolated bushes. The location under an old nest is an important one according to Sopyev because of the protection given against the sun. Measurements of the temperature in such a nest throughout a 24-hour period varied from 30–40.6°C (range 10.6°) compared with 24.8°–42.6° (range 17.8°) in the vicinity of the nest. The open nest in a tree is an irregular mass of dry grass and small twigs, domed over, the entrance sloping upwards at one end. Sopyev gives the following measurements for 15 nests in Turkmenistan: length 136–250 mm (mean 213 mm), height 135–180 mm (mean 148 mm). The nest cavity is lined with soft plant fibres and feathers. These tree nests are built round small branches, which in thorny trees afford some protection against predators. Heights range from 1.5 m in thick bushes to 4 m in trees.

Little has been published about courtship, though Bundy and Morgan (1969) describe a display near a nest site in Tripolitania in January, where breeding does not occur until April, with the male hopping after a female with outstretched, shivering wings.

Both sexes take part in building and continue to add material through the incubation period up to the hatching of the first young. For instance, at one nest watched by Sopyev that had a fresh clutch on 11th June, 57 visits with nesting material were seen in 12 hours on 4th June, and another 32 visits with material in 13 hours on 8th June, five days before the young hatched.

The Desert Sparrow is not a social breeder and frequently nests in isolated pairs. Heim de Balsac (1929) mentions a large number of nests in the understorey of a nest of a Lappet-faced Vulture (*Aegypius tracheliotus*), but it is not clear whether more than one of these was in occupation. Sopyev, in fact, remarks that the durability of the nest may give a misleading impression of colonial breeding, with 4–5 nests from previous seasons remaining in one bush in addition to the nest actually in use. On occasions he found a nest from previous seasons to be repaired and re-used.

Published data on the breeding season are summarised in Fig. 40. The clutch size in North Africa ranges from 2–5 eggs (Hartert 1913), in Turkestan from 2–6, exceptionally, 7–8 eggs. In Turkestan two clutches may be laid each year, with an average of 5.5 eggs for the first clutch and 4.4 for the second. Eggs are normally laid at daily intervals.

Detailed observations on the nesting activity of the Desert Sparrow have been made by Sopyev in Turkmenistan. The following summary is based mainly on his work. During the incubation period both sexes spend some time in the nest, though whether they actually incubate the eggs during the day is somewhat doubtful in view of the high ambient temperatures. The fact that hatching of the clutch may last from 4–5 days suggests that egg development starts as soon as it is laid and that the eggs

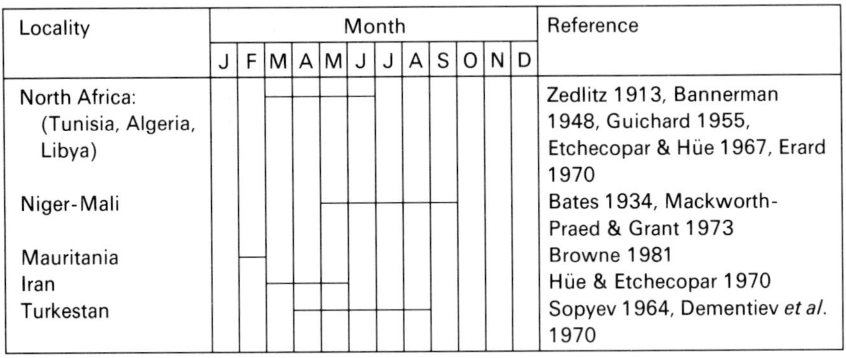

112 *Desert Sparrow*

Locality	Month												Reference
	J	F	M	A	M	J	J	A	S	O	N	D	
North Africa: (Tunisia, Algeria, Libya)													Zedlitz 1913, Bannerman 1948, Guichard 1955, Etchecopar & Hüe 1967, Erard 1970
Niger-Mali													Bates 1934, Mackworth-Praed & Grant 1973
Mauritania													Browne 1981
Iran													Hüe & Etchecopar 1970
Turkestan													Sopyev 1964, Dementiev et al. 1970

Fig. 40   Breeding season data for Desert Sparrow

may only be truly incubated at night. Time spent in the nest depends on the ambient temperature, as shown in Table 23, and is probably influenced also by the state of development of the eggs. It will be seen that the female devotes about twice as much time in this activity as the male; the female also spends longer periods in the nest: female 1–35 minutes (mean 8.7 minutes), male 1–23 minutes (mean 5.5 minutes). Both sexes roost in the nest at night. The incubation period is 12–13 days.

Table 23: Activity of Desert Sparrow at nest during incubation period

Day temperature range °C	Period in nest			Stage of incubation period
	Male	Female	Total	
20–30	161 (28%)	409 (72%)	570	middle
25–36	105 (33%)	211 (67%)	316	beginning
31–42.6	35 (34%)	68 (66%)	103	end

Both sexes feed the nestlings, taking almost equal parts, as observed by Sopyev: at one nest watched for 13 hours (07.00–20.00), when the chicks were 5–6 days old, the birds paid 98 feeding visits (mean 7.5 visits per hour), 43 by the male and 55 by the female. The feeding rate ranged from 8–9 visits per hour in the morning and evening, falling to 3.5 visits per hour from 11.00–17.00, the hot time of the day when the ambient temperature rose to 46°C. The nestling period ranged from 12–14 days. Sopyev found an overall success rate of 33%, with a hatching success of 44% (19 out of 43 eggs) and a fledging success of 76% (22 out of 29 young).

MOULT

Dementiev et al (1970) state that the moult in Turkestan begins in late July or early August and ends in late August or early September. The only reports of moulting in the African population come from Algeria: a young bird collected by Spatz (Zedlitz 1913) and two seen by Berg and Roever (1984) in 1982; both of these observations were in July.

VOICE

No detailed description of the voice has been published. The birds I saw in the Sahara were silent, though other observers state that they call frequently. The call is said to be softer and more musical than that of the Willow Sparrow (Bundy & Morgan 1969); Berg and Roever (1984) describe it as a subdued *chu*. In addition, Dementiev *et al.* (1970) state that the bird has a song similar to that of the Goldfinch *Carduelis carduelis*; while Hüe and Etchecopar (1970) say it is similar to that of the Greenfinch *Carduelis chloris*.

FOOD

The main food is the seed of *Aristida pungens* in the Sahara and the related *Aristida pennata* in Turkestan. The bird also takes seeds of the 'acheb' flora, the ground vegetation that springs up in the desert after rain, from seeds that may have lain dormant for years. For example, Bates (1934) recorded the seeds of *Panicum turgidum*. According to Guichard (1955), the bird can become a pest by attacking the green heads of barley and Bates describes how it fed on scattered grain around houses, just like many other members of the genus. In the Sahara it attacks the flowers of Retama.

Food also includes invertebrates: Coleoptera (beetles, both larval and adult forms), Lepidoptera (caterpillars) and Arachnida (spiders) being mentioned.

The Desert Sparrow, like many other desert species, is able to live without direct access to water, obtaining the water it requires from its food.

9: House Sparrow *Passer domesticus*

NOMENCLATURE
Fringilla domestica Linnaeus, Syst. Nat. 1758 Edn. 10: 183.
　　Sweden.
Passer domesticus (Linnaeus) 1758.

Vaurie (1949, 1956) has studied the geographical variation of the House Sparrow and has reduced the populations that justify racial separation to twelve. He further suggests that these can be divided into two groups: a *domesticus*-group, in which the birds are typically larger, with grey cheeks and underparts, that occurs in the Palaearctic region, and an *indicus*-group, generally smaller (bill always smaller) with white cheeks and underparts, and generally a richer colour on the upperparts, that is basically located in the oriental region. These characteristics follow the ecogeographical rules of Bergmann and Gloger, and the groups taken as a whole are sufficiently distinct to suggest that they have followed two separate evolutionary lines; though with secondary expansion further adaptive modifications have occurred so that the most southerly of the races of the *domesticus*-group occurring in Egypt is smaller than the largest race of the *indicus*-group. For convenience I shall refer to these as the Palaearctic and Oriental groups respectively.

　A number of authorities have gone further than Vaurie and have placed *indicus* as a separate species from *domesticus*, mainly on morphological grounds. The situation in Kazakhstan is particularly relevant in this connection. Here, according to Gavrilov (1965) and Gavrilov and Korelov (1968), the race *bactrianus*, a summer visitor that winters in India, overlaps the sedentary nominate race of the House Sparrow with negligible interbreeding. These authors show that there are also differences in behaviour and physiology, as well as in morphology, between the two taxa. Further, Yakobi (1979) has cited additional behavioural differences in support of Gavrilov's hypothesis. These are strong arguments for recognising the 'Indian Sparrow' as a distinct species. There are, however, difficulties. There is an extensive zone of intergradation between the Indian Sparrows of the race *indicus* and the *persicus* subspecies of *Passer domesticus* lying to the west. This implies that *indicus* must be recognised as a race of *domesticus* and hence the same must apply to *bactrianus*, unless it is to be considered a species on its own, a course that is not suggested by Gavrilov. Clearly we have here a difficult classification problem. For the present I prefer to retain *bactrianus* as a subspecies of *Passer domesticus*, arguing that the morphological and physiological differences are not sufficient to justify specific identity and that the behavioural differences are no more than a consequence of the migratory habits of *bactrianus*. The absence of interbreeding could merely be the result of the displacement of the breeding seasons of the two populations that would reduce the chances of mixing. The spread of sedentary *domesticus* into the breeding range of the migratory *bactrianus* is a comparatively recent event and it may well be that a stable situation has not yet had time to develop.

This is one of the taxonomic problems that are found with *Passer domesticus* (see later for the relationship between *P. domesticus* and *P. hispaniolensis*), but in view of the extensive intergradation that occurs between birds of the two groups where they come into contact, particularly in Iran, but also to a lesser extent in Arabia and the border between Egypt and Sudan, I consider that they have not yet evolved sufficiently to be separated in the way that is clearly the case with *domesticus* and *hispaniolensis*. It may be relevant that the bird described by Zarudnyi as *Passer enigmaticus* (Zarudnyi 1903), type locality Hurmuk (Hormak) and Kamschar (near Sarbāz), Persian Baluchestan, and subsequently identified as an intersex by Mayr (1949), together with eight other birds (four males, four females) showing intermediate plumage, were all obtained in eastern Iran and western Afghanistan, the transition zone between *P. d. indicus* and *P. d. persicus*, that belong respectively to Vaurie's Oriental and Palaearctic Groups.

I think that Vaurie's division into two groups is helpful in interpreting the evolutionary situation and follow the races he considers warrant separation with one exception: *italiae*, which I feel is better placed as a race of *P. hispaniolensis*. The birds occurring in Italy and Crete are intermediate between *P. domesticus* and *P. hispaniolensis*; they are usually treated as a subspecies of *domesticus* but, for the reasons given in Chapter 10, I consider they are more appropriately described as a race of *hispaniolensis*.

The following races are recognised.

PALAEARCTIC GROUP

Passer domesticus domesticus (Linnaeus) 1758.
    Synonyms: *Passer domesticus hostilis* Kleinschmidt, Falco 1915: 19.
        Tring, Herts, Great Britain.
        Passer domesticus semiretchiensis Zarudny & Kudashev, Nasha Okhota 1916: 37.
        Verny [Alma-Ata], Djarkent [Panifilov] and
        Przhevelsk, Russian Turkestan.
    Passer domesticus balearoibericus v. Jordans, Falco 19, Sonderhaft: 4.
        Valldemosa, Mallorca.
    Passer domesticus baicalicus Keve, Anz. Akad. Wiss. Wien Math. Naturwiss. 80: 20.
        Kultuk, southern Lake Baikal.

Passer domesticus tingitanus Loche, Expl. Sci. Algeria Ois. 1867: 132.
   Algeria.
Passer domesticus biblicus Hartert, Vög. pal. Fauna 1904 1: 149.
   Sueme, Palestine.
Passer domesticus niloticus Nicoll & Bonhote, Bull. Br. Orn. Cl. 1909 23: 101.
   El Faiyum, Egypt.
   Synonym: *Passer domesticus halfae* Meinertzhagen, Bull. Br. Orn. Cl. 1921 41: 67.
      Wadi Halfa, northern Sudan
      (now considered to be intergrades between *P. d. niloticus*
      and *P. d. rufidorsalis*).
Passer domesticus persicus Zarudny & Kudashev, Nasha Okhota 1916: 37.
   Arabistan [Khuzetsān], southwestern Iran.

ORIENTAL GROUP

Passer domesticus indicus Jardine & Selby, Ill. Orn, 1835 3: 118.
   India; restricted to Bangalore by Kinnear
      Ibis 1925: 751.
   Synonym: *Passer domesticus soror* Ripley, Spolia Zeylandica 1946 24: 241.
      Nikawella State Farm, Rattota, Matale, Ceylon [Sri Lanka].
Passer domesticus rufidorsalis Brehm, Naumannia 1855: 37.
      northeast Africa; restricted to Khartoum, Sudan, by Vaurie, *Birds of the Palearctic Fauna* 1959: 570.
Passer domesticus hyrcanus Zarudny & Kudashev, Nasha Okhota 1916: 37.
      Astrabad [Gorgān], Gilan and Mazandaran, northern Iran.
Passer domesticus bactrianus Zarudny & Kudashev, Nasha Okhota 1916: 37.
      Merv [Mary], Tashkent; the type is from Tashkent according to Meinertzhagen, Ibis 1938: 507.
Passer domesticus parkini Whistler, Bull. Br. Orn. Cl. 1920 41: 13.
      Srinagar, Kashmere [Kashmir].
Passer domesticus hufufae Ticehurst & Cheesman, Bull. Br. Orn. Cl. 1924 45: 19.
      Hufuf town, Hasa Province, eastern Arabia.

DESCRIPTION

The House Sparrow belongs to the 'black-bibbed' group of sparrows. The sexes are dimorphic. It is one of the larger sparrows, length typically 160–165 mm, up to 180 mm in the north and down to 140 mm in the smaller, southern races.

Passer domesticus domesticus
Male. Crown and nape dark grey; black round lores and eye, becoming chestnut behind eye, outlining cheek and joining across neck; small white spot at rear of eye; mantle and scapulars boldly streaked black, chestnut and buff, with white wing bar formed by tips of lesser coverts; back and rump grey-brown; tail dark brown; cheeks grey; black bib from bill, fanning sideways over breast; underparts pale grey. White tips of black feathers of bib and breast patch, renewed at moult, gradually abraded so that by beginning of breeding season bib becomes uniformly black and more extensive on breast.
   Variants in which chestnut-brown feathers occur in bib and grey of crown not uncommon and some have bib or crown completely chestnut-brown.
   Bill black in breeding season, otherwise horn.
   Female. Uniform dull brown above, except for some darker streaking on mantle and scapulars and a buff streak behind eye; underparts pale greyish brown, tending

to whitish on belly. Bill becomes darker in breeding season and a few birds have completely black bills.

Juvenile. Similar to adult female, but upperparts buffy and underparts browner. A tendency for throat to be grey in juvenile males and whitish in juvenile females. Harrison (1961) was able to sex about 95% of juveniles on basis of throat colour and Johnston (1967a) 85–90%. Males also tend to show a light postocular spot immediately behind and above eye. This is a less consistent sexual character than colour of throat; it was present in Johnston's sample in 70% of males, but only in 12% of females.

Immature. According to Selander & Johnston (1967), first-year males highly variable in degree to which plumage departs in colour from cryptic juvenile plumage and approaches full adult male pattern. It appears likely that males from later broods show least development of fully mature plumage.

Passer domesticus tingitanus
Male. Not well differentiated from nominate *domesticus*, but in worn spring plumage head speckled with black, cheeks, ear coverts and underparts paler.

Passer domesticus biblicus
Male. Similar to nominate *domesticus*, but paler on cheeks and underparts.

Passer domesticus persicus
Male. Paler than *biblicus*, but cheeks still grey and not white; slightly smaller than *biblicus*.

Passer domesticus niloticus
Male. Similar to *biblicus*, but significantly smaller.

Passer domesticus indicus
Male. More richly coloured on upperparts than *domesticus*; cheeks and underparts white; smaller than nominate *domesticus*.

Passer domesticus rufidorsalis
Male. Similar to *indicus*, but chestnut on upper back much more extensive; smaller than *indicus* and *niloticus*.

Passer domesticus hyrcanus
Male. Similar to *indicus*, but larger.

Passer domesticus bactrianus
Male. Upperparts paler than *indicus*, but with white cheeks and underparts, and small bill of the Oriental-group.

Passer domesticus parkini
Male. Generally darker and larger than *indicus*, with more extensive black on breast.

Passer domesticus hufufae
Male. Paler than *indicus*, but similar in size.

BIOMETRICS

It is not easy to give a balanced picture of the biometric data for the House Sparrow. A mass of information is available for the nominate race, not only from its extensive range in Europe, but also from the introduced populations in America. In contrast, data available for the other races are relatively rather sparse. Typical values are given in Table 24.

The spread of dimensions for the nominate race is very large; this is partly a consequence of the very large number of specimens measured, increasing the incidence of exceptionally small and large specimens in the samples, but there are also considerable clinal variations. It is apparent from the European data that significant differences exist between populations located at no great distance from each other, possibly the result of partial genetic isolation. However, overriding this there is a marked cline of increasing size (in weight and wing length) from west to east. Separate clines exist for the Palaearctic and oriental groups of races and Fig. 41 shows clearly how the birds of the latter group are generally smaller, despite the fact that they contain the two migratory races living at high altitudes, both effects that tend to result in increased size.

The relationship between size and latitude in the House Sparrow is confusing. In the Palaearctic group of subspecies the most southerly race, *niloticus*, is the smallest, but apart from that the extensive data shows no correlation, neither within the nominate race nor within the group. There is a positive correlation of size and latitude for the Oriental group, with the most southerly race, *rufidorsalis*, the smallest, but the most northerly races, *parkini* and *bactrianus*, occur at high altitudes, where birds tend to be larger, and there are insufficient data to be able to differentiate between the effects of latitude and altitude.

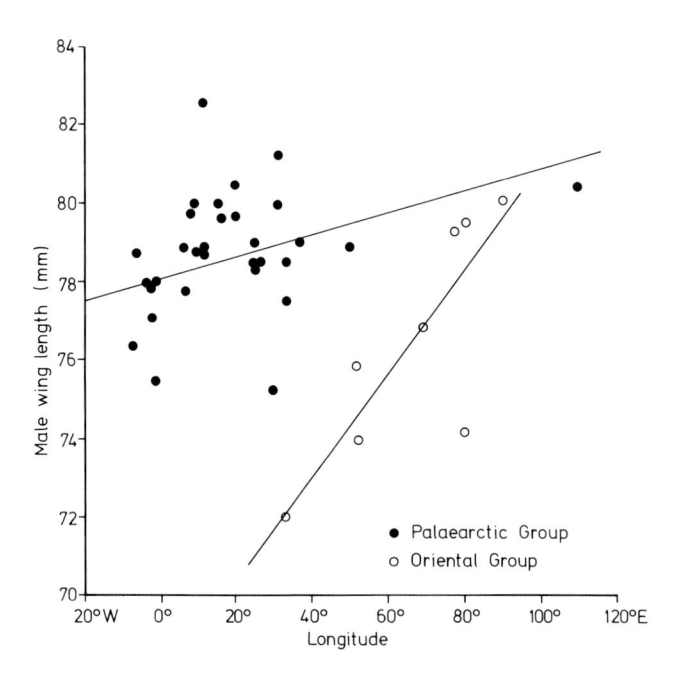

Fig. 41 Variation of wing length of male House Sparrows with longitude

Table 24: Biometric data for House Sparrow

Feature	Subspecies	Locality	Males Range	Males Mean (Median)	Females Range	Females Mean (Median)
weight	domesticus	UK		28.6		28.5
		European Continent	27.7–39.5	30.2	24.0–39.5	30.5
	tingitanus	Morocco	23–30.5	26.7	19–35	26.7
	indicus		22–33	(27.5)		
	bactrianus		21–28	(24.5)		
	parkini		24–30*	26.5*		
wing	domesticus	UK	71–82	76.0	71–76	(73.5)
		European Continent	67–89	80.4	70–86	77.7
		Asia	76.5–83	(80)		
	tingitanus	Morocco	76–81	78.8	74–80	77.0
	biblicus		75–85	80.9	72–82	75.3
	niloticus		71–80	(75)		
	persicus		73–85	78.3		
	indicus		70–83	74.2	70–80	(75)
	hyrcanus		71–81	75.9	69–75	71.9
	bactrianus		70–80	76.9	71–78	74.3
	parkini		71.5–85	80.8	73–80	77.8
	hufufae		71–77	(74)	70.5–74	(72.3)
	rufidosalis		69–75	72		
tail	domesticus	Europe	55–64.5	59.6	52–65	57.5
	tingitanus	Morocco	55–65	59.5	54–60	57.6
	indicus		49–61	(55)	51–57	(54)
	parkini		55–62.5	(58.8)	52–61	(56.5)
tarsus	domesticus	Europe	17–24	19.9	16–25	19.5
	tingitanus	Morocco	16–18.5	17.6	16.3–18.9	18.0
	indicus		16.5–20	(18.3)	17–20	(18.5)
	parkini		18.5–20	(19.3)	19–20	(19.5)
culmen	domesticus		10–14	12.4	10.5–14.5	12.4
	tingitanus		12.0–12.9	12.5	12.7–13.8	13.3
	biblicus		13–15	14.0		
	persicus		11–14.5	12.8		
	indicus		11.5–15	12.5	13–15	(14)
	hyrcanus		12.0–14.0	13.1		
	bactrianus		13–14	13.3		
	parkini		12.5–15	13.5	13–15	(14)
	hufufae		13.3–14.5	(13.9)	12.5–13.5	(13)

*not sexed

In contrast, Johnston and Selander (1973) found that a positive correlation of size with latitude had already developed in House Sparrows introduced to North America, and this has subsequently been confirmed by Blem (1975) for a 31° range of latitudes (28–59°N), his most northerly birds weighing 6 g (>20%) more than those in the extreme south. The cline in weight was matched by a cline in wing length. These results were obtained from birds in mid winter, when at high latitudes they have to carry additional fat and food reserves to withstand the long, cold nights. Blem was concerned with wing-loading and he found that his most northerly birds were at the maximum of wing-loading (0.33–0.35 g/cm^2) for effective flight, setting a physiological limit to the winter range, though this can be overcome to some extent by the fact that such birds spend much of their time in the winter inside buildings (grain stores, cattle byres) where they are insulated from the extreme climatic conditions of these high latitudes.

The data in Table 24 show that there is a large variation in weight, even within birds of the one race. There are, in fact, considerable daily and seasonal fluctuations. For example, Dexter (1949) found variations of 5–10% in individual House Sparrows retrapped on the same day, increases of 10–20% in birds retrapped 1–14 days apart and decreases of up to about 30% in birds retrapped 8–11 days apart. These weight changes were recorded in juvenile birds and the large decreases could have involved individuals that were not going to survive; this cannot, however, apply to those birds that showed weight increases. It is not known if the weight of adult birds fluctuates to the same extent over short periods, though there are considerable seasonal variations resulting from changes in body fat content. Investigations by O'Connor (1972) in England and by Folk and Novotny (1970) in Czechoslovakia showed fluctuations in monthly means of approximately 5% about the overall means. These studies (Fig. 42) do not show a very clear pattern, though there is a tendency for males to be heavier in winter months, and females in the breeding season, possibly through an increase in the size of the ovaries. According to O'Connor the variation of weight in the English birds in winter months is largely inversely correlated with temperature. First-year birds are generally lighter than full adults.

These large samples, which illustrate the large differences in weight that can occur within a population (the heaviest birds weighing over 50% or so more than the lightest at all times of the year), emphasise the need for caution in making comparisons based on biometric data when only small samples are available, as is generally the case.

Significant seasonal changes also occur in the length of the bill. Bill length is determined by a balance between the rates of growth and wear. Wear rate is deter-

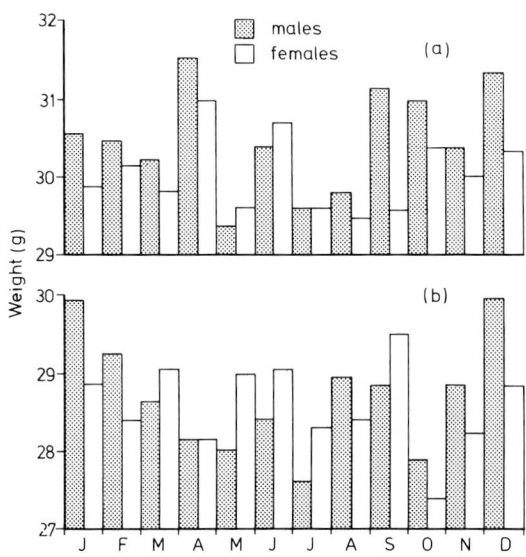

Fig. 42   Seasonal pattern of mean weights of House Sparrows in (a) Czechoslovakia, (b) England

Table 25: Seasonal changes in the bill length of the House Sparrow

Locality	Decrease in bill length in winter compared with summer	Period of comparison	Reference
Germany	12%	July:December	Steinbacher 1952a
Berkeley, California, USA	8%	May/June: December/January	Davis 1954
Pasadena, California, USA	3.5%	ditto	ditto
Colorado, USA			
Males	1.6%	August:October	Packard 1967a
females	4.4%	ditto	ditto
Kansas, USA			
Males	6.7%	March:October	Rising 1973
females	5.4%	ditto	ditto

mined by diet and is lower in summer when the birds feed on a mixture of seeds and soft-bodied insects, than in winter when the diet is almost exclusively hard seeds. The magnitude of the difference must obviously depend on the local dietary regime. Five studies in the USA and Germany show average bill length 1.6–12% greater (mean 5.5%) in summer than in winter (Table 25).

DISTRIBUTION

The House Sparrow has a very extensive natural distribution in the Palaearctic and Oriental regions that has been greatly increased by successful introductions to other parts of the world, beginning in the middle of the 19th century, making it now one of the most widely distributed species of land birds in the world. The most recent study of the ranges occupied by the different subspecies has been carried out by Vaurie (1949, 1956); in some cases there is quite a wide zone of intergradation with birds intermediate in character, so that it is not always possible to define the ranges of the different races with precision.

Passer domesticus domesticus

The nominate race extends from the Iberian peninsula and Ireland in the west, through the remainder of Europe from the Mediterranean north to about 70°N in Norway, with the exception of Italy and some of the Mediterranean islands, notably Corsica, Sicily, Malta and Crete, where it is replaced by birds intermediate between *P. domesticus* and *P. hispaniolensis*, that I allocate to *hispaniolensis*. *P. domesticus* does not occur on Sardinia, where the indigenous sparrow is *P. hispaniolensis*, though the island has recently been colonised by *P. montanus*.

The occupation of northern Europe is comparatively recent, the bird spreading north of the Arctic Circle in the middle of the 19th century and to Nordland, Finnmark, Lapland, the Kola peninsula and the northern shore of Arkhangel'skaya Oblast at the mouth of the Mezen' and Pechora rivers in the 20th century. Stragglers have been recorded as far north as Novaya Zemlya, though without evidence of breeding.

At the beginning of the 19th century this race was confined almost entirely to Europe, with the boundary about the Ural Mountains in the east and the Caucasus in the south. Since then, however, there has been a remarkable extension of range following the growth of cultivation in Siberia and the building of the trans-Siberian railway, so that the nominate race now extends in a narrowing belt across Siberia to the east coast, where it reached Nikolayevsk-na-Amure, at the mouth of the River Amur in 1929. The northern limit reaches Salekhard on the River Ob' and along the same latitude on the River Taz, falling south to Turukhansk on the Yenisey, along the course of the Nizhnyaya Tunguska to Suntar and Nyurba on the Vilyuy river, Yakutsk on the Lena and then southeast through Amga, to Nel'kan on the Maya and to Chumikan at the mouth of the Uda. The northern limit is not clearly defined, partly because of the scarcity of authenticated records from northern Siberia, but also since colonies in the extreme north may come and go.

From Nikolayevsk-na-Amure, the southern limit extends to Khabarovsk on the Amur river, through Heilungkiang Province in Manchuria, in extreme northeast China, northern Mongolia south to the Gobi desert, the Tuvinskaya Autonomous Region of the USSR, along the northern shore of Lake Balkhash, west to Aral'sk, on the Aral Sea, and to the mouth of the Emba river on the Caspian Sea. Recently there has been a spread through Semireche, south of Lake Balkhash, to the northern slopes of the Tien Shan at about 40°N in Kirghizia and east to about Tashkent, in Uzbekistan, where it comes into contact with the migratory race, *P. d. bactrianus*. According to Gavrilov and Korelov (1968), there is little intergradation between the two subspecies in the area of overlap and this is, in fact, given as justification for the specific separation of *bactrianus* (and hence of *Passer indicus* of which it is a subspecies). The situation between the Aral and Caspian Seas has not been studied in detail, so that whether the nominate race is spreading there into the range of *P. d. bactrianus* or whether the two remain separated is not known. The only other race with which contact could occur is *P. d. biblicus*, but the latter appears to be separated by physical barriers: the Caucusus at the northeast limit of its range in Turkey and by the Sea of Marmara in the north west.

The other boundary of some interest is that between *P. d. domesticus* and the Italian Sparrow, that I place as a subspecies of *P. hispaniolensis* for the reasons given under that species. This boundary was investigated by Meise (1936) using museum skins and more recently by Johnston (1969a) with up-to-date material collected specifically for the purpose. Both of these investigators used a 'hybrid index' based on a number of characters; in addition, the transition zone, where the pure phenotypes *P. d. domesticus* and *P. hispaniolensis italiae* as well as intermediates occur, has been determined in a number of Alpine regions in Switzerland, Austria and Italy under field conditions, using the crown colour of the male as a criterion (Wallis 1887, Ris 1957, Niethammer 1958, von Wettstein 1959, Schweiger 1959, Schöll 1959, 1960, Niethammer & Bauer 1960, Löhrl 1963, Schifferli & Schifferli 1980, and the author). In broad terms the transition zone runs from the Franco-Italian border on the Mediterranean coast, roughly along the arc of the Alps, near the Italian border, to the Adriatic coast near Trieste; in detail, however, the exact situation is quite complex.

In the west the hybrid zone extends, according to Johnston, for about 40 km from Menton in France to Imperia in Italy. The position in the Alpes Maritimes does not appear to have been studied, but in the Alpes Cottiennes in 1985 I found a very narrow hybrid zone with pure *italiae* in Montgenèvre, 1,854 m, in France to the west of the Col de Mont Genèvre, pure *domesticus* in Briançon and hybrids only in la Vachette, 1,321 m, 3.5 km to the east of Briançon. To the north there appears to be complete separation in the area of the Massif du Mont Cenis, with *domesticus* in Lanslebourg, 1,399 m, and Lanslevillard, 1,479 m, to the north of the Col du Mont Cenis, 2,083 m, with no sparrows at Bar Cenisio, 1,475 m, south of the col and only *italiae* in Cesana Torinese, 1,344 m, Oulx, 1,121 m, Salbertrand, 1,054 m, Gaglione, *ca* 750 m, and Susa, 503 m.

Continuing to the north, *domesticus* extends to la Rosière du Montvalescan, 1,850 m, on the French side of the Col du Picco St Bernard, 2,188 m, and there is an extensive hybrid zone in the Valle d'Aosta on the Italian side, with hybrids occurring from la Thuile, 1,441 m, to Aosta, 583 m, a distance of over 30 km, though at Courmayeur, 1,278 m, at the head of the valley on the Italian side just below Mont Blanc, there were only pure *domesticus*. North of Aosta the situation is reversed. Here not only is the hybrid zone in the Rhône valley in Canton Valais (Wallis), from Martigny-Ville, 476 m, to Brig, 713 m, north of the Grand St Bernard, 2,473 m, and Simplon, 2,009 m, passes, but there is even a band of pure *italiae* in the south Valais valleys, *eg* at Evolène, 1,378 m, Stalden, 795 m, Zermatt, 1,616 m, and Sass Fee, 1,790 m, with no penetration of *domesticus* south of the passes (Ris 1957). In the Rhône valley the hybrid zone is about 20 km wide.

The hybrid zone remains in Switzerland to the east, though at the St Gotthard Pass, 2,112 m, it lies to the south of the col in the Valle Leventina, from Airolo, 1,100 m, in the north, to south of Biasca, 309 m, in the south, with an extent of about 35 km (Niethammer & Bauer 1960, Schifferli & Schifferli 1980). The situation in the Ober Engadin does not appear to have been studied, though to the east the hybrid zone once again penetrates to the north of the watershed at Nauders, 1,365 m, extending south over the Reschen Scheideck Pass, 1,510 m, as far as Lasa (Laas), 800 m, giving a width of about 35 km. In contrast, the situation in the south Tirol has been well studied, though there is some divergence in the findings: Niethammer (1958) found hybrids from San Leonardo, 680 m, to Merano (Meran), 320 m, in the Val Passiria (Passeiertal), as I did in 1985, and from Mules (Mauls), 889 m, to Chiusa (Klausen), 523 m, in the Val Isarco (Eisacktal). Schöll (1959) found pure *italiae* in Colle Isarco (Gossensass), 1,065 m, 12 km north of Mules, and in 1985 I found hybrids there, though we both recorded only pure *domesticus* from Brenner, 1,375 m. On the other hand, however, von Wettstein (1959) reported *italiae* to the north of the Brenner in Trins and Steinach, *italiae* and hybrids in Neustift. It is difficult to resolve this divergence of view as both von Wettstein's and Neithammer's observations were made in the same year (1958) and, although Niethammer later implied (Niethammer & Bauer 1960) that the birds seen by von Wettstein on the Austrian side of the Brenner could have been out of breeding season wanderers, this seems unlikely as von Wettstein was there in July. Niethammer specifically mentioned pure *domesticus* in Gries am Brenner, 5 km north of Brenner. Males that I saw in that area (Gschnitz, Trins, Steinach, Gries and Brenner) in July 1985 had a broader chestnut band on the side of the head, restricting the grey on the crown to a narrower band than is normal for typical *domesticus*, but this was a constant feature and, moreover, one I have seen elsewhere well away from any possible influence of *italiae*, *eg* in the Auvergne region of central France; so I do not feel that it necessarily represents an infusion of *italiae* genes; Wallis actually mentions such a bird in the hybrid zone at Dobbiáco (Toblach). Thus I tend to agree with Niethammer and Schöll that the hybrid zone lies to the south of the col with an extent of about 30 km.

The hybrid zone continues south of the mountain range, along the Val Pusteria (Puster-tal) in the east Tirol from Vandoies (Ober-Vintl) to Dobbiáco (Toblach), 1,240 m, (Schöll 1960), with *domesticus* penetrating to Cadipietra (Steinhaus), 1,054 m, between San Pietro and San Giovanni in the Val Aurino (Ahrntal). The same situation obtains in the Karnische Alpen, with the hybrids south of the watershed; for example, at Paluzza, 600 m, and Arta, *domesticus* at the Plöcken Pass (Passo di Carnico), 1,362 m, and *italiae* at Tolmezzo, 323 m, giving a hybrid zone probably no more than 10 km wide (Schweiger 1959). The hybrid zone then runs south to the Gulf of Venice, where it extends from Monfalcone to Trieste.

The complete transition zone is shown in Fig. 43.

There is some evidence of wandering outside the breeding season by both species into the breeding range of the other. Visual observations and ringing recoveries show that *domesticus* moves south over the high Alpine passes and equally, *italiae* moves

Fig. 43 Zone of intergradation between P. d. domesticus *and* P. hispaniolensis italiae

north beyond the transition zone; for example, Schweiger found five *italiae* males in Pressenger in Untergailtal, in August 1952, 30 km to the northeast of the hybrid zone; these appeared to be wanderers as they disappeared after two weeks and no *italiae* could be found there in the following year. In the same area a single male has been collected at Klagenfurt, 80 km from the transition zone south of the Karnische Alpen (von Wettstein 1959). In the west there are reports of *italiae* from the ornithological research station at la Tour du Valat in the Camargue between October and December (Hoffmann 1955), more than 200 km to the west of Menton, and Meise reported an early (1865) specimen from Nice, though whether this was a wanderer or an indication that the transition zone lay further to the west of its present position is not known. Again, how far these observations imply mere dispersal, with the birds subsequently becoming absorbed into the local population of the other species, or a true migration, has not been determined, though there have been recoveries of *italiae* in Italy of birds ringed in the winter at la Tour du Valat.

It is difficult to define the ecological conditions that determine the limits of the two species and the transition zone. Some authors have suggested that altitude may be an influencing factor, with *domesticus* more tolerant of high altitudes than *italiae,* but this is clearly an over-simplification as can be seen from the summary in Table 26; it is clear that *domesticus* does show a tendency to go to higher altitudes, but both taxa occur at the maximum recorded altitude for sparrows in the Alps, 2,063 m at San Bernardino. In the mountain range the hybrid zone is mainly found at altitudes between 500–1,500 m.

It is possible that the penetration of the high mountain valleys by these sparrows has been comparatively recent; for example, according to von Wettstein (1941), Trins in Gschnitztal, 1,210 m, was only colonised between 1924 and 1940, and Gschnitz,

House Sparrow 125

Table 26: Limits of the transition zone between P. d. domesticus and P. hispaniolensis italiae and maximum altitude for the pure phenotypes

Region	Height of Pass m	Maximum Altitude for *domesticus* Phenotype, m	Hybrid Zone Location	Approx. Extent, km	Maximum Altitude for *italiae* Phenotype, m
Côte d'Azur	—	—	Menton to Imperia	40	—
Alpes des Cottienes	Montgenèvre 1,845	la Vachette 1,321	Western (French) side	5	Montgenèvre 1,854
Massif du Mont Cenis	Col du Mont Cenis, 2,083	Lanslevillard 1,479	Complete separation	0	Gaglione, ca. 750
	Col du Picco St Bernard 2,478	la Rosière du Montvalescan, 1,850	Eastern (Italian) side of col: la Thuile, 1,441, to Aosta, 583	30	la Thuile, 1,441
Mont Blanc		Courmayeur, 1,228 Chamonix, 1,037			
Canton Valais (Wallis) Switzerland	Grand St Bernard, 2,478 Simplon Pass, 2,009	?	Rhône valley, Canton Valais, north of cols: Martigny-Ville, 476, to Brig, 713, south to Mauvoisin, 1,900	20	Saas Fee, 1,790
Canton Ticino (Tessin) Switzerland	St Gotthard, 2,112	S. Bernardino, 2,063	Canton Ticino south of col from Airolo, 1,100 to Biasca, 309	30–35	S. Bernardino, 2,063
Ober Inntal	Reschen Scheideck Pass, 1,510	Reschen, 1,494	Spanning pass from Nauders, Austria, 1,365, to Lasa, Italy, 800	35	S. Valentino alla Muta, 1,470
Tirol	Brenner, 1,375	Brenner, 1,375	South of col from Colle Isarco, 1,065, to Chiusa, 523	30	Colle Isarco, 1,065, Dobbiáco, 1,250
Karnische Alpen	Plöcken Pass, 1,363	Plöcken, 1,363	South side of col, ca 600	5	Paluzza, 600
Trieste	—	—	Monfalcone to Trieste	30	—

1,242 m, further up the valley, apparently some time later as von Wettstein did not mention sparrows there in 1940, though I found them in 1985. The recent penetration is confirmed by a statement by Ticehurst and Whistler (1927) that it seemed possible that the spread into the hills had only occurred in the last 40 years, that is, at the end of the last century. The position actually in Brenner is less clear; Schöll found no sparrows on the Italian side in 1959, though he saw 10 on the Austrian side, where according to a customs official they had been present for some time. Both Niethammer and von Wettstein failed to find sparrows in 1958, despite searching for them. This would suggest that colonisation did not take place until the 1950s. It may well be a matter of chance which species arrived first and now dominates. The major factor in maintaining the separation would appear to be the high mountains – Johnston points out that this barrier would have been even more effective during the 'little ice age', 500–600 years ago, when a persistent period of cold weather completely closed the Alpine passes throughout most of the year – with incursions of both phenotypes maintaining the hybrid zone, but neither able to make any significant penetration into the range of the other.

The main point of interest is that not only is the transition zone sharply limited in extent (usually no more than 30 km wide, see Table 26), but it appears to be stable in time as far as can be judged from the historical records, even when not controlled by the physical barrier of the high mountain passes. This implies that there are subtle ecological, and perhaps behavioural, factors operating to maintain separation that override the incursions of the two phenotypes into each other's range.

Passer domesticus tingitanus

This is a completely allopatric race that occurs in two discrete populations. The major range is in northwest Africa, where it occurs through most of Morocco, northern Algeria (south to the oasis towns of Beni-Abbès, Adrar and Ain Salah), northern Tunisia and extreme northwest Libya, though in eastern Algeria, most of Tunisia and northwest Libya it hybridises with *P. hispaniolensis*, forming a complete range of intermediates, better described as a 'hybrid swarm' to distinguish it from the more uniform 'hybrid' phenotype that occurs in Italy and on Crete. The distribution of this population and the extent of the hybrid region is shown in Fig. 44 (Summers-

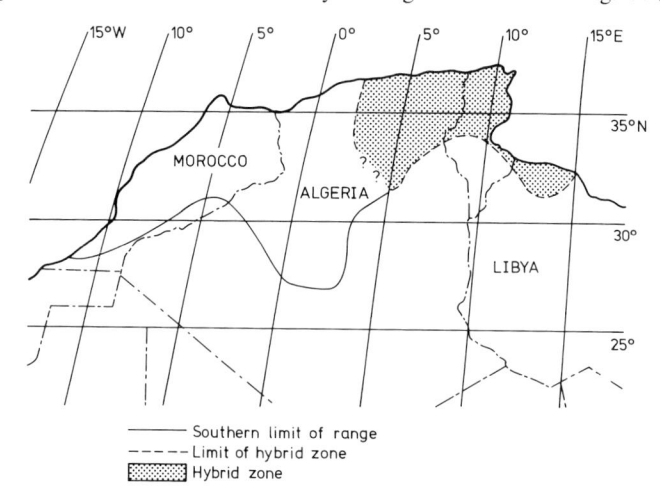

Fig. 44 Range of House Sparrow P. d. tingitanus *and the zone of hybridisation of House and Willow Sparrows in northwest Africa*

Smith & Vernon 1972). In addition, there is an isolated population to the east of Cyrenaica, where, according to Bundy (1976), it occupies most towns and villages east of Ajdābiyah to the Egyptian border. There is evidence to suggest that this population also shows some hybridisation with *P. hispaniolensis* (Hartert 1923, Stanford 1954).

Passer domesticus niloticus
This race occurs along the Mediterranean coast of Egypt from Alexandria to El 'Arish in Sinai, then south along the towns and villages of the Nile valley to Wadi Halfa, and south along the Suez Canal and the Gulf of Suez to Abu Zenima on the Sinai coast. According to Vaurie (1956) where *P. d. niloticus* meets *P. d. rufidorsalis*, a member of the Oriental (*indicus*) group, and very distinct from *niloticus*, the two forms do not intergrade smoothly, but form a narrow transition zone where the population is very variable.

Passer domesticus biblicus
Occurs in Cyprus, Asia Minor and the Near East. As already mentioned, it is separated from the nominate race at the Continental divide between Europe and Asia and by the Caucasus massif in the northeast. The boundary with *persicus* is less clear, but probably lies from the southwest corner of the Elburz mountains to the head of the Persian Gulf. According to Meinertzhagen (1954), it 'merges' into *indicus* in northern Arabia (Tabūk, Al Jawf and further north), though no detailed examination or specimens from that area has been reported. It is separated from *niloticus* by the Sinai Desert.

Passer domesticus persicus
The final race of the Palaearctic group adjoins *biblicus* in the west and extends from south of the Elburz mountains in the north to the Persian Gulf. To the east, there is a wide zone of intergradation with members of the Oriental group, running from eastern Iran in Khorāsān and Kermān to the borders of Afghanistan and Pakistan.

Passer domesticus indicus
This race has a very wide distribution in the Oriental region from Burma (south through Arakan to Rangoon, but not in Tenasserim), west across the Indian subcontinent into west Pakistan, where it breeds up to altitudes of 3,000 m in Baluchistan, and into southeastern Iran, where it intergrades with *persicus* as already mentioned. In the north, it extends to the foothills of the Himalayas, reaching up to altitudes of about 2,000 m in the west and to about 1,500 m in Sikkim, where it is replaced by *Passer montanus* at higher altitudes. It extends to Sri Lanka in the south, reaching its southern limit at Matara.

Further to the west it occurs over most of Arabia south of *biblicus*, except for a band stretching along the western shore of the Persian Gulf to Oman, where it is replaced by another race *hufufae*.

At higher altitudes in the Himalayas and Baluchistan it is only a summer visitor, withdrawing to the foot of the mountains in the non-breeding season.

Passer domesticus hufufae
Occupies the eastern side of Arabia from about Al Hufuf in the north to Oman in the south. It is almost completely separated from the *indicus* population of the Arabian peninsula by the arid central plateau in which House Sparrows do not occur.

Passer domesticus rufidorsalis
Occurs along the Nile valley from about Wadi Halfa, where there is a narrow zone

of intergradation with *niloticus*, as already described, south to about Renk on the White Nile.

Passer domesticus hyrcanus

This is a race of restricted distribution, occurring from the south shore of the Caspian Sea to the Elburz mountains, which separate it from *persicus* and *indicus*, the subspecies lying to the south of the Elburz range.

Passer domesticus bactrianus

Occurs in Russian Turkestan from eastern shore of the Caspian Sea eastwards to Semireche and north to Karaganda, east of the Aral Sea and south to about 40°N, where as already described its range overlaps with that of the nominate race. To the south the breeding range extends over much of Afghanistan (excluding the lowlands of the south and west) and northern Baluchistan. It intergrades with *persicus* in western Afghanistan.

Unlike the races considered so far it is almost completely migratory, withdrawing to the plains of northwest Pakistan and India in the winter months.

Passer domesticus parkini

The final subspecies of the Oriental group occurs in the high Karakoram and Himalayan ranges north of *indicus* at altitudes above about 1,500 m from Kashmir and Ladakh in the west to southern Tibet and southern Nepal in the east. It mostly occurs above 2,000 m and breeds up to 4,500 m in Ladakh. It is separated from *bactrianus* by the Pamirs and is largely migratory, descending to lower ground in the winter, though in milder winters some birds can remain in the breeding areas (Scully 1881). The birds of the Punjab Himalayas, which are sedentary, are intergrades between *parkini* and *indicus*.

The distribution in the Palaearctic and Oriental regions of the above eleven races of the House Sparrow is shown in Fig. 45. Different hatching is used to distinguish between the races of the two groups; boundaries are approximate.

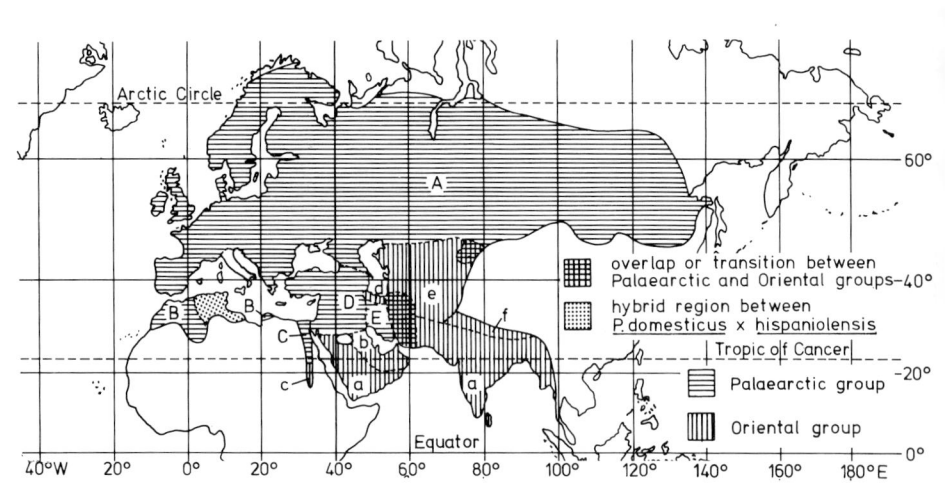

Fig. 45 Distribution of races of House Sparrow. *Palaearctic group:* A domesticus; B tingitanus; C niloticus; D biblicus; E persicus. *Oriental group:* (a) indicus; (b) hufufae; (c) rufidorsalis; (d) hyrcanus; (e) bactrianus; (f) parkini

Grey-headed Sparrow
P. griseus

Swainson's Sparrow
P. swainsonii

Parrot-billed Sparrow
P. gongonensis

Swahili Sparrow
P. suahelicus

Southern Grey-headed Sparrow
P. diffusus

Golden Sparrow
P. l. euchlorus

♂ ♀

Golden Sparrow
P. l. luteus

♂ ♀

♂ ♀

Chestnut Sparrow
P. eminibey

Great Sparrow
P. motitensis

White Nile Rufous Sparrow
P. m. cordofanicus

Kenya Rufous Sparrow
P. m. rufocinctus

Iago Sparrow
P. iagoensis

Willow Sparrow
P. h. hispaniolensis

Italian Sparrow
P. h. italiae

House Sparrow
P. d. domesticus

House Sparrow
P. d. indicus

Dead Sea Sparrow
P. moabiticus

♀ ♂

Sind Jungle Sparrow
P. pyrrhonotus

♀ ♂

Pegu Sparrow
P. flaveolus

♀

♂

Tree Sparrow
P. montanus

♂

♀

Cinnamon Sparrow
P. rutilans

♂

♀

Somali Sparrow
P. castanopterus

Cape Sparrow
P. melanurus

♂

♀

RG

Desert Sparrow
P. simplex

Saxaul Sparrow
P. ammodendri

The 'natural' range of the House Sparrow has, however, almost been doubled since the middle of the 19th century by deliberate introductions to other parts of the world by acclimisation societies set up by colonists to introduce familiar plants and animals from their original homelands. This was done either on sentimental grounds or on the assumption, usually mistaken in the case of the House Sparrow, that they would control some insect pest. Most of these introductions ceased by the end of the century, but the spread is continuing in some of the areas of introduction and has also been aided by a few more recent, accidental, or at least rather less deliberate, introductions.

The introductions have involved mostly birds of the nominate race from western Europe, or of the subspecies *indicus* from the Indian subcontinent, though *rufidorsalis* has been introduced to the Comores Archipelago and in a few cases the affinity of the introduced birds has not been clearly established. Some evolutionary differentiation of the introduced birds has been found, notably in North America by Johnston, Selander and their co-workers (Johnston & Selander 1964, 1971, 1973a, 1973b, Selander & Johnston 1967, Johnston 1969c, 1973, 1976, Packard 1967b, Hamilton & Johnston 1978), but also in New Zealand (Baker 1980) and in South Africa (Crowe, Brooke & Siegfried 1980), though as yet this has not led to the description of new races.

I have described in some detail the introductions and the course of the spread up to 1960 (Summers-Smith 1963) and it would be out of place to repeat this here. Much has, however, occurred since then and it is appropriate to bring the situation up to date.

NORTH AND CENTRAL AMERICA

The first successful introduction was from Europe to New York City in 1852 (birds liberated in 1851 apparently did not survive) and fresh introductions continued until 1880 at least, mainly in New England and the northeastern states, but also further afield in Wisconsin, Iowa, Utah and Texas, with the spread stimulated by trans-shipments of American reared birds to other parts of the country. By the end of the century it was present in all States of the Union (see Barrows 1889, Bent 1965 and Robbins 1973 for further details), and had penetrated to the limits of cultivation in Canada. It is now resident in scattered pockets in southern Yukon, North West Territories (*eg* Fort Smith, Fort Simpson), and the northern parts of the remaining Provinces, *viz* Lake Athabasca in Saskatchewan, Churchill in Manitoba, Fort Severn in Ontario, Chibougamau in Quebec, St Anthony in Newfoundland (Godfrey 1966, American Ornithologists' Union 1983). There is a recent sighting of 4–5 birds at Anchorage Airport, Alaska, in June 1981 (C. R. Cole, pers. comm.), though no evidence that it is established there as a breeding bird.

Surveys showing the relative abundance of the House Sparrow in North America have been reported by Robbins (1973), based on roadside counts during the breeding season, and by Wiens & Johnston (1977) using the Audubon Christmas bird count data from Hailbrun & Arbib (1974). While there are differences in detail, both surveys show the maximum concentrations in the central and midwest wheatlands, on the one hand, and the populous northeast on the other. These maps are reproduced in Fig. 46.

The bird reached Mexico at the beginning of the present century and the spread has continued into Central America, with records from Quezaltenango, Guatemala, in 1966, the Department of La Paz, El Salvador, in 1972 and from Cartago, Costa Rica, in 1975 (Reynolds & Stiles 1982). It is presumably also present in Belize, Honduras and Nicaragua, though I know of no published records from these countries. It is now widespread in Costa Rica, has been reported from Concepción in Chiriqui, the westernmost province of Panama (Reynolds & Stiles 1982) and more recently from eastern Panama Province (Monroe 1983).

The House Sparrow is present in five Caribbean island groups. In the Bahamas, it

130 *House Sparrow*

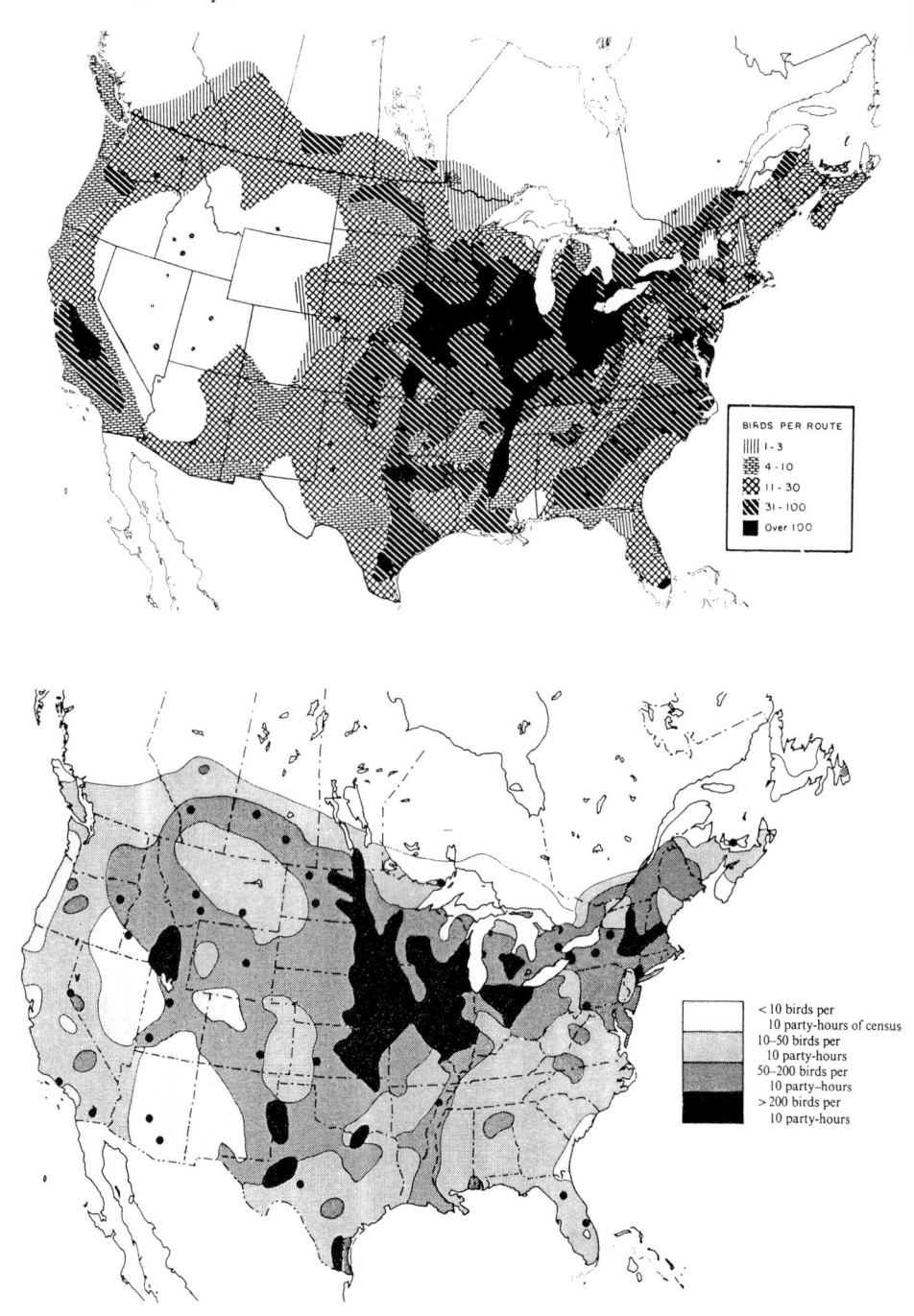

Fig. 46 Distribution and relative abundance of House Sparrow in North America during breeding season (upper map) *and in mid winter* (lower)

was introduced to Nassau, New Providence Island, in 1875, but is said to have been exterminated by a hurricane in 1909 (Gebhardt 1959); according to the *Check List of North American Birds* (American Ornithologists' Union 1983) and B. L. Monroe, Jr. (*in litt.*), however, it was present on Grand Bahama and New Providence Islands in 1971 and presumably it is still there. It was introduced to Cuba in the 1890s and is present chiefly in the large towns (Bond 1960). The *Check List of North American Birds* (American Ornithologists' Union 1983) gives it as present on Hispaniola and Puerto Rico in 1978. Finally, it has been present in Charlotte Amalie, St Thomas Island in the Virgin Islands since 1954. It was introduced to Jamaica about 1903, near Annotte Bay in the north, and spread from there before apparently decreasing in numbers; Lack (1976) stated that the last published record was in 1966 and he failed to find any in 1971.

SOUTH AMERICA

The original introductions were to Buenos Aires, Argentina, in 1872, Santiago, Chile, in 1904 and Rio de Janeiro, Brazil, in 1905, with known transplantations to Punta Arenas, Chile, in 1918 and Callao, Peru, in 1953. When I reviewed the position in 1960, the distribution was continuous in inhabited parts from Ushuaia, Tierra del Fuego, to about 12–15°S right across the continent. Since then it has spread north, reaching Ecuador in 1969 and is now north of the equator at Esmeraldas, where it was found in 1977 (Ortiz-Crespo 1977) and in February 1986 at Puerta Merizaldo, 60 km south of Buenaventura, Columbia (C. Hinkelmann, pers. comm.). In the centre it has consolidated the position in Bolivia, where it goes as far north as Magdalena (Dott 1986).

In Brazil, House Sparrows have almost reached to the equator in Pará Province. Fig. 47 shows recent reports for northeastern Brazil south of the Amazon (Müller 1967, Sick 1968, 1979, Smith 1973, 1980), together with tentative lines showing the timing of the spread. Smith postulates that there have been two lines of spread: one from Brasilia, where the bird was first reported in 1959, north along the Brasilia-Belém highway to Maraba (1964) and Imperatriz (1965), with more recent extensions east to Floriano (1964), north to Itinga (1973), Itupiranga (1974) and Vila Rondon (1979); the other along the coast from Recife to where they were introduced in 1963, with records from Fortaleza (1968), São Luis, Maranhão (1979), Salinópolis (1976), São Caetano de Oliveira (São Caetano de Odivelas) (1979), Mosqueiro (1978), Belém (1978) and Castanhal (1978), with extensions inland to Teresina (1979) and Bacabal (1979). This is not an area subject to close ornithological study and the first recorded dates are not necessarily those of first arrival, though sparrows were not found in Belém in 1967. With that proviso, Smith's hypothesis seems consistent with two major lines of penetration along the newly developed highways, particularly as he was unable to find any sparrows in Paragominas in 1977 and 1978, and São Miguel do Guamá in 1979, towns on the Brasilia-Belém highway north of Itinga. Sick (1968) attributes much of the recent extension in range to the development of the road network in central Brazil that occurred after 1957, suggesting that birds were carried deliberately by road to Brasilia and also to parts further north on the route to Belém, whereas previously the spread could only take place along the rivers that up to then had provided the principal communications. The probability that much of the spread has been the result of human agency is shown by the large gap between Teófila Otôni, 17°25'S, and Recife, 8°06'S, in central coastal Brazil. There is a possible record for Maceió, 9°40'S (Sick 1968), but even if this is confirmed there is still a long stretch where House Sparrows are lacking.

The record from Belém is a particularly interesting one, as House Sparrows were previously reported there in 1927 (most probably ship-borne immigrants), though they disappeared after two years. Sick, however, was unable to find any in 1979, the

132 *House Sparrow*

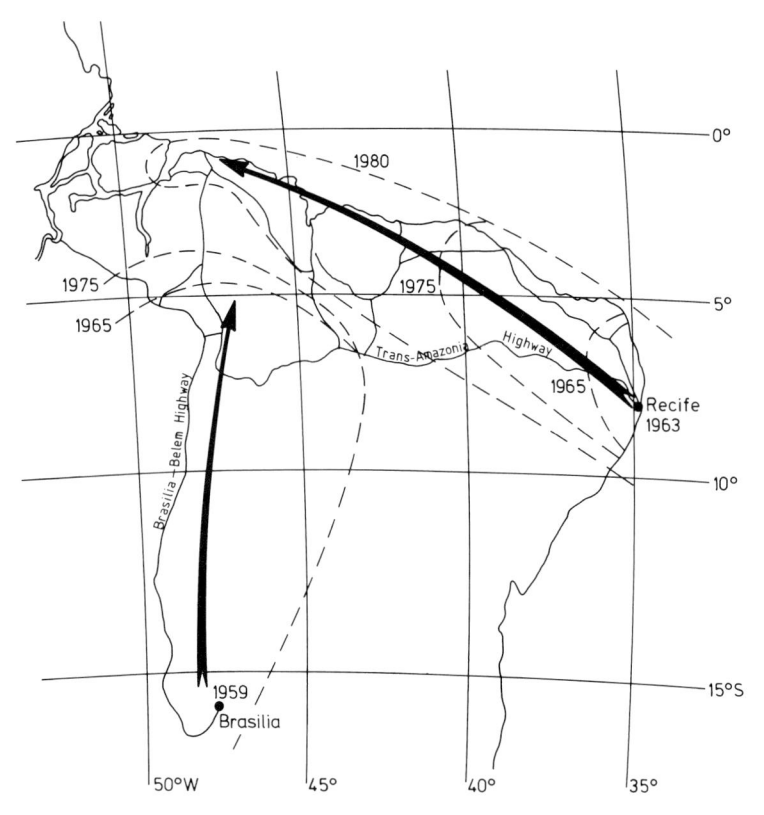

Fig. 47 Spread of House Sparrow in northeastern Brazil south of the Amazon

year following the sighting by Smith. With an annual rainfall of 3,000 mm, this could well be a marginal location for the somewhat arid-adapted House Sparrow.

It seems inevitable that the birds from Central America and South America will join up in Venezuela in the near future. I searched for House Sparrows in Maracaibo and Caracas in 1983, but without success.

In my previous review I was unaware that House Sparrows had been introduced to Easter Island from Chile in 1928. Despite the island being uninhabited, the sparrows have apparently thrived, breeding in sea caves in the absence of human habitations and trees (Sick 1968, Johnston et al. 1970).

About 20 House Sparrows arrived at Stanley in the Falkland Islands in November 1919, having travelled in four whaling vessels from Montevideo, Uruguay (Bennett 1926). By 1961 they had spread to Teal Inlet, Fitzroy, Green Patch and Port Darwin in East Falkland (Cawkell & Hamilton 1961). Three colonies have been established in West Falkland: in October 1959 three pairs arrived on Carcass Island and about the same time a small colony was formed on West Point Island; more recently, another colony has been reported from Port Stephens (Woods 1975, 1982). The first two of these lie off the west coast of West Falkland, the third in the southwest; all are over 150 km from the nearest colonies in East Falkland. It seems unlikely that the birds have made the journeys unaided and more probable that they have travelled from Stanley aboard local coastal vessels.

AUSTRALIA

The House Sparrow was introduced to Victoria in 1863. From this and subsequent introductions up to 1870, the bird has colonised most of eastern and southeastern Australia, the islands of the Bass Strait and Tasmania (Slater 1975, Long 1981). According to Slater, the spread north and westwards is continuing with the birds now north of the tropic of Capricorn. Birds occasionally penetrate to Western Australia, where there were five positive sightings near Fremantle between September and December 1962 (Gooding & Walton 1963), though nothing further has been reported of these birds and Western Australia remains uncolonised through vigilant action by the human inhabitants.

There is also a record for Port Moresby, Papua New Guinea, where five were seen on 30.12.76. Sightings continued up to 7.1.77, when the birds disappeared. These birds were presumed to be ship-borne immigrants; they do not appear to have survived.

AFRICA

The greatest change in the recent status of the House Sparrow has taken place in the Afrotropical region. The details of the introductions to South Africa are by no means clear. The first introduction was of birds of the Indian race, *P. d. indicus*, to Durban, Natal, almost certainly at the turn of the century, though the precise date is in doubt: 1893 and 1897 according to Mackworth-Praed and Grant (1963), *ca* 1902 (Gebhardt 1959). Birds from Europe, *P. d. domesticus*, were first liberated at East London, Cape Province, with various authorities quoting dates between 1907 and 1930 (Courtney-Latimer 1955, Mackworth-Praed & Grant 1963, Winterbottom 1959, Harwin & Irwin 1966, Vierke 1970). Other reported introductions of the nominate race were to Lourenço Marques, Mozambique (from Portugal) in 1955 (Pinto 1959) and Harare, Zimbabwe, in 1957 (Harwin & Irwin 1966). There is no evidence, however, of much spread of birds of the nominate race from any of these points of release; most of the expansion of range has been of *P. d. indicus* from the original point of introduction in Natal, and for about 50 years even this was on a very modest scale.

The situation began to change dramatically about 1950 and there was a sudden extension of range that took the bird all over the Republic of South Africa and northwards into Namibia, Botswana, Zambia, Malawi, Swaziland, Mozambique and the extreme south of Zaire and Tanzania (Harwin & Irwin 1966). Fig. 48 shows the expansion in 5-year intervals from 1950.

While there is no clear evidence, it seems possible that the more southerly *indicus* race is better adapted physiologically to a high temperature environment than the nominate race and this could have given it an advantage in southern Africa. There is circumstantial evidence, however, from a study of introduced birds of the nominate race in North America by Blem (1973) over a latitudinal range going from 28°N in Florida to 59°N in Manitoba, that demonstrates such an adaptation. This showed that the more northerly birds were able to withstand significantly lower winter temperatures, these birds apparently being better insulated against heat loss through having greater amounts of subcutaneous fat. In addition to the direct ability to cope with lower environmental temperatures, the increasing insulation also enables the birds to survive the longer winter nights without food. Blem takes pains to point out that there is no direct evidence that these adaptations are under genetic control. To some extent the ability of House Sparrows to survive in such a wide range of environmental conditions is the result of behavioural adaptations – living in towns where there are more favourable microclimates, living inside buildings in the far north (grain elevators in northern Canada, cattle byres in northern Norway), the use of

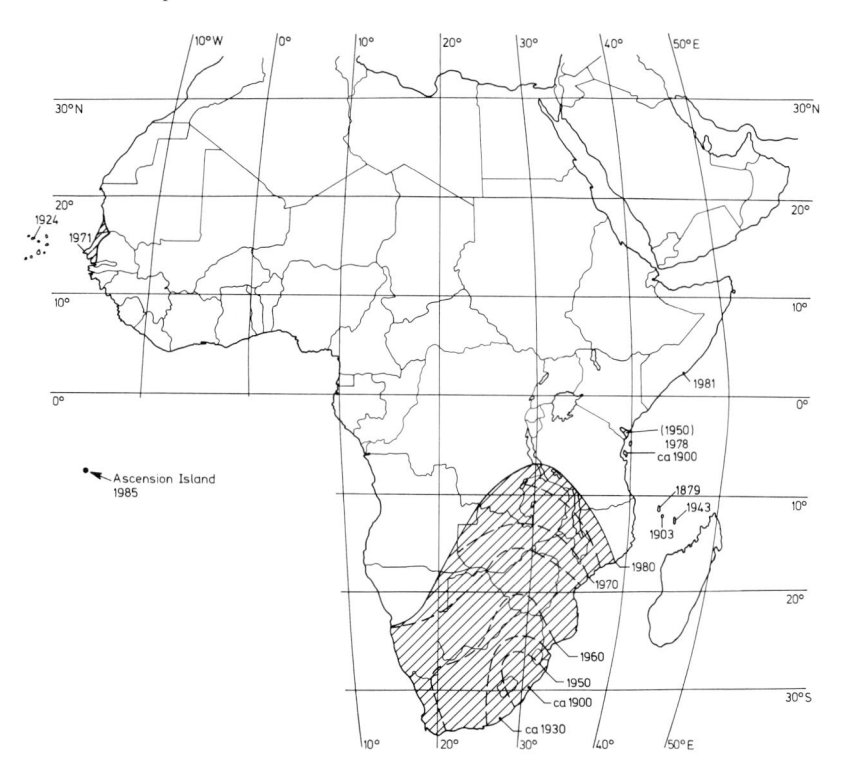

Fig. 48 Dates of introduction of House Sparrow in Afrotropical region and timing of spread in southern Africa

winter roost nests – but it is plausible that there are also genetically controlled adaptations, even more so when the different races are considered. A gradual adaptation by nominate *domesticus* to more tropical conditions might also account for the recent success of the House Sparrow in Amazonia, compared with its original failure to colonise Belém, and also the recent colonisation of Columbia, where a previous introduction had failed (Sick 1968).

There have been no recent changes in the situation off the east coast of Africa, where it is present on three of the islands of the Comores archipelago (Grande Comore since about 1879; Moheli since 1903 – birds of the race *rufidorsalis*; Mayotte since 1943) and Zanzibar, where birds of the race *indicus* were introduced about 1900 but, though well established, are still confined to the port area of the town (Pakenham 1979). At last the bird seems to be establishing itself in Kenya. There have been odd reports of House Sparrows in Mombasa dating back at least to 1950 (Gebhardt 1959, Mackworth-Praed & Grant 1960), though Williams and Arlott (1980) stated that its present status was unknown and Pakenham (1979) reported that it was not now known to occur. In February 1978, I saw a single female in Mombasa and P. L. Britton found a pair in April 1979. Subsequently, a small colony was discovered near the railway station; this was thought to be a fresh introduction and not of birds remaining from the earlier reported dates (Ash & Colston 1981). It is now sufficiently common to have become a nuisance at Moi Airport in Mombasa and is spreading along the coast and inland along the Mombasa-Nairobi highway, where it has reached Mtito Andei, 200 km from Nairobi (G. R. Cunningham-van Someren, *in litt.*) Further north, three birds (1 male, 2 females) appeared in Mogadiscio, Somalia, in December 1981, but nothing is known of their fate since then (Ash & Miskell 1983).

House Sparrows are present in two locations in the west of Africa. I overlooked, in my previous survey, that they had arrived at Mindelo, São Vicente, Cape Verde Islands, between 1922 and 1924 (Bourne 1966). They were still present when I visited in 1983, but restricted to Mindelo and the impoverished farmland to the south of the town (Summers-Smith 1984b). An examination of the specimens collected on the 1923–24 *Blossom* expedition, by courtesy of the Curator of the Peabody Museum, New Haven, Connecticut, USA, where they are now lodged, showed that these birds are of the nominate race.

A more recent expansion in range followed its arrival in Dakar, Senegal, about 1970 (Ndao 1980). Since then it has spread north along the coast to Nouakchott, Mauritania, south to Banjul, The Gambia, and about 100 km inland to Podor on the Senegal river, as well as to Diourbel and Kaolack. So far its occurrence is limited to the European style towns (M.-Y. Morel, in prep. and pers. comm.). A preliminary examination of a few skins suggests that the source of the introduction was South Africa; a ship-borne immigration seems most likely.

Even more recently a House Sparrow turned up on Ascension Island at Christmas 1985, again most probably by ship as there is currently considerable sea traffic there, and since then 'half a dozen' more have been introduced 'to keep it company'! (W. R. P. Bourne, *in litt.*). It remains to be seen if it will be any more successful than those on St Helena, where 26 House Sparrows introduced in 1820 died out after a few years (Melliss 1870).

It has been noted in the Afrotropical region that the House Sparrow only occurs in European style towns and has not become established in settlements composed only of African huts (for example, see Cole 1962).

The range of the House Sparrow in the Afrotropical region is shown in Fig. 48.

EUROPE

The only recent change in distribution of significance in Europe has been the successful colonisation of the Azores. According to Agostinho (1963), 'some tens of sparrows, brought from Portugal, were set free at Lajes Airport on Terceira in 1960'; by 1982 the bird had overrun the island (Le Grand 1983). All the other inhabited islands have now been occupied with the exception of Santa Maria and Corvo; it was established as a breeding species on Graciosa, São Jorge, Pico and Faial by 1970; it had arrived on São Miguel by 1972 or 1973 and on Flores in 1982 or 1983 (Le Grand 1977, 1983 and pers. comm.). I estimated the breeding population on the archipelago in 1984 as 50,000–60,000 birds.

ASIA

In addition to the Mascerenes, which were colonised in the last century, the House Sparrow has been introduced to a number of other Indian Ocean island groups. They were reported as common on two of the islands of the Chagos archipelago in 1962 (Loustau-Lalanne 1962) and in the same year some captive birds were liberated on Malé in the Maldives (Ash & Shafeeg in prep.). According to the latter workers these birds were reinforced by about a hundred more that had arrived on a grain ship from Sri Lanka in 1980 and, despite attempts at extermination by a rodent control team the following year, the population has survived and is now thriving.

The situation on the Andaman Islands is slightly confused. Gurney (1866) reported that there had been an introduction of House Sparrows from Calcutta, India, some time prior to his visit, though there were doubts about its success as he was unable to find any in 1866. Butler (1899) stated that about half-a-dozen were released on Ross Island, South Andaman, in 1882; they remained there for some time and then

136 *House Sparrow*

Fig. 49 Distribution of House Sparrow

disappeared, apparently moving to the main island. A further 20 were imported to South Andaman in 1895. Butler found them in 1899 and according to Abdulali (1964) they have thrived as he found them common at Port Blair and also in other localities in South Andaman. Abdulali found the birds on South Andaman to be closer to Burmese birds (originally separated as *P. d. confucius*) than to typical *indicus* from India and suggested that they originally came from Burma aboard the mail steamer from Rangoon.

The current world distribution of the House Sparrow is shown in Fig. 49.

HABITAT

The House Sparrow is primarily associated with human habitations, farms and villages, including the surrounding land with cereal cultivation, and also in more extensively settled areas, where it occurs from the suburbs with their gardens to the more built-up centres, particularly where there are squares and parks, and industrial areas with waste ground. Heij (1985) in Rotterdam, The Netherlands, and Earlé (in press) in Bloemfontein, South Africa, in comparative studies of House Sparrows in urban, suburban and surrounding rural areas, both showed that the bird did best in the suburban area in terms of breeding density, an earlier start to breeding, higher clutch size, greater breeding success and a higher survival rate. Furthermore, they found that the suburban population attained its stable level with juvenile recruitment in the autumn, whereas this did not occur in the rural population until the spring, the surplus from the suburban population making up the deficit in the rural one. These findings suggest that the optimum habitat for the House Sparrow, in a temperate region at least, is a combination of buildings with holes under tiles or eaves to provide suitable nest sites and sufficient green areas to provide insect food for the young. Urban zoos, with availability of food supplied to the animals, together with pickings to be obtained from open-air cafes, provide particularly favourable conditions for House Sparrows. The bird is not, however, entirely confined to human habitations or areas of cereal cultivation; it can occur in almost any habitat, from sea level, searching for food along the tide line and in marram grass in sand dunes, up to altitudes of 4,500 m, though it does not penetrate into denser woodland or forest, or tundra or high open country, such as mountain grassland and maquis.

Although, with its broad geographical distribution, the House Sparrow is clearly tolerant of a wide range of climate, its preference is for dry conditions; for example, Reynolds and Stiles (1982), in a study of the colonisation of Costa Rica by House Sparrows, showed a negative correlation between the density of the sparrow population and the rainfall. It is even able to survive in areas without open water by feeding on berries (Walsberg 1975). It also shows a high salt tolerance (Minock 1969), another feature suggesting adaptation to arid conditions.

In areas with long dry summers and open country, it becomes increasingly frequent away from towns and villages. This is the case in southern Europe; for example, in Cyprus, it is an exceedingly common bird in the low lying cultivated land, where it nests in trees, telegraph poles and holes in earth embankments, as well as in towns and villages, and this is also taking place with the introduced birds in the Azores (G. Le Grand, pers. comm.). A similar situation obtains in many parts of Asia. Sladen (1919) reported that in Palestine it nested in caves and rocky cliffs far from houses; Philby (quoted by Bates 1936) found it breeding in a desert valley in central Arabia, where Meinertzhagen (1958) and Jennings (1981a) state that it breeds many kilometres from buildings and farmland, becoming a bush bird; the same was found by Gallacher and Rogers (1978) in Bahrain. In southern Iran, Desfayes and Praz (1975) found House Sparrows well away from human habitations, and even cultivation, breeding in loose colonies in cliffs; the same is reported for Tamil Nadu in southern India (Davison 1883). Hobbs (1955b) found the introduced House Sparrow breeding beside dams in grazing country in the Riverina district of New South Wales, Australia, more than 15 km from the nearest habitations, and the bird has even colonised some uninhabited islands – Easter Island, as already mentioned, and White Island, 50 km off the mainland of New Zealand in the Bay of Plenty (Wodzicki 1956), where in both cases they nest in the cliffs. In northern Iran, Afghanistan and southern Kazakhstan, where the sedentary Tree Sparrow is the sparrow of the inhabited areas, the migratory House Sparrows of the race *bactrianus* are almost exclusively open country birds and can be found well away from cultivated areas.

In arid and desert regions the House Sparrow is restricted to oasis towns and villages; this is also the case in regions of high rainfall with extensive forest, as, for example, in the 'wet zone' of Sri Lanka, where the House Sparrow is restricted to towns and the clearings round villages. In these hot areas the birds spend most of the time in shade trees.

Where conditions are more severe distribution becomes more patchy; not all villages have their populations of House Sparrows. This is the situation in the Himalayas, the high Swiss alpine villages, the higher villages in the Black Forest (Schwarzwald) in Germany and in villages and towns in northern Europe and Siberia.

Its penetration of the built-up environment is quite remarkable, not only in the open where they have been observed feeding around the observation floor of the Empire State Building in New York City (80 stories up!) at night (Brooke 1973), but many House Sparrows appear to spend much of their time indoors. This applies not only to birds living at high latitudes, that spend their lives inside cattle byres as they do in northern Norway; elsewhere they readily come inside large buildings, such as railway stations and factories. And not only buildings with large doors or ready access to the open; a number of House Sparrows lived in the Queen's Building of London Heathrow Airport when this was used as a passenger terminal, feeding on scraps at the cafeteria and resting in the security of concealed roof lights. This was an air-conditioned building and it is unlikely that the birds could move freely in and out. Airport terminals in the tropics provide excellent locations for House Sparrows; the newly arrived birds in Senegal nest commonly in buildings at Dakar airport. Even more extraordinary, however, are House Sparrows that lived for several years in a coal mine in South Yorkshire, England, 640 m below the surface, where they survived on food provided by the miners and even bred successfully in this alien environment (Summers-Smith 1980a).

BEHAVIOUR

The House Sparrow is a social species; this is shown in a number of ways. In large areas of apparently suitable habitat it tends to breed in small colonies, usually of 10-20 pairs, rather than spreading uniformly, though probably better described as a 'clumped' rather than a colonial breeder. Outside the breeding season it normally is to be found in flocks that associate in many activities, ranging from communal roosting to feeding, dust and water bathing, and what I term 'social singing', when the birds collect in bushes and call together; this can occur when they emerge from their roosting sites prior to searching for food and regularly on dark winter afternoons.

As soon as the fledged young become independent of their parents, they group together where there is a suitable food supply, such as in waste areas with grass and weed seeds; in grain-growing areas they gravitate to the ripening cereal fields, where they are joined by adults that have finished with breeding. These grain-field flocks can amount to several thousand birds and can do considerable damage, not only by feeding on the grain, but also by breaking off the ears on which they perch. In general they prefer to feed close to cover, for shelter in the event of danger. This means that fields with hedges and hedgerow trees tend to suffer the most damage and particularly a strip about 5 m from the nearest cover. However, according to Barnard, who has extensively studied flocking behaviour in House Sparrows (Barnard 1979, 1980a,b,c,d), larger flocks (30 or more birds), in which individuals need to spend less time scanning for predators, will move further from cover, perhaps up to 10–15 m (Barnard 1983). This movement to the grain fields is, in general, limited to birds living in country districts or on the outskirts of towns (see, for example, Preiser 1957), though, as Fallet (1958) has shown from her study of the House Sparrow population in Kiel, West Germany, some decrease in numbers can occur even in quite large

threat

towns. According to her results, about half the birds of the year moved out into the countryside, 2–4 km round Kiel, during the period of ripening grain, returning about the middle of November.

Once the grain has been harvested and the fallen grain seeds gleaned, the birds of the non-migratory populations return to the breeding areas.

The House Sparrow shows a wide range of comfort/cleaning activities. Dust bathing and water bathing involve similar movements and are invariably social activities. Dust bathing creates little hollows in the ground and these are actively defended against competitors, using the same threat display as in disputes over food. The bird crouches with the tail erect, the wings held out and rotated to the front, showing off the white wing bars; from this position it may lunge forward with opened bill towards the opponent. Feeding and bathing bouts are followed by preening, and then spells when the birds rest together in cover, calling conversationally (social singing); this is particularly prevalent on dull winter afternoons, suggesting that even at this time of year they do not have great difficulty in obtaining their food. In addition, House Sparrows occasionally sunbathe, with feathers ruffled, and there has been one report of 'anting' (Common 1956). As well as normal preening activities, bill wiping is particularly frequent; to such an extent that it may have some ritualistic function, possibly an indication of mild anxiety – for example, the male frequently bill wipes during coition between successive treadings of the female.

Roosting is another communal activity, particularly with young birds that have not yet adopted a permanent breeding site. Large roosts are formed in dense trees by the post-breeding flocks at the grain fields, sometimes reaching spectacular proportions: C. Harrison described one on the outskirts of London that, on 29th August 1949, was estimated to contain 19,000 birds (see Summers-Smith 1963), and Moreau (1931) recorded a late summer roost in Egypt that at its peak contained about 100,000 birds. In urban areas in particular, the birds assemble in small groups in trees before flying to the main roost. These gathering places are so familiar that they have been given the name 'chapels', no doubt because of the social singing that again takes place. Communal roosts are also formed in the colony areas when the birds return after the summer grain-field flocking; sheltered sites, such as creepers on walls being particularly favoured. As the leaves fall in temperate regions and the roosts in trees and creepers become more exposed, the birds tend to forsake them for more sheltered sites in holes in buildings if these are available. In places where the winter climate is severe, special roost nests are built to reduce heat losses during long, cold nights (Mayes 1926–27; Kalmbach 1940, at 3,100 m altitude at Leadville, Colorado, USA; Janssen 1983, Minnesota, USA). The House Sparrow has also been recorded roosting on street lights, no doubt taking advantage of the heat from the lamps (Wynne-Edwards 1927–28).

House Sparrows, with the exception of the two migratory races, *bactrianus* and *parkini*, are among the most sedentary of wild birds, though in some of the nominally sedentary races there may be small scale withdrawals, during the winter months, of local populations living at high altitudes to the lower lying foothills, *eg* birds of the race *domesticus* in parts of the Swiss Alps, *indicus* in the Himalayas and Baluchistan. Observations on colour-ringed birds in England suggests that, once the birds have bred, they remain faithful to the breeding area, rarely moving away more than a kilometre or two (Summers-Smith 1956). A certain amount of movement, however, does occur, mostly involving birds of the year (for example, see Rademacher 1951). Most of this appears to be an undirected wandering or dispersal that takes place, first in autumn, after the break-up of the grain-field flocks and, secondly, at the beginning of the breeding season by birds that have been unsuccessful in establishing a breeding site. Preiser (1957), in his study in Germany, found that only 25% of the surviving young returned to their natal colony, whereas Lowther (1979b) and Fleischer *et al.* (1984) in a study in Kansas, USA, estimated that 25–50% returned after the autumn dispersal. According to the latter workers, dispersal was not related to the time of year that the young had fledged, but proportionately more females than males did not return to the colony where they had been born. The extent of movement is, however, limited and the majority of juveniles settle within a kilometre of their place of birth (Cheke 1973).

In addition, more directed movements can take place in the autumn, as evidenced by reports of House Sparrows on the North Sea coasts of England and the European continent (Summers-Smith 1956, 1963, Williamson & Spencer 1958), the east coast of North America (Broun 1972) and through the high passes of the Swiss Alps (Jenni & Schaffner 1984). Although these movements follow the normal autumn migration direction for the area, their irregular occurrence suggests they are in response to over-population in locally unfavourable conditions, *ie* an irruption type of movement rather than a true migration. The extent of the movement and the subsequent fate of the participants are not known, but ringing recoveries would suggest that the distance moved is not great. Apart from a few isolated recoveries up to 500 km, the vast majority are less than 10 km from the place of ringing.

To a limited extent some true migration occurs in the essentially non-migratory races where the winter conditions are extremely unfavourable for sparrows; for example, populations living in high Swiss alpine valleys disappear outside the breeding season (Jenni & Schaffner 1984); though in Norway, north of the Arctic Circle, the birds remain during the winter period without proper daylight by moving into the sheds with domestic animals.

The dispersal of juveniles is important in a sedentary species in preventing excessive inbreeding in local populations. Preiser (1957) showed this dispersal dramatically in an experiment in which, after a period of study, a substantial proportion of the young were ringed. The sparrow population in selected areas was then significantly reduced by poisoning. These 'sparrow-free' areas were quickly restocked by immigrants from the immediately surrounding areas. New areas that become suitable for House Sparrows, *eg* new housing estates, and of course new parts of the world to which the bird is introduced, are quickly colonised in this way. In addition, the small proportion that moves greater distances (4.3% of Preiser's birds moved more than 10 km from the place of ringing) also helps to give a wider spread that is subsequently infilled by the smaller scale dispersal mechanism.

As already mentioned, the races *bactrianus* and *parkini* are almost completely migratory, withdrawing from their breeding areas in Turkestan and the high mountain ranges to winter in the plains of northwest Pakistan and India, reaching as far south as Bahawalpur in Sind and Bharatpur in Rajasthan (birds ringed in the latter place have subsequently been recovered on their breeding grounds in Tadzhikistan), where they mix not only with each other, but with the sedentary Indian population of the race, *P. d. Indicus*. The migration is concentrated by geographical features. Gistsov and Gavrilov (1984) found House Sparrows (*P. d. bactrianus*), mixed with Willow Sparrows, funnelling through the Chokpakskii pass in the western Tien Shan on their way to and from the breeding grounds in Kazakhstan. Ringing recoveries showed that the birds tended to use the same route in spring as in autumn and, moreover, they were very consistent in the timing of the spring movement, with repeats in subsequent years within a few days of the original capture. Spring migration takes place from the third week in April and lasts until the end of May. *P. d. parkini* is absent from its breeding areas between October/November and March/April.

Dolnik (1973) found, by collecting *P. d. bactrianus* in southern Tadzhikistan from migrating flocks in autumn and spring, that the fat content was higher than normal with weight increases of up to 20–30% (including higher water as well as fat). In contrast, the weights of migrants of the nominate race passing through the Col de Bretolet, Canton Valais, in the Swiss Alps were no different from those of non-migrating birds in central Switzerland, thus showing no adaptation for migration and confirming that the nominate race is basically sedentary (Jenni & Schaffner 1984).

Despite females being smaller, they tend to be more aggressive towards males at feeding stations than *vice versa* (Johnston 1969b, Kalinoski 1975); at the nest site the female is dominant over her mate. Johnston suggests that high-level aggression by males outside the breeding season has been suppressed to avoid disruption of the pair bond.

Although in special circumstances, such as the London parks where the birds are regularly fed and come freely to people's hands for food, House Sparrows, despite their close association with man, perhaps even because of this, tend to be extremely wary and cautious. Even where they are regularly fed, a change in the method of putting out the food tends to put the sparrows off for a few days. Birds that are closely watched tend to show their nervousness by tail flicking and bill wiping, behaviour that is elicited by the presence of obvious enemies, such as cats.

BREEDING BIOLOGY

According to a study of individually marked House Sparrows that I carried out over a period of seven years in two locations in England, pairs remain faithful to each other and to their nest sites for life. This pattern is different from that of many other small birds, such as finches, that only form pairs for breeding. A few cases of bigamy occurred when a male took over an additional neighbouring site and its owner female after she had lost her mate. Bigamy has also been found by Deckert (1969) in Germany and North (1980) in the USA, though it was not reported, in either study, how the male acquired a second mate. The retention of the nest site by one pair is not, however, an invariable pattern. Weaver (1939) in a study in Ithaca,

New York, USA, found that as soon as the young of one pair left a nest site, it was immediately taken over by another pair. Naik and Mistry (1980) in India, in contrast, found that some pairs changed their nest sites for successive broods; there was no suggestion, however, from this study that pair faithfulness was not maintained. In yet further contrast, Sappington (1977), in another study on marked birds in Missouri, USA, found quite a different pattern: only 60.5% of pairs remained faithful during a particular breeding season, with 86% of males retaining a nest site for the whole breeding season, and only 10% using the same site in consecutive seasons. Females were less faithful, with only 45% remaining at a nest site for the whole of the breeding season. Sappington suggested that this may have been the result of excessive disturbance (old nests torn out), but the differences between populations may possibly reflect differences in availability of nest sites and the pressure for suitable places. Latitude could also be an influence as less protected sites are increasingly used in the warmer parts of the bird's range, with presumably less competition for particular sites.

Dr T. Burke (*in litt.*), in contrast to my observations of colour-ringed birds, found, through enzyme analysis of blood samples from adults and *pulli*, a high level of promiscuity in two populations studied near Nottingham, England. There is no obvious explanation for this difference: whether promiscuity occurs more frequently in large and dense populations (as Burke's were) or coition also occurs away from the nest sites, where it would have been more likely to have passed unnoticed in my studies, is not known. In fact, coition away from the nest site by two different males has been observed by Brackbill (1969) at the conclusion of a group display.

How then are the pairs formed? My observations suggest that the nest site provides the focus. Where the nest is retained throughout the year, the normal method of pair formation is by replacement of a lost mate. This can occur in sedentary populations at any time of the year, except for the immediate post-breeding period when all sexual activities are at a low ebb.

During the breeding season a lost mate of either sex is quickly replaced, indicating that there is a reserve of non-breeding birds available to fill the gaps. Weaver (1939), in fact, noted small flocks of immature birds throughout the breeding season. First-year birds tend to start breeding later than adults, and Selander and Johnston (1967) have even suggested that birds from the late broods of the previous year may delay breeding until their second year of life. Except in years of high adult mortality, or in the case of birds forming new colonies, there is no competition from mature adults with these birds; such birds do not develop full male plumage after post-juvenile moult, and Selander and Johnston argue that this could be an adaptation to avoid unnecessary conflict with mature adults. A dramatic example of mate replacement is given by Clark (1903) where in a period of two months (25th March to 19th May) he shot four successive females at one nest and even then the male obtained a fifth mate; the status of these birds was, of course, not known. Nelson (1907) relates a converse case from Yorkshire, England. In order to control the birds, bounties were paid, 1 penny for males, $\frac{1}{2}$ penny for females. Males were thus shot in preference and in one case a female acquired no less than seven mates, one after the other as the incumbent male was shot. In tropical countries with a prolonged breeding season it has been found that the early young of the year are able to breed before they are twelve months old. (The age can be determined in collected specimens from the state of pneumatisation of the skull; according to Nero (1951) this is complete in most House Sparrows after 180 days and, in all he examined, after 220 days at most.) Ali and Whistler (1933) reported precocious breeding for birds in India and Dr M.-Y. Morel has found the same for introduced birds in Senegal (pers. comm.).

After the post-nuptial sexual regression, which coincides with the autumn flock formation at the ripening grain fields, adult pairs return to their nesting sites. If one of the pair fails to return it can be replaced by a bird of the year, or a previously unmated bird; pairs of this type continue to be formed at any time up to the breeding season, as well as during the breeding season itself, when mated birds disappear, though in higher latitudes little pair formation takes place when winter weather is severe and days are short. With the coming of spring, unmated juvenile males adopt suitable nest sites and call to proclaim ownership and attract a mate; pairs of juvenile birds can be formed in this way.

The holding of a nest site thus appears to be the key to pair formation, though once formed the members of a pair clearly recognise each other away from the nest. It is not known if the members of migratory populations also mate for life, but this may well be so in the case of birds returning to a fixed nest in a hole in a cliff or earth bank.

A very wide variety of nest sites is used, ranging from holes and crevices in man-made structures, or in trees, in cliffs and earth banks, to other covered but open sites under roofs, and free-standing nests built among the branches of trees or the tops of telegraph poles. A favoured site that I have seen all over the world is a hole in a street light; perhaps the additional warmth at night is an attraction. Cink (1976) showed that unmated females chose males calling at enclosed nest sites in preference to those at open sites. This would tend to favour the use of hole sites, except in colonies where the number of holes is limited. In my experience, open, unprotected sites are more commonly used in the southern parts of the House Sparrow's range in the northern hemisphere; this is consistent with the earlier observation that population density is higher in areas with long dry summers, high enough to exceed the number of available holes.

courtship

The adaptability to the use of man-made structures is quite extraordinary. Meinertzhagen (1949) records a pair that nested on a house boat on the Nile, staying with the boat as it made journeys of up to 30 km from base. Another pair nested in the coke oven ram machines at a steel works in south Wales, tolerating not only the continuous movement, but also the high temperature when the coke oven was 'pushed' (Daily Telegraph, London, 5.8.64). Finally, several nests with young were located on 'nodding donkeys' in an oilfield in Kansas, USA, where they were subjected to an up and down movement of about 60 cm every 3–4 seconds (Tatschl 1968).

The nest is built of dry grass or straw, loosely woven into a globular structure, domed over with an entrance at the side. The external dimensions are typically 200×300 mm. In a hole, the cavity is completely filled with nesting material, provided it is not much larger than the normal size of a free-standing nest; again it is domed over with the entrance in the side. In smaller spaces, such as the old nests of House Martins *Delichon urbica,* the roof may be omitted. The nest cup is lined with feathers and other soft material. Nest spacing is usually determined by the spacing of suitable holes; neighbouring pairs apparently recognise and tolerate each other, even when the hole entrances are only 0.5 m apart. Nests built openly in trees are usually further apart, though McGillivray (1980), in a study in Calgary, Alberta, Canada, found cases where nests were actually joined together in a communal structure.

An unmated juvenile male establishing ownership of a nest site will begin to add nest material, and once he has atttracted a mate, the female joins in and both birds take an active part in completing the structure. No lining material is added until a pair is formed.

In addition to the formation of new pairs during the period of sexual recrudescence that follows the post-breeding moult, there is a certain amount of desultory nest building, and occasionally attempted breeding, by old pairs that have returned to their nest sites.

The male proclaims his ownership of a nest site by regular chirrup-calling beside it. In the case of an unmated bird this calling can be almost incessant at the beginning of the breeding season; if a female approaches, the calling rate is speeded up excitedly

threat

and the male displays with wings held out, drooped slightly and shivered. In a more intense form the head is pushed up vertically, displaying the black bib, and the tail is raised and spread; this is alternated with deep bows.

The male becomes sexually active before the female and attempts to mate with her at that time result in aggressive display by the female, which may attack the male before flying off. She is then pursued by the male and this pursuit flight attracts other sexually mature males in the vicinity that hurry to the scene and the group of males display together in front of the female (communal or group display). Such displays rarely lead to successful coition. Once the female is ready she crouches in front of her mate and invites copulation by shivering her wings and calling a soft *dee-dee-dee*. Copulation is frequent before and during the egg-laying period, with the male mounting many times on each occasion. In contrast to the display before an unresponsive female, copulation attempts do not attract neighbouring males to a group display, though they may stimulate other sexually ready pairs to copulation.

With such a wide distribution from the tropics to the Arctic Circle in the north, and south to the limits of the land masses in the southern hemisphere, there is a great variation in the breeding season. Even in one locality, where studies have been carried out over a number of years, there are considerable differences from year to year: for example, 4th April–3rd May (33 days) for the start of the breeding season in the 14-year study by Pinowska and Pinowski (1977) in Dziekanów Leśny, Poland; 27th April–6th May (10 days) in the 10-year study by Rassi in Kangasala, Finland (quoted by Pinowska & Pinowski 1977). The date of the start of breeding is plotted in Fig. 50 as a function of latitude. The correlation is good (correlation coefficient $r = 0.94$). No allowance is made in the figure for the effect of altitude, though there is evidence that the start of laying is delayed by increasing altitude (Dyer *et al.* 1977). The start of breeding is clearly influenced by temperature, and temperature differences may be a factor affecting the annual variations at the same locality. It is more probably temperature change rather than absolute temperature that is the trigger for breeding;

solicitation

as Murphy (1978) and Naik and Mistry (1980) have pointed out, breeding starts at a much lower temperature at 52°N in Oxford, England (10°C) than at 22°N in Baroda, India (24–25°C). The start of breeding probably correlates with the availability of sufficient invertebrate food to enable the female to build up the necessary protein for egg-formation (female body weight increases by about 20% prior to egg-laying, Pinowska 1979); this will vary locally and will be affected by annual variations in temperature rather than by the absolute temperature.

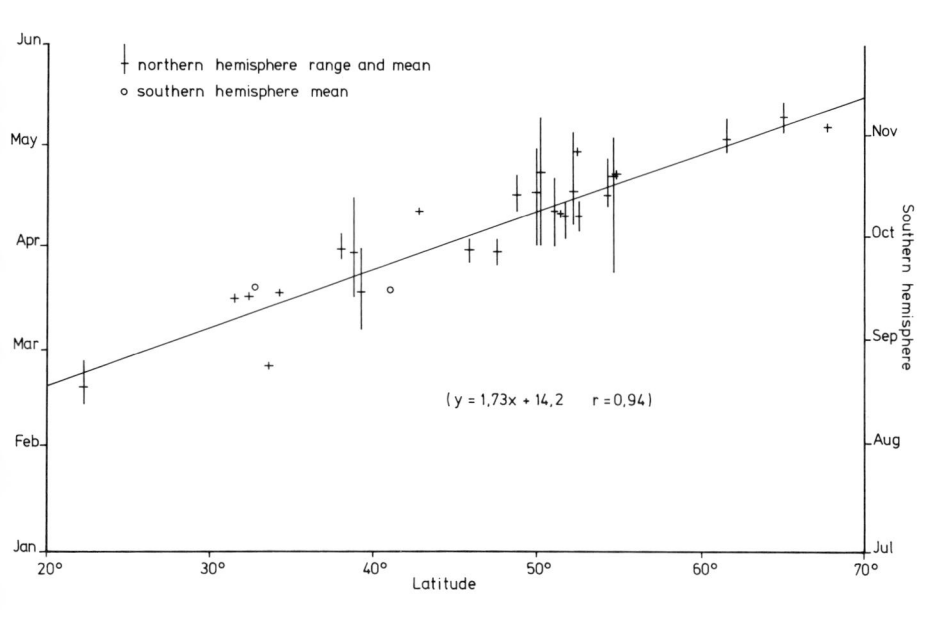

Fig. 50 Relationship between start of breeding season and latitude in House Sparrow

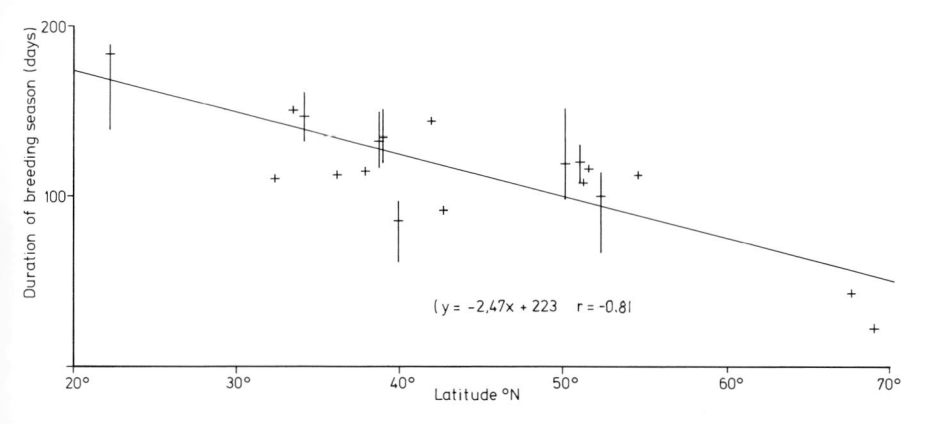

Fig. 51 Duration of breeding season of House Sparrow as a function of latitude

The duration of the breeding season, defined as the interval between the first egg of the first clutch and the first egg of the last (Moreau 1950), is plotted against latitude in Fig. 51. Again there is a distinct negative correlation ($r = -0.81$), duration decreasing with latitude. Breeding can be terminated by high temperatures and arid conditions, as has been found in Israel, 32°N (R. Singer & Y. Yom-Tov, pers. comm.) and Spain, 40°N (Alonso 1984a), or be interrupted by spells of bad weather – heavy rain and a drop in temperature – as shown for Baroda, India, by Naik and Mistry (1980); these effects make for considerable annual variations. These data refer to the normal pattern, though in some places breeding is stated to take place throughout the year. No doubt confusion has arisen through observations of nest building, that can occur at all times, being mistaken for actual breeding, but there is good evidence that in some favoured localities breeding can occur in all months; eg Burma, 10–25°N (Smythies 1953) and the Azores, 35°30′N (Le Grand 1983), though in the latter place Le Grand gave the start of the main breeding season as March, consistent with Fig. 50. Rather more abnormally, egg-laying has been recorded in all months of the year in Europe, America (Summers-Smith 1963) and India (Naik & Mistry 1980).

In the temperate regions of the southern hemisphere the breeding season is displaced by about six months from that at the equivalent latitude in the northern hemisphere: for example, mid September in South Africa (32–34°30′S), compared with mid March in the north at the same latitude; in New Zealand (35–47°S) the duration is about 120 days, the value on the regression line for the northern hemisphere at 41°N. In the tropics, however, it may be controlled more by local climate than by latitude: in Brazil the House Sparrow breeds in the dry season, March–October, the same time as in the northern hemisphere.

The number of clutches laid per female each breeding season is clearly related to its duration, though this is difficult to determine unless the birds have been individually marked. More data are available on the number of clutches per nest site per year; these are plotted against latitude in Fig. 52. There is a reasonable negative correlation ($r = -0.58$). The few results available from marked females (Naik & Mistry 1980, Baroda, India, 23.3°N; R. Singer & Y. Yom-Tov, pers. comm., Israel, 32.4°N; Sappington 1977, Missouri, USA, 32.5°N; Summers-Smith 1963, England, 51.4°N; Craggs 1967, England, 53.4°N) are reasonably close to the regression line derived from the number per nest, though as has already been pointed out, for at least some populations, more than one pair may use a nest during the same season and hence the regression line probably overestimates the number of clutches per pair, though the fact that in some populations pairs change nests between clutches may compensate for this to some extent. Individual females may lay up to four clutches in temperate

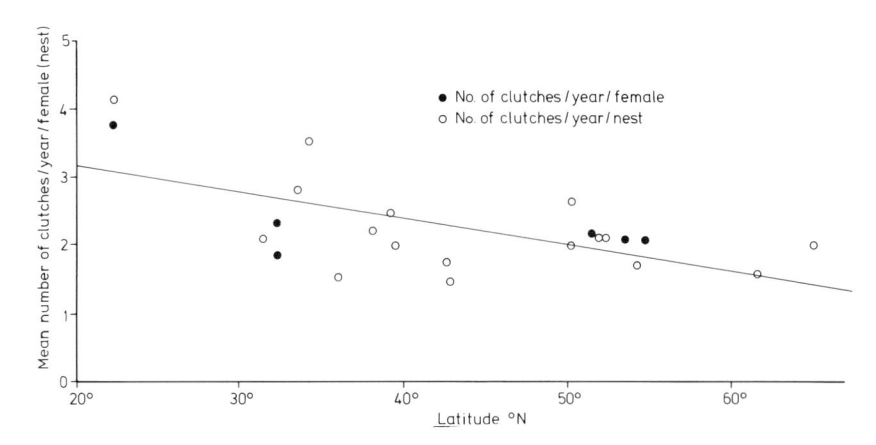

Fig. 52 Number of clutches laid per year by House Sparrows as a function of latitude

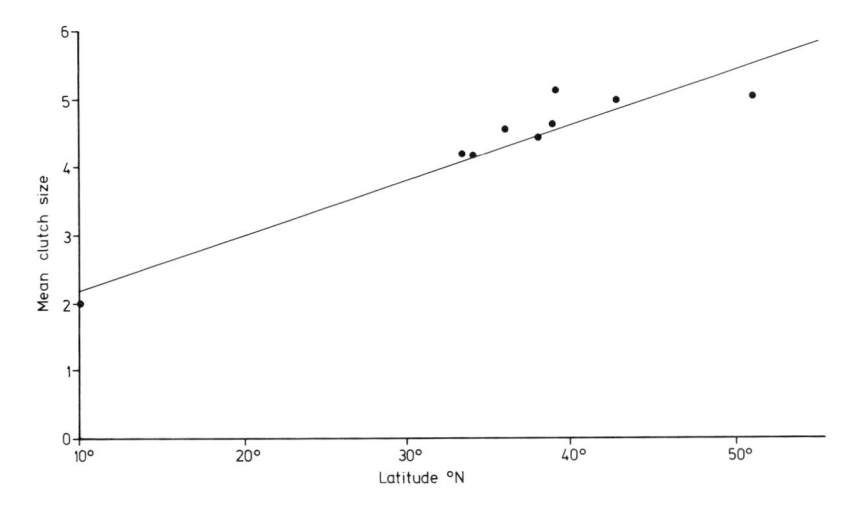

Fig. 53 Mean clutch size of House Sparrow in North America as a function of latitude

regions (England and USA, for example) and even more in the tropics – in Baroda, where the breeding season lasts from February to October, Naik and Mistry recorded instances of some females attempting up to seven clutches per year.

Clutch size normally ranges from 2–5 eggs (exceptionally up to 10 eggs, though this could involve egg dumping by another female), with 4 the modal value in the UK and 5 in continental Europe and North America. Clutch size appears to increase with latitude, but this is not easy to quantify as there is some variation throughout the breeding season, initially rising and then declining towards the end of the season, and also a tendency for clutch size to increase from west to east across the Eurasian continental land mass. However, if one limits the analysis to North America there is a strong correlation between clutch size and latitude ($r = 0.95$) as shown in Fig. 53, with data coming from over 40° of latitude from Costa Rica, 10°N, (Fleischer 1982) in the south, to Calgary, Alberta, 51°N (Murphy 1978).

Both sexes take part in incubation. The female develops a brood patch and this remains bare until the feathers are replaced at the moult. As the male does not develop a brood patch, he can only cover the eggs rather than truly incubate, but the role played by the male is important. Schifferli (1978) showed that hatching success was lower in nests from which the male had been removed than in those in which both sexes were present. The female takes the major share, spending on average about 10 minutes per spell in the nest, compared with about 7 minutes for the male, and she roosts in the nest overnight. Novotny (1970) found that, in general, incubation began in half of the clutches he studied (286 clutches in total) before the laying of the last egg. The time spent in incubation is influenced by ambient temperature: North (1973) recorded an average of 58.7% of daylight hours in Wisconsin, USA, temperature 25–40°C, compared with 87% in England, temperature 10–25°C (Summers-Smith 1963).

The incubation period, defined as the interval between the laying of the last egg of the clutch and the hatching of the first young, ranges from about 10–17 days, with average values from different studies typically 11–14 days. There is some evidence to show that the incubation period decreases with increasing ambient temperature; for example, Lowther (1979b) found a decrease from about 13 days at the beginning of the breeding season to about 11 days at the end, the incubation period showing a negative correlation with the seasonal increase in temperature, and R. Singer and Y. Yom-Tov (Yom-Tov pers. comm.) found the mean period decreased from 15.8 days at 16°C to 13.9 days at 24.5°C mean daily air temperature.

The nestlings are fed by both adults. Most studies show that the sexes play an almost equal role (Seel 1960, 1969; Summers-Smith 1963; North 1980) though McGillivray (1984) in Calgary, Canada, found that the females contributed almost twice as much as their mates. A day or two before fledging the male's share drops as he spends an increasing amount of time calling and displaying outside the nest (Summers-Smith 1963, Seel 1969). Some brooding takes place during the first eight days of life of the chicks until they develop sufficient feather coverage to become homoiothermic, but with low ambient temperatures brooding may be continued until the young fledge. Cottam (1929) describes an exceptional nest in Utah, USA, that on 1st January 1929 contained five recently hatched young. During the 18 days until the young fledged, one or both parents were almost constantly in the nest, with both staying in the nest overnight. During this period the ambient temperature was close to 0°F (-18°C) and fell as low as -14°F (-25°C). Brooding is shared by the sexes.

The overall average feeding rate during the nestling period in six studies ranged from approximately 15 feeding visits (England, Summers-Smith 1963) to 21 visits per hour (Australia, Marples & Gurr 1943; England, Seel 1969), with intermediate values of 16.5 (Missouri, USA, Sappington 1977), 17.5 (Wisconsin, USA, North 1973) 18.7 (Calgary, Canada, McGillivray 1984). In the colony in Missouri studied by Sappington, already referred to above as unusual for the unfaithfulness of the birds to their nest sites, there was a further unusual feature in that 'helpers' joined in the feeding of the young at 161 (63.4%) of the 254 nests watched. None of the numerous other observers who have published on this species have reported the presence of helpers at the nest and I did not record a single instance in my seven-year study; Dr T. Burke informs me (*in litt.*) that when he was working on the genetics of House Sparrow populations near Nottingham, England, he recorded a few instances that fitted in with 'helping', though his observations were not sufficiently complete to be sure.

The nestlings remain in the nest for periods ranging from 12–18 days (the shorter periods probably caused by observer disturbance), with averages typically 14–16 days. R. Singer & Y. Yom-Tov (Yom-Tov pers. comm.) found that the fledging period increased with the number of young in the brood, but decreased with increasing mean daily air temperature, falling from $16\frac{1}{2}$ days at 17°C to $14\frac{1}{2}$ days at 25°C; the latter effect was also noted by Lowther (1979b).

In the event of the death of the male during the incubation or nestling period, the female may still be able to rear the young, the later in the breeding cycle that this occurs, the greater the chance of a successful outcome.

The complete cycle of nest occupancy from the beginning of egg laying to fledging of the young amounts to 28–31 days, with an interval between successive clutches typically 36–40 days (Seel 1968b, Anderson 1978, Naik & Mistry 1980), though Lowther (1979a) recorded a number of cases of overlap in which the first egg of the new clutch was laid before departure of the young.

Published information on the fecundity of the House Sparrow in different parts of its range is summarised in Table 27. This shows a breeding success rate (percentage of young raised from eggs laid) that ranges from 25% (mean of a 4-year study in Baroda, India) to 85% (mean of a 4-year study in Nowy Targ, Poland) with an overall average from 33 different studies of 48.5% (allowing a weighting for the number of years duration of the different studies). This is significantly lower than the average figure of 66.0% given by Nice (1957) for hole-nesting altricial birds.

Despite the fact that breeding success varies from year to year, depending on climatic and other random effects, the results of 30 widely spread studies show that

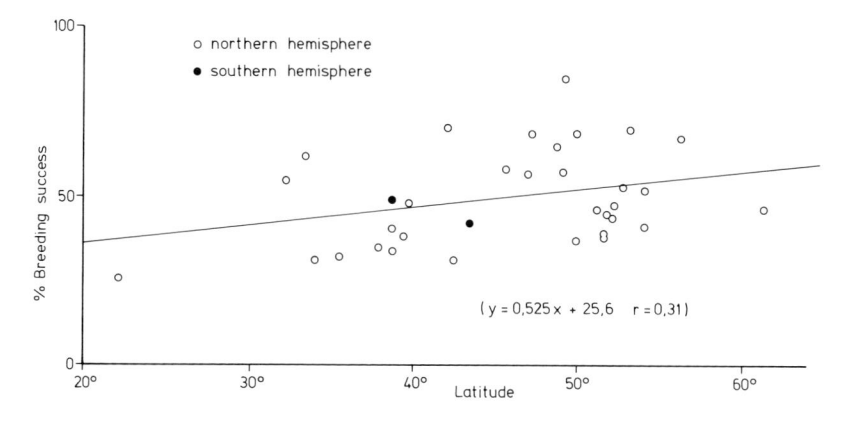

Fig. 54   Breeding success of House Sparrow as a function of latitude

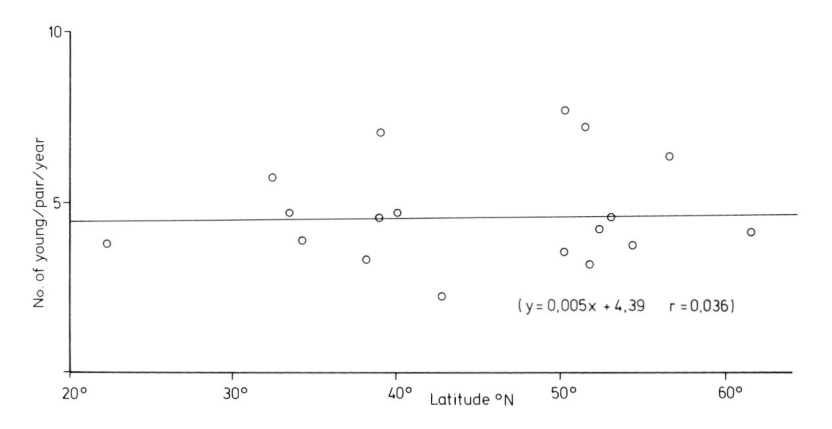

(y = 0.525x + 25.6   r = 0.31)

Fig. 55   Number of young House Sparrows per pair per year as a function of latitude

(y = 0.005x + 4.39   r = 0.036)

it probably increases with latitude, Fig. 54, the regression line increasing from 35% at 20°N to 60% at 60°N, though the correlation is poor (r=0.31). (Two points for the southern hemisphere included in the figure, though not in the calculation of the regression, generally agree with the results from the north.) As far as productivity is concerned, however, the effect of increasing breeding success and clutch size with latitude is cancelled by the decreased duration of breeding and number of clutches laid, so that the number of young produced per year is almost independent of latitude, Fig. 55, and closely approximates a value of 5.

House Sparrows nesting in protected sites – nest boxes, holes in buildings – tend to start breeding earlier and are more successful than those nesting among tree branches (McGillivray 1981). First-year birds begin later than adults and no doubt have to make use of inferior sites as well as being less experienced. When I started working on my Hampshire colony, several pairs were using tree sites; but when I put up additional nest boxes the tree sites were abandoned in favour of the boxes, again suggesting that a 'hole' is the preferred nest site. The decreased heat loss in enclosed sites gives an advantage in cooler, temperate regions, and it is perhaps not surprising

Table 27: House Sparrow breeding statistics

Locality	No. of years	No. of eggs	% hatched	No. of young	% fledged	Eggs/pair/year	Young/pair/year	Breeding success %	Reference
United Kingdom									
BTO records (ca 53°N) sample 1		2,774	71	1,401	74	8.61	4.56	53	Summers-Smith 1963
sample 2		2,109						50(3)	ditto
Oxford (51.7°N)	4	3,008	86	2,446	45			38.3	Seel 1968a, 1970
Oxford	1	915(1)	80(1)		48(1)	8.4(2)	3.22(2)	38.4	Dawson 1972
Hilbre (53.4°N)	5	53						60.3	Craggs 1967
Continental Europe									
Valencia, Spain (39.7°N)	1	271	79.7	216	48.1			38.3	Pardo 1980
Spain (40.0°N)	2	1,684	75	1,255	64.1	9.79	4.71	48.1	Alonso 1984a
Berlin, GDR (52.5°N)	1	407(1)	88(1)		54(1)			47.5	Encke 1965
Bulow, GDR (53.5°N)	1	495(1)	84(1)		83(1)			69.7	ditto
Slezké Rudoltice, Czechoslovakia (50.2°N)	3	1,903	80.5	1,308	85.4	11.18(2)	7.7(2)	68.7	Novotny 1970
Veverská Bitýška, Czechoslovakia (49.1°N)	3	940	82.6	776	78.9			65.1	Balat 1974b
Zakopane, Poland (49.5°N)	2	207(1)	73(1)		79(1)			57.7	Mackinowicz et al. 1970
Nowy Targ, Poland (49.5°N)	4	162(1)	95(1)		89(1)			84.6	ditto
Kraków, Poland (50.1°N)	7	897(1)	61(1)		61(1)	9.6(2)	3.58(2)	37.2	ditto
Gdánsk Poland (54.3°N)	5	787(1)	73(1)		71(1)	7.31(2)	3.79(2)	51.8	Strawinski & Wieloch 1972
Dziekanów Lesny, Poland (52.3°N)	14	6,239	68.4		80.5	9.87(2)	4.29(2)	43.5	Pinowski & Kendeigh 1977
								55.1(4)	
Wieniec, Poland (54.3°N)	3	6,239	55(1)		79.1(1)	9.87(2)	4.29(2)	43.5	ditto
Bujor Târg, Roumania (45.9°N)	2	2,561(1)	59(1)		70(1)			41.3	Pinowska 1975
Cimpulung Meldovenesc, Roumania (47.5°N)	2	159(1)	79(1)		74(1)			58	Ion 1973
	2	132(1)	80(1)		86(1)			68.8	ditto
Iasi, Roumania (47.2°N)	2	316(1)	68(1)		83(1)			56.4	ditto
Kangasala, Finland (61.5°N)	5	637(1)	85(1)		54(1)	8.96(2)	4.12(2)	45.9	Rassi (unpub.) quoted by Pinowski & Kendeigh 1977
Asia									
Israel (32.4°N)	2	197	70.1			11.6	5.75	52.3	R. Singer & Y. Yom-Tov (Yom-Tov pers. comm.)
Baroda, India (22.3°N)	4	5,460	61.7		41	15.0	3.78	25.3	Naik 1974
Lahore, Pakistan (31.5°N)			92						Mirza 1973

House Sparrow 153

Location								Reference
Australasia								
Christchurch, New Zealand (43.7°S)	9		71		60			42.6	Dawson 1972
Hawke Bay (Hawke's Bay), New Zealand (39°S)	9		82		60			49.2	ditto
North America (6)								
Plainview, Texas (34.27°N)	1	397	57	(225)	28	16.36	2.62	16.0	Mitchell *et al.* 1973
Cloverlake, Texas (34.34°N)	1	690	61	(419)	67	12.6	5.15	40.8	ditto
Stillwater, Oklahoma (35.8°N)	1	219	50.7		64.9			32.5	Ables 1960 (quoted by North, 1968)
		215	50						North 1968
McLeansboro, Illinois (38.1°N)	3	1,502	66		53	10.12	3.54	35.0	Will 1969
	3	1,502	65.8		53.3	8.94	3.14	35.1	Will 1973
Portage des Sioux, Missouri,	3	2,142	64		66	10.8	4.57	42.2	Anderson 1973
USA (38.9°N)	6	2,652	64.2		61.9			39.7	Anderson 1975
	6	2,894	64.6		62.8			40.6	Anderson 1978
Coldspring, Wisconsin (42.9°N)	1	370	50.8	188	61.2	7.35	(2.29)	31.1	North 1973
Oktibbeha Co., Mississippi (32.5°N)	3	(910)	83.2	758	77	7.46	(4.78)	64.1	Sappington 1977
Calgary, Alberta (51.4°N) 1975	1		68		76	15.6	7.22	52	Murphy 1978
1976									ditto
Leavenworth Co., Kansas (39°N) 1975	1		65		61			40	ditto
1976	1		50		66	15.48	7.11	33(5)	ditto
	1		56		62			35(5)	ditto
Ithaca, New York (42.4°N) 1975–78(7)	4	7,316	62.5	4,750	64.6			40.4	Lowther 1979b
	4	180						70.5	Weaver 1942
Glenn Dale, Maryland (39°N)	4	114	85	95	(90)			78.5	McAtee 1940

(1) Pinowski & Kendeigh 1977, Appendix 3.5
(2) Pinowski & Kendeigh 1977, Appendix 3.7
(3) According to North (1973) this has been recalculated by Beimborn (1967) as 37.7%, but I have been unable to consult this reference.
(4) % hatched × % fledged does not agree with $\frac{\text{young/pair}}{\text{eggs/pair}} \times 100$
(5) Obtained from No. of eggs/pair/year and No. of young/pair/year.
(6) Figures in parenthesis have been calculated from other data given in the paper.
(7) Includes the 1975 and 1976 data.

154 *House Sparrow*

that open tree-sites become commoner in the more southerly parts of the breeding range in Europe, where heat loss is of less significance, though of course such sites are more vulnerable to violent, summer thunderstorms. McGillivray, in his study of tree-nesting birds in Canada, found that early nesters chose the most protected sites and were more successful in raising young, again suggesting a climatic influence, with the older, more experienced birds nesting first and choosing the best sites.

SURVIVAL

Little is known about the fate of the fledglings once they leave the nest. The adults continue to feed them for up to 14 days after fledging as they gradually become independent and begin to look after themselves. This is clearly a dangerous time for young inexperienced birds and ringing records show that they suffer a high mortality (Fig. 56). Observations on a colour-ringed population, where the adults remained faithful to their nest sites and thus absence could be equated to mortality, together with fledging success based on the British Trust for Ornithology ringing records, allowed the construction of an hypothetical seasonal composition of a suburban House Sparrow population in Stockton on Tees, England (Summers-Smith 1958). This is shown in Fig. 57 for an initial population of 100 adults, starting on 1st April; also included in this figure are the results of a similar exercise based on sparrow studies in McLeansboro, a small town in Illinois, USA (Will 1973). It will be seen that each population follows a similar pattern with a maximum in Stockton in August, in McLeansboro a month later. The high production of young (Table 27) is balanced by the high mortality of juveniles, so that at its maximum the population never reaches twice that at the beginning of the breeding season. One point worth noting is that the highest mortality of adults occurs during the breeding season (April–August) – 5% per month, compared with only 2–3% per month for the remainder of the year, suggesting that the former is the time of greatest stress for the House Sparrow, not the problems of survival in the winter months. In a study of a suburban, colour-ringed population in the Netherlands, Heij (1985) also found that mortality was at its highest during the breeding season and the post-breeding moult, least during the winter months. Even in severe winters in Britain there appears to be little increase in mortality (Jourdain & Witherby 1918–19).

Estimation of the survival rates of adults are based on recoveries of ringed birds, but the calculations depend on the statistical analytical techniques used to allow for the incompleteness of recoveries resulting from some of the birds still being alive.

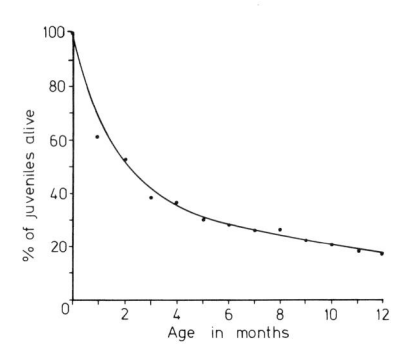

Fig. 56   Percentage mortality curve for young House Sparrows

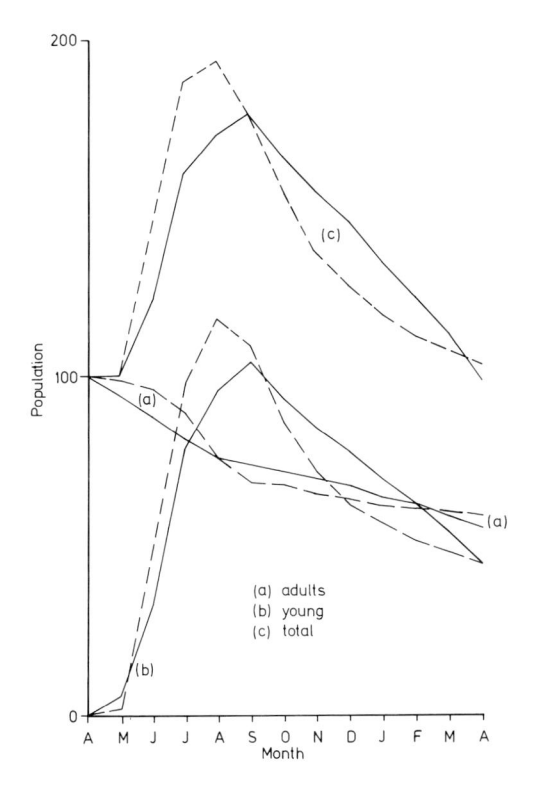

Fig. 57 Seasonal composition of two hypothetical House Sparrow populations. Dashed lines England; solid lines Illinois, USA.

The values derived for the annual survival rates of adults range from 45–65%, depending on the source of the data and method of analysis used (Dyer et al. 1977). Gramet (1973), in a more sophisticated analysis, found a slight difference between the survival rate (February to February) of full adults and first-year birds, 65% compared with 57%. Individual birds have been known to survive for 13 years in the wild (Dexter 1959) and I have had under observation a pair of marked birds that remained together until both were at least six years old. There are several records of birds in captivity surviving for 12–14 years and a remarkable record of a bird that lived for 23 years (Flower 1938). Heij (1985) constructed survival curves from observations on colour-ringed birds for suburban and rural populations in the Netherlands; these showed that the oldest suburban birds survived until their eighth year, compared with rural ones that survived only until their sixth year.

There is evidence from trapping and ringing data (Nichols 1934, Beimborn 1976, Heij 1985), and analysis of large samples collected after poisoning campaigns (Neithammer 1953, Piechocki 1954, Löhrl & Böhringer 1957, Geiler 1959), that males tend to outnumber females (on average 53:47), even after making allowance for obvious biases. This appears to be the result of a higher fledging success for males compared with females (perhaps the larger male is more successful in competition for food in the nest), coupled with a higher survival rate of males after fledging. Blem (1975) has shown that wing-loading is greater for females than males (the sexes have similar weights, but the females on average smaller wings); this could contribute to reduced survival for the latter, particularly at high latitudes where wing-loading tends to be at its highest, and could thus affect the ability of the birds to obtain enough food in a competitive situation for overnight survival during the long and cold winter nights.

BREEDING DENSITY AND NUMBERS

Pinowski and Kendeigh (1977, Appendix 3.1) summarise over 150 estimates of House Sparrow breeding densities in the Palaearctic region, identifying these with particular habitat types. These have to be interpreted with some caution, particularly with a colonial species that is not uniformly dispersed, as the density is greatly influenced by the size of the area in which the census has been carried out. There is also considerable difficulty in allocating the censuses to a precise habitat classification. With these reservations I have arranged in Table 28 the densities for the eighteen habitat types proposed by Pinowski and Kendeigh in order of decreasing population density. (Seven biotopes in which no census figures for House Sparrows have been reported – and in which House Sparrows seldom if ever breed – are included for comparison with the Tree Sparrow, see Chapter 15, that was also covered in the study.) This shows a clear relationship between House Sparrow density and the human population, the highest densities occurring in towns, villages and the surrounding cultivated land. Other studies suggest that, in temperate areas at least, the House Sparrow population roughly correlates with the human one at a ratio of 0.2–0.3:1. There is also some indication, however, that numbers tend to be greater in the warmer, more southerly parts of the range in the northern hemisphere. In view of the caveats expressed above about the validity of the density it is perhaps not surprising that correlation of density with latitude in all of the biotopes in Table 27 tends to be rather weak. Nevertheless for all of the town habitats there is a decrease in density (y) with latitude (x):

	regression equation	correlation coefficient
commercial and shopping areas	$y = -41x + 2990$	$r = -0.47$
residential areas with appartments	$y = -43x + 3350$	$r = -0.23$
suburban areas with gardens	$y = -52x + 3399$	$r = -0.37$
old parks in larger towns	$y = -78x + 4499$	$r = -0.43$

But, more importantly, there is an extension of habitat tolerance in warmer climates; for example, in the Mediterranean area, House Sparrows spread into cultivated areas surrounding villages, nesting in holes in earth banks, openly in trees and telegraph poles.

Although there have been few specific studies of the density of the introduced House Sparrow in the United States, there is good evidence to show that numbers peaked there in the early years of this century and since then have decreased dramatically. This has been particularly noted for urban populations and has been attributed to the displacement of the horse by the internal combustion engine, that not only denied the sparrows the food that had come from the spillage of grain from nose-bags and undigested grain in droppings, but also made life in the streets much more hazardous (Bergtold 1921). Counts made in cities suggest that the sparrow populations decreased by a half to two-thirds between 1900 and 1925 as the horse virtually disappeared from the streets. (For a summary of the US data see Bent 1965.) There could also have been a backlash after the dramatic expansion that followed the introduction. Introduced birds may flourish initially if diseases are not introduced with them, until such time as local diseases and predators become adapted to the new species.

Table 28: House Sparrow densities in different habitat types (Based on Pinowski & Kendeigh, 1977, Appendix 3.1)

Biotope	No. of census results	Breeding density (birds/km^2)	
		Mean	Range
Family dwellings with livestock	3	3952	1714–5428
Towns: residential areas with apartment complexes	23	1145	10–3020
Towns: commercial and shopping areas	12	909	268–1670
Villages	21	724	14–3810
Towns: surburban areas with gardens	19	611	44–1884
Old parks in larger towns	29	394	14–3600
Old parks in villages and open fields with trees	18	338	20–1096
Old orchards	1	160	160
Riparian areas and cemetaries	11	124	24–360
Towns: small allotments and gardens	6	64	10–354
Deciduous forests: 1–50 years old	3	68	38–90
Deciduous forests: 50–100 years old	0		
Deciduous forests: over 100 years old	0		
Deciduous forests: over 50 years old with nest boxes	0		
Pine forests: 1–50 years old	0		
Pine forests: 50–100 years old	0		
Pine forests: 1–50 years old with nest boxes	0		
Young orchards with nest boxes	0		

Presumably the effect of the replacement of the horse must also have operated in the old-established parts of the range in Europe, though this has attracted less attention than in the New World. A study in Kensington Gardens, London, suggested that the population there had fallen by two-thirds between 1925/26 and 1948/49. It would be expected that overall House Sparrow numbers are on the increase as man continues to modify the environment, increasing the biotope more favourable to House Sparrows (Table 28). As far as the British Isles are concerned, however, I believe the population has recently shown a slight decline. It is not easy to quantify this, but I have noticed that the House Sparrow appears to have disappeared from some of the higher Yorkshire villages and my impression is that densities in both urban and suburban areas have decreased. More quantitative evidence for this decrease has been provided by my friend L. R. Lewis, who has carried out regular counts for a number of years on the North Hampshire and Berkshire Downs, involving breeding colonies and autumn passage; he found both a retraction in breeding numbers and also an almost complete disappearance of autumn passage – in approximately 200 hours of observation, 181 passage birds were seen in 1966 and 206 in 1967, whereas in 1981 none was observed and only one in 1982 (Lewis 1984). Lewis ascribes this to more intensive cereal cultivation, reducing the insect population on which the House Sparrow depends for rearing its young.

Based on censuses carried out in different habitat types, I estimated the House Sparrow breeding population in Great Britain to be about 9.5 million birds in the 1950s (Summers-Smith 1959). Using more recent data, including the British Trust for Ornithology Winter Atlas Enquiry, 1981–84 (Lack 1986), I believe the population reached a plateau in the early 1970s and has since then declined slightly. My estimates (Summers-Smith 1986) for the winter population over the period are:

1959	early 1970s	1984
11.5 million	13 million	12 million

The reasons for this are not clear, though it would seem most likely that it is tied to a decrease in the availability of invertebrate food that is required for successfully rearing the nestlings. Modern agricultural practice, with cleaner fields and more careful control of food for livestock, could be having an effect in rural areas, increased use of garden insecticides a similar one in the suburbs. The recent increase in the numbers of Sparrowhawks *Accipiter nisus,* that are now to be seen regularly on the outskirts of towns and villages, could perhaps also be a contributory factor. In a study in the Netherlands by Tinbergen (1946), it was shown that Sparrowhawks took about 8% of the House Sparrow population in May, about 80% of the House Sparrows that died in that period; as their name implies they can obviously be a significant predator of the House Sparrow. Domestic cats also take their toll; Churcher and Lawton (1987), in a study in an English village, found that cats accounted for at least 30% of the sparrow deaths that occurred in the year studied.

MOULT

Four to six weeks after they leave the nest young House Sparrows begin a moult of the juvenile plumage. After a further four weeks the signs of sexual dimorphism begin to be apparent: first the chestnut patches on the scapulars of males, followed by the development of a black bib and then the distinctive head pattern. The moult is a complete one, including wing and tail feathers. It lasts about two to three months, so that in temperate northern hemisphere latitudes young in moult can be found from the end of May until the end of October or middle of November. In temperate regions adults begin to moult slightly before, or immediately after, they have stopped breeding: at the end of July or beginning of August. The duration in individual birds is about two months and adults in moult can be found in the northern hemisphere from the beginning of July until late October (Ziedler 1966; Haukioja & Reponen 1968; Ginn & Melville 1983; Alonso 1984b), with the moult in earlier moulters lasting about 80 days, reducing to about 60 days in the later ones.

In tropical regions, where there is a prolonged breeding season – in India at 22°N, for example, birds may be occupied with breeding duties from February until November (Naik & Mistry 1980), though possibly with intermissions in bad weather during the monsoon between June and September – the regime of moult of breeding birds is contemporaneous with the breeding season, starting earlier than in temperate populations. Unlike the temperate population, however, moult is not continuous, but arrests during the intensive periods of breeding and restarts during intermissions. This means that moult in individual birds may last for four or more months in contrast to the two months in temperate latitude birds, with birds in moult from May until December (Mathew & Naik 1986). Arrested moulting has also been recorded for House Sparrows in Texas, USA (Ginn & Melville 1983), and in Senegal (M.-Y. Morel pers. comm.).

VOICE

I have previously described in some detail the calls used by the House Sparrow (Summers-Smith 1963). The most basic of these and the circumstances in which they are used are summarised here so that comparisons can be drawn with those used by other sparrow species.

Chirping is almost synonymous with sparrow calling and there can be no more familiar bird sound to those that live where there are House Sparrows than the *chirrup* call of the male. This is usually disyllabic, though at times it can sound like a monosyllabic *chirp*. It is used basically as a nest ownership proclamation call, and as such it can also function as an invitation to pairing by an unmated male. Just prior to and during the breeding season calling by the unmated male takes on a greater urgency, becoming higher pitched and speeded up into a rhythmic sequence; I have named this ecstatic calling, and it is the nearest approach to House Sparrow song. Excited *chirrup* calls are also used by males taking part in group displays.

The *chirrup* is frequently uttered away from the nest as a basic communication call by both sexes in a variety of social situations: when emerging from or prior to entering the roost, in a social group in a tree or hedgerow when the birds are resting between bouts of feeding (social singing).

In more aggressive situations the *chirrup* changes to a trilled chur-*chur-r-r-it-it-it-it*. The female becomes dominant over the male during the breeding cycle and this call is used when she arrives at the nest, either to take over incubation or with food for the young, and she wants to displace her mate. The male, on the other hand, churrs against intruders at the nest, both of its own and of other species.

In less aggressive situations the mated pair use a soft call that has a characteristic long *ee* sound, *eg, dee, quee*. This appears to have the function of inhibiting aggression between two birds in close contact. I have described it as an appeasement call. It is used, for example, by the female as a prelude to and during coition.

Finally there are a number of calls that signal alarm. These have a distinct nasal quality, with the basic sound *quer*. This can be given as a single note or repeated *quer-quer-quer*, with variants *quer-it, ki-quer* and *ki-quer-kit*. These calls are used most frequently in ambivalent situations when the bird's motivation is to fly away from danger, but it is inhibited from doing this by wanting to feed or protect the young. In extreme alarm the bird uses a shrill *chree*.

FOOD

The House Sparrow is primarily a seed eater, specialising on the seeds of man's cultivated grain crops – oats, wheat, barley, corn, sorghum, maize, millet and rice are all taken, depending on availability, with a preference for oats and wheat when there is a choice. Large flocks feed on ripening grain and can cause considerable damage, not only through the amount of grain they consume, but also, as already mentioned, by the damage to seed heads as they perch on them, though most damage is confined to within a few metres of a suitable place of safety, such as a hedge or tree where the birds can take refuge in the event of threatening danger. The other major food source is the seeds of annual herbs, such as grasses (Graminae), rushes (Juncidae), goosefoot (*Chenopodium* spp.), docks (Polygonacaea, *eg* Knotgrass, *Polygonum avicular*) and chickweed (*Stellaria* spp.). Birds living in inhabited areas supplement their diet of natural vegetable matter with a variety of household scraps, spillage from luncheon packs in parks and open-air restaurants, in addition to deliberate feeding, largely of bread. When offered a choice, urban and suburban House Sparrows will take seeds in preference to bread and other cooked food.

The composition of the diet reflects mostly the availability of different types of food and is subject to considerable variation with locality and season. A wide range of studies suggests that throughout the year vegetable matter makes up 85–90% of the food, with the balance being of animal origin (Wiens & Dyer 1977). Of the vegetable matter, the cereal grain content ranges from 40–75%, being highest in grain-growing and rural areas, where it comes from growing crops, spilt seeds and animal foodstuffs, with weed seeds, leaves, buds and household scraps accounting

160   *House Sparrow*

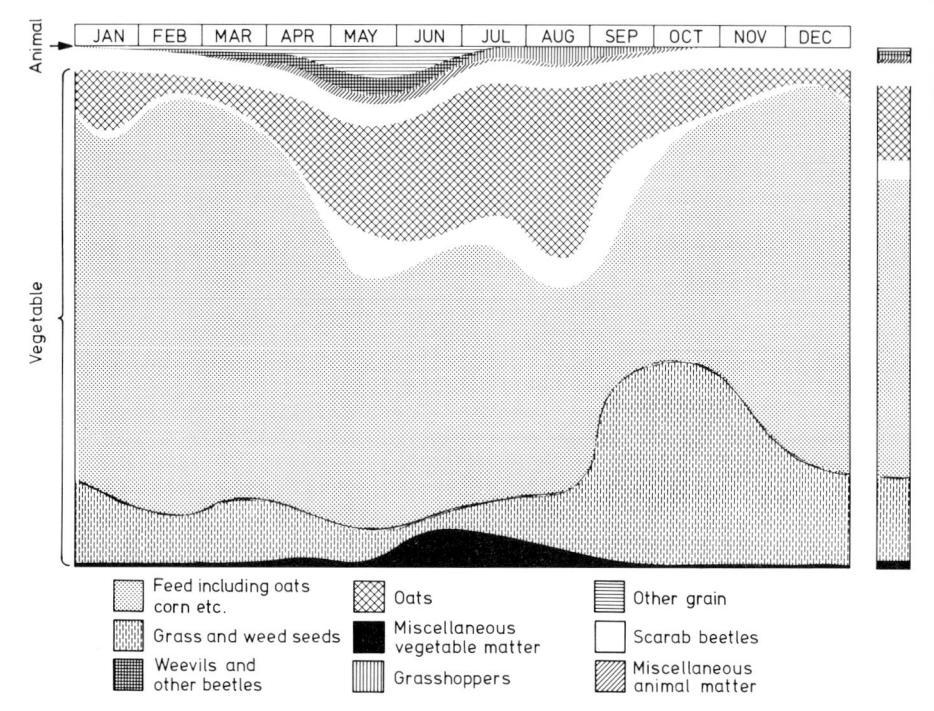

Fig. 58   Seasonal variation in type of food taken by House Sparrows in North America

for 20–50%, the latter, particularly bread, replacing cereal grains in urban and suburban areas, though a considerable amount of pilfering of grain can occur in towns and markets where there are open-fronted shops and stalls. The proportion of animal food taken is seasonally dependent; it can amount to 15–30% in spring and summer, continuing through to the time of the moult, falling to 5% in autumn, when cereal grains are plentiful, and to a negligible proportion in winter. As with all seed eaters a small amount of grit is also regularly taken, House Sparrows being frequently seen pecking at the mortar on walls.

Fig. 58 gives a diagrammatic representation of the food taken by adult House Sparrows in North America throughout the year. This is based on the analysis of 4,848 birds carried out by Kalmbach (1940). The types of food taken will obviously vary in different parts of the bird's range, but the proportions of animal matter, grain, grass seeds, weed seeds and other vegetable matter is probably generally typical of non-urban House Sparrows. The consumption of animal matter according to this study was limited to the summer months, reaching a maximum of 11.6% in May, but only amounting to just over 3% on an annual basis.

In contrast, nestlings are fed initially almost exclusively on insects and other invertebrates, both larval and adult forms: aphids (Aphidoidea), weevils (Curculionidae), grasshoppers (Orthoptera) and caterpillars (Lepidoptera) being the most important. As the young develop, the insect diet is increasingly supplemented by vegetable matter, and by the time they have fledged the latter forms by far the most important part. Clearly the precise composition of the diet will be affected by local circumstances and conditions, but three studies from the Holoarctic region (England, Germany, USA), summarised in Table 29, show a considerable similarity in the way

Table 29: Percentage of animal food given to nestling House Sparrows as a function of their age

Locality	Age (days) 1 2 3 4 5 6 7 8 9 10 11 12 13 14 15	Reference
England	← 84 → ← 67 → ← 50 → ← 40 → ← 33 →	Summers-Smith 1963
Germany	← 80 → 50 20–25	Mansfeld 1939
USA	← 90 → ← 65 → ←———— 49 ————→	Kalmbach 1940
India	← 90 → ← 93 → 74 ← 80 →	Simwat 1977

that vegetable food becomes increasingly important as the chicks approach fledging; this is also shown by the Oriental region study from northern India, though here the proportion of animal food is much greater.

The House Sparrow is, however, a complete opportunist as far as food is concerned and the basic diet summarised above can be supplemented by such a wide variety of foodstuffs that the bird can almost be described as omnivorous. The adults freely take animal food when it is available – small molluscs and crustaceans from the sea shore, defoliating caterpillars from trees, swarming ants from the air – and where fruit is cultivated they can cause damage by eating the buds as well as the ripe fruit, with grapes, cherries and dates receiving most mention. Wild fruits are also taken; Howells (1956) mentions the berries of the Christ Thorn, *Poterium spinosum,* which he says he has not seen eaten by any other bird. The habit of tearing the petals of flowers, particularly yellow ones like crocuses, in spring is well known; the reason for this is not clear and it is not known whether it forms part of the diet, though this is certainly the case with the young vegetable foliage that it also attacks in the spring. I have an impression that petal tearing is more prevalent in dry springs, so perhaps it is connected with moisture content.

10: Willow (Spanish) Sparrow *Passer hispaniolensis*

NOMENCLATURE
Fringilla hispaniolensis Temminck, Man. d'Orn. 2nd Edn. Part 1 1820: 353.
    Algeciras, southern Spain.
Passer hispaniolensis (Temminck) 1820.
    Synonyms: *Fringilla Italiae* Vieillot, Nouv. Dict. Hist. Nat., nouv. ed. 12 1817:
        190. Italy.
    Fringilla salicicola Vieillot 1825.
    Pyrgita salicaria Bonaparte 1838.
    Subspecies: *Passer hispaniolensis hispaniolensis* (Temminck) 1820.
    Passer hispaniolensis transcaspicus Tschusi, Orn. Monatsb. 1902 10: 92.
        Transcaspia (Lenkoran) eastwards; the type is from Iolotan according to Tschusi, Orn. Jahrb. 1903: 10.
    Passer hispaniolensis italiae (Vieillot) 1817.

The sparrows that occur in Italy and Crete are intermediate in plumage between the House Sparrow, *P. domesticus*, and the nominate Willow Sparrow, *P. h. hispaniolensis*. They were originally given specific status, with the name *Fringilla* (*Passer*) *italiae*, but are now recognised as a hybrid population *domesticus* × *hispaniolensis* and most recent authorities (*eg* Vaurie 1959, Moreau 1962) place them as a subspecies of the House Sparrow, *Passer domesticus italiae*, though Johnston (1969a) has suggested that, because the hybrid population has resulted from events long past, and is currently stable with little if any inflow from the parent species, they might well be given specific rank, for which the name *Passer italiae* (Vieillot) 1917 is available. This does not, however, appear to me to be a sound argument, as surely it is the definition of a race or subspecies. In my opinion it is more sensible for the following reasons to put *italiae* as a subspecies of *hispaniolensis* rather than of domesticus.

1. The birds in northern Italy, south to about Naples and Foggia, show little variation and are almost intermediate between the two parent phenotypes. Moreover, there is a sharp transition from *italiae* to *domesticus*, with few intermediates, at its northern limit as we have already seen in Chapter 9, whereas to the south the transition zone extends through southern Italy to Sicily and even Malta, where the birds are still not pure *hispaniolensis*, assuming more and more of the *hispaniolensis* characteristics, so that it is not possible to define a boundary with any precision. This is much more typical of the intergradation that takes place between subspecies, than the sharp transition at the northern boundary, and suggests a closer relationship to *hispaniolensis*.

2. Although *hispaniolensis*, as we shall see later in this chapter, penetrates the urban habitat in the absence of *domesticus*, it is more typically found away from human habitations. This is shown by its displacement from the urban habitat on the east side of Sardinia (*domesticus* does not occur on Sardinia) by *Passer montanus*, a recent colonist. *Passer montanus* is much more adapted to built-up environments than *hispaniolensis*, being almost exclusively found in this environment when *domesticus* is lacking, though in most parts of its range giving way to *domesticus* where the two species occur together (see Chapter 15). In Naples, a similar situation to that in Sardinia exists between *montanus* and *italiae*, with *montanus* in the centre and *italiae* displaced to the outskirts. The situation is less clear cut in northern Italy. Giglioli (1865) reported that both species bred in company in the roof of his house in Pisa, and in Turin and Biella I found *montanus* sharing the built-up area with *italiae*, though here it tended to be in the more open parts. Wallis (1887) also noted *montanus* as an urban bird in Ponte Gardena (Waidbruch). On balance this interrelationship suggests that *italiae* is closer to *hispaniolensis*, as elsewhere in the southern parts of Europe occupied by *domesticus*, *montanus* tends to be displaced into the countryside.

3. Harrison (1961) after an investigation of 500 specimens of juveniles and adults of both sexes of *domesticus*, *italiae* and *hispaniolensis*, while not going as far as to remove *italiae* from *P. domesticus* to *P. hispaniolensis*, nevertheless on the basis of head plumage, both the chestnut colour and the white post-ocular spot that is vestigial in *domesticus*, considered *italiae* to be very close to *hispaniolensis*, and, by implication, closer to *hispaniolensis* than to *domesticus*.

The familiar English name, Spanish Sparrow, derives from the species having been first reported from Spain. It is not very appropriate as the bird is by no means as common there as in other parts of its range. The alternative, Willow Sparrow, is to be preferred in view of the bird's frequent association with damp river valleys, often where there are willows; it is also in line with the German *Weidensperling*.

The Willow Sparrow is very closely related to the House Sparrow. Over much of its range the two species are sympatric; they remain separate, breeding as good species, but, in addition to the stabilised hybrid population, *P. hispaniolensis italiae*, described above, the Willow Sparrow hybridises freely with the House Sparrow in other parts of the Mediterranean basin, particularly in northeastern Algeria and Tunisia. Meinertzhagen (1921) has also described a hybrid from Palestine, apparently between *P. domesticus biblicus* and *P. hispaniolensis transcaspicus*.

DESCRIPTION

A typical black-bibbed sparrow, rather similar in appearance to House Sparrow, length about 160 mm.

Passer hispaniolensis hispaniolensis

Male. Crown, nape and sides of neck chestnut red; white supercilium from bill to just behind eye, giving clear white pest-ocular spot; black stripe from bill through eye joining chestnut at back of head; cheeks white. Upper parts grey, heavily streaked with black and white on mantle. Black bib more extensive than in House Sparrow, spreading sideways on breast and going over into heavy black streaks on flanks. Belly greyish white. Wings dark brown with chestnut lesser coverts, white tips of feathers forming conspicuous wing bar.

After the moult, chestnut of crown is obscured by brown feather tips and black of bib, breast and flank streaks by light feather tips; the tips of feathers are removed by abrasion so that, by spring, full breeding plumage is revealed. Bill black in breeding season, otherwise horn.

Female. Very similar to female House Sparrow, but rather more boldly marked; pale creamy supercilium; broad creamy streaks on upper back; necklace of darker streaks that continue down flanks. Bill horn.

Juvenile. Very similar to juvenile House Sparrow.

Passer hispaniolensis transcaspicus

In worn breeding plumage, according to Vaurie (1949), this race is so similar to the nominate race that it cannot be separated with certainty. In fresh plumage, however, both sexes of *transcaspicus*, but particularly the female, are significantly paler than the nominate race and the chestnut tones are less pronounced.

Passer hispaniolensis italiae

The so-called Italian (or Cisalpine) Sparrow is intermediate between the House and Willow Sparrows.

Male. Head pattern is similar to that of nominate *hispaniolensis*, with crown, nape and sides of head chestnut red and cheeks white; upper parts similar to nominate *hispaniolensis* and brighter than those of *domesticus*; in contrast, underparts resemble those of House Sparrow, with a small black bib that spreads out sideways over breast, but is less extensive than black bib of nominate race, and black streaking on flanks of latter are entirely lacking.

In south of its range in Italy, a cline to almost pure *P. hispaniolensis hispaniolensis*.

After moult, clear chestnut of crown obscured by greyish yellow borders to feathers, giving greyish tone.

Female. Very similar to the female of nominate race, but lacking any sign of darker markings on underparts.

BIOMETRICS

Typical biometric data are given in Table 30. The measurements indicate that the Asian race, *transcaspicus*, is slightly larger than the nominate race.

Gavrilov (1963) gives mean monthly weights for the Asian birds on their breeding grounds in Kazakhstan (Fig. 59) and Ali and Ripley (1974) typical winter weights from India of 25 g for males and 23.5 g for females. These seasonal weights show the same general trends as for the House Sparrow, with males heavier than females, except for the beginning of the breeding season, and an increase in weight in August, just prior to migration. Unlike the nominate race of the House Sparrow, however, the birds are significantly lighter in winter. This is presumably an adaptation to the higher temperatures in their winter quarters, whereas sedentary House Sparrows tend to have increased weight in the colder winter months.

Table 30: Typical Biometric data for Willow Sparrow

Feature	Subspecies	Locality	Males Range	Males Mean (Median)	Females Range	Females Mean (Median)
weight	hispaniolensis			28.1		29.1
	transcaspicus	Kazakhstan (summer)	24.5–27.5	28.9	23.7–37.8	28.8
		India (winter)	20–28	24.9	18–28	23.5
wing	hispaniolensis	Europe	73–81	77.2	74.5–77.5	(76)
		NW Africa	74–83	78.5		
		Canaries	76–80	77.5		
		—	74–83	78.2	72–77.5	(74.8)
	italiae	N Italy	73.5–82	77.5		
		Crete	76–82	77.9		
	transcaspicus	India	73–87	80	73–82	77.1
		USSR	76–83	78.9	72–80	75.5
tail	hispaniolensis	—	49–56	52	49–53	51
	transcaspicus	—	49–62	54.5	48–55	51.6
tarsus	hispaniolensis	—	19–21	(20)		
	transcaspicus	—	19–20	(19.5)		
culmen	hispaniolensis	—	13–14	(13.5)	13–14	(13.5)
	transcaspicus	—	13–15	13.9	13–14	13.4

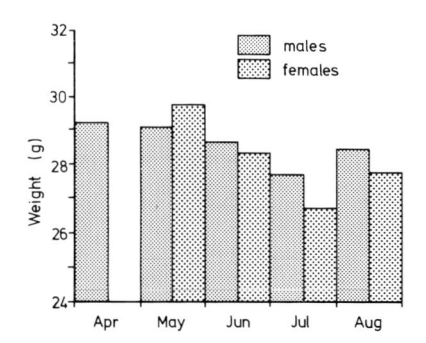

Fig. 59   Seasonal variation in mean weight of Willow Sparrow

DISTRIBUTION

The Willow Sparrow extends in a rather narrow band of latitude from southern Europe and north Africa eastwards to Russian and Chinese Turkestan, with the northern limit approaching the July 25°C isotherm (Voous 1960). There are three races: the nominate race in the west, except for Italy and Crete, where *italiae* occurs, *transcaspicus* in the east, with the boundary between the nominate race and *transcaspicus* occurring about Asia Minor, though it is not clearly defined.

Passer hispaniolensis hispaniolensis
The nominate race of the Willow Sparrow has a rather confined breeding range in

166 *Willow (Spanish) Sparrow*

the Iberian peninsula; it occurs most commonly in the Tagus valley in east-central Portugal and west-central Spain, but more sporadically in the valleys of the Guadiana and Guadalquivir and on the east coast of Spain (Sacarrão & Soares 1975, Alonso 1976, R. F. Johnston, pers. comm.). Despite the specific name it is by no means common in this part of its range and Alonso is, moreover, of the opinion that it is decreasing in Spain. It is absent from mainland Italy and Malta, where it is replaced by the subspecies *italiae*, which forms a broad zone of intergradation with it through southern Italy, Sicily and Malta, but on Sardinia the birds are of the nominate race, with only one record of the Italian subspecies from Olbia on the northeast of the island on 29.3.55 (Bezzel 1957).

Further east it is found from southern Yugoslavia—Montenegro east of Boka Kotorska to Skadarsko Jezero (Lake Scutari) and inland to Titograd, where it is rather patchily distributed. Since about 1950 it has spread north to the Vojvodina Region of Yugoslavia, at about 45°N, and the Danube basin in southern Roumania north to the Moldavian SSR at about 46°N. It is more frequent, though still rather local in Albania, Macedonia, the Greek mainland and Bulgaria, but is absent from Corfu, the Peleponese and the Aegean Islands. It does not breed on the Balearic Islands, but is present on Pantelleria.

On mainland North Africa it is found from Morocco to Tripolitania in Libya, though with an extensive zone of hybridisation with *P. domesticus* in eastern Algeria and Tunisia (to be discussed later). The breeding range in northwest Africa is shown in Fig. 60 (Summers-Smith & Vernon 1972). There is also an isolated population in Libya further to the east in Cyrenaica; this is of very limited extent, being confined to the coastal region from about Benghazi to the Egyptian border, south to about 31°30'N.

It also occurs on the Atlantic archipelagos of the Cape Verdes, Canaries and Madeira. The populations on the Atlantic islands are of particular interest as they represent recent extensions of range. This is known to have occurred naturally on Madeira and Porto Santo, where the arrival was recorded in May 1935 after a period of persistent easterly winds (Bannerman & Bannerman 1965). It is usually said to have been introduced to the Canary Islands, but if the first recorded dates for the

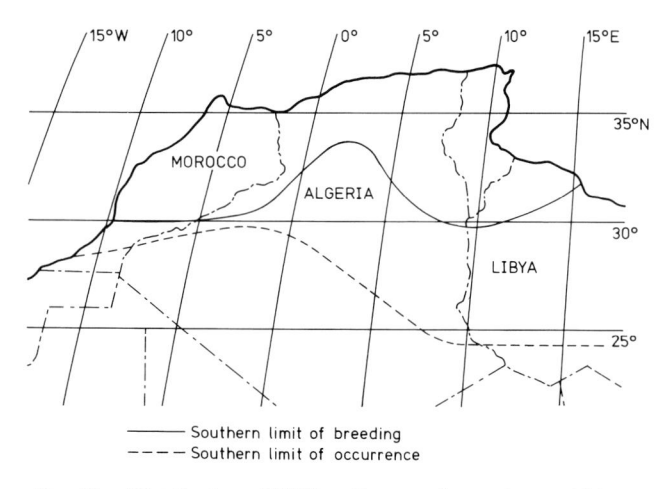

Fig. 60   Distribution of Willow Sparrow in northwest Africa

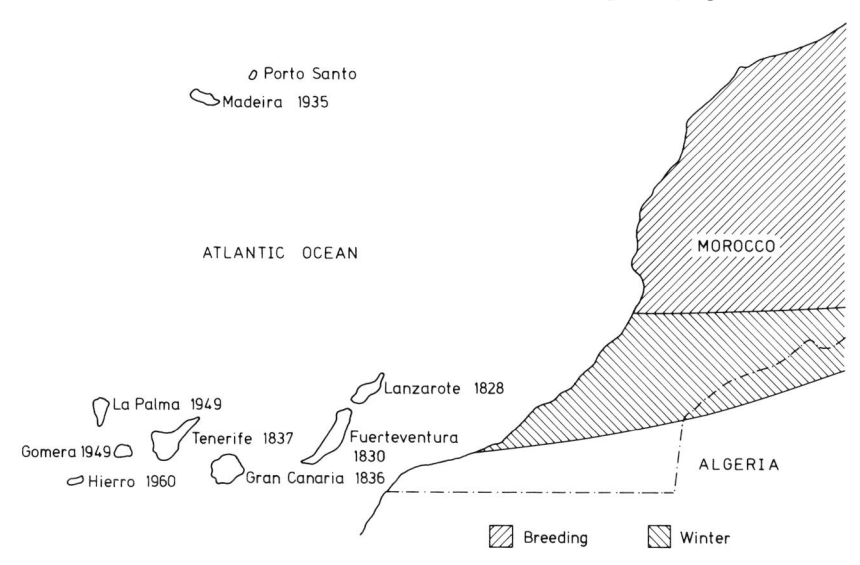

Fig. 61   Distribution of Willow Sparrow in northwest Africa and first-recorded dates on Madeira and the Canary Islands

different islands are plotted, as they are in Fig. 61, it will be seen that they show a gradual spread to the west and this seems likely to be the consequence of a natural extension of range. Fuerteventura, where they were first recorded in 1830, lies only 200 km from the wintering range of the species on the African mainland, much less than the 650 km to Madeira.

The situation is much less clear for the Cape Verdes, where Darwin found it on São Tiago in 1832. Fig. 62 shows the first-recorded dates of the Willow Sparrow on the different islands of the Cape Verdes Archipelago together with the date (in parenthesis) of the first visit by an ornithologist who published his records. The dates, which show that the Willow Sparrow was not recorded on several of the islands, particularly the more westerly ones, until later than the original record for São Tiago, imply that the birds were still colonising the islands in the second half of the 19th century, and did not even reach the northwesternmost islands of São Vicente and Santo Antão until the second half of the present century. This suggests that the Willow Sparrow was a fairly recent colonist at the time of Darwin's visit.

As there are no recorded cases of deliberate introductions of Willow Sparrows anywhere else (in contradistinction to the House and Tree Sparrows) I have argued that the birds probably arrived naturally early in the 19th century (Summers-Smith 1984b). The extensive autumn migration of the Willow Sparrow, more pronounced than for most *Passer* species, to the south and west in North Africa makes such an extension of range by no means improbable.

The Willow Sparrow has had a somewhat chequered career on the Cape Verdes, where certainly in recent years the prolonged drought conditions have made the islands rather unsuitable for this, one of the less arid-adapted sparrows. Alexander (1898a) mentioned it as abundant on Brava in 1897; this was apparently still the case at the time of the *Blossom* expedition in 1924, but by 1951 Bourne (1955) was unable to find it and reported that the Willow Sparrow had died out in the droughts of the 1940s. Sometime between 1963 and 1982 it evidently recolonised the island, as Nørrevang and den Hartog (1984) found it present in the latter year. On the other

168 *Willow (Spanish) Sparrow*

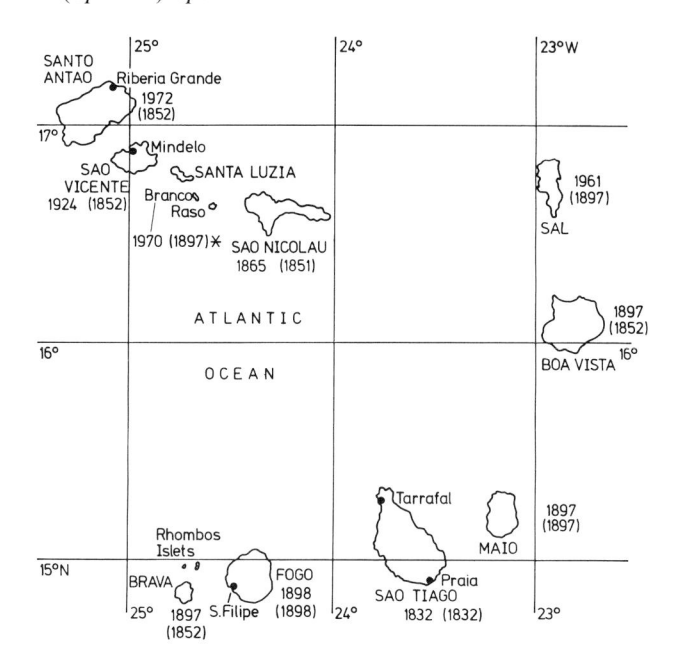

Fig. 62 First-recorded dates of Willow Sparrow on different islands of Cape Verde Archipelago. Date of first visit by an ornithologist in parenthesis

hand, on neighbouring Fogo it is the only sparrow, and is present in all habitats. On São Tiago, where it and the Iago Sparrow are both present, the Willow Sparrow is confined to the town of Praia, the larger villages and the more fertile cultivated land. Dohrn (1871) described the Willow Sparrow as common on São Nicolau in 1865; it was recorded there again in 1897, but was not collected on the *Blossom* expedition in 1923. Frade (1976) obtained it in 1970, but Nørrevang and den Hartog (1984) could only find it in one locality on the island in 1982 and Hazevoet (1986) failed to see any on his visit in 1986. It probably did not arrive on São Vicente until the present century – Murphy (1924) makes no mention of it in 1912 and Correia did not collect it in 1922, though the *Blossom* expedition did in 1924. I have examined the sparrows from the *Blossom* collection, by courtesy of the Curator of the Peabody Museum, New Haven, Connecticut, USA, and was able to confirm both House and Willow Sparrows from São Vicente. In the 1960s de Naurois (1969) found hybrids between Willow and House Sparrows on São Vicente as well as the parent species. According to Nørrevang and den Hartog (1984), Willow Sparrows were still present in 1982, but on my visit eighteen months later in 1983 I could find only a single hybrid *hispaniolensis* × *domesticus* on the outskirts of the town of Mindelo, though in March 1985 M. A. S. Beaman (pers. comm.) reported a significant population of hybrids in the hundred or so birds examined in this area. In addition to the islands already discussed, Willow Sparrows are also present on São Nicolau, Sal, Boa Vista, Maio and Santo Antão, where it was first definitely reported in 1972. It has, however, never been able to colonise the small uninhabited islands – the Rhombos Islets and the Desertas, though there is a single record for Branco, one of the Desertas, in 1970 (Frade 1976).

If, as I suggest, the Willow Sparrow is in the first place a natural invader of the

Cape Verdes, colonisation, extinction and recolonisation of the different islands is likely to occur as conditions change from extreme drought to a rather damper regime.

The birds of the western race have a complex pattern of movements. Extensive ringing studies in northwest Africa (100,000 birds ringed in Morocco) have shown the birds to be nomadic with considerable fluctuations in numbers in any one place, no doubt in response to variations in annual rainfall affecting the ecological conditions. For example, some of the young birds ringed at Berkane, Morocco, in May, moved in autumn to Oran, Algeria, and bred there for the following two years; whereas others were found the following spring in the region of Fès-Meknès; a young bird ringed in Morocco was subsequently recovered near Malaga, Spain (Bourdelle & Gabin 1950–51; Heim de Balsac & Mayaud 1962).

At least part of the population has a more normal pattern of migratory movement. The birds breeding in central Spain move to the south coast, and even across the Strait of Gibraltar into northwest Africa (Alonso 1986), and those that winter in the south of France probably come from the Spanish population. Many of the birds breeding in north Africa move south in winter into the arid regions to the north of the Sahara (Fig. 60), where they range widely searching for seeds; for example, I saw a flock of about 50 birds in the Algerian oasis of Djanet in April 1971.

At the eastern end of the range, movements appear to be rather more directed, with the birds moving with the advancing spring for successive broods as conditions for breeding deteriorate in the south and become progressively more favourable further north (Baumgart 1980). Willow Sparrows winter commonly in the Nile valley as far south as the Sudan, at least some of which are birds of the nominate race (Meinertzhagen 1921; Nicolls 1922). I saw a flock of about 100 near Chersónisos, in Crete, in October 1980; these were probably migrants pausing on their way from Continental Europe to winter in northeast Africa (Summers-Smith 1980b).

The extensive wanderings of the Willow Sparrow lend support to the possibility of a natural extension of range to the Canaries and Cape Verde Islands.

In addition, there are a few records of extralimital wandering following the breeding season. A flock 10 birds was seen on Majorca on 14.8.71; there are four records for England (Nottinghamshire autumn 1900, Whitaker 1907; Lundy Island, 9.6.66, Waller 1981; Isles of Scilly 21.10.72, Charlwood 1981, and 22–24.10.77, Britton 1981), and a male turned up on the Dalmation island of Pag in autumn 1959.

Passer hispaniolensis italiae

The Italian Sparrow occurs in northern and central Italy. In the north it forms a narrow zone of hybridisation with *P. domesticus domesticus*, from about Imperia in the west, following a broad arc along the Alps at about the 500–1,000 m contour on the southern side, to Monfalcone in the east, as has already been described in Chapter 9 and is shown in Fig. 43. South of about the latitude of Naples there is an extensive clinal zone of intergradation with nominate *hispaniolensis*, through southern Italy to Sicily and Malta. A small amount of movement of Italian Sparrows north of the breeding range into southern France and Austria also occurs, as has already been described in Chapter 9.

A population that has been described as 'pure' *italiae* occurs on Corsica, separated by the 10 km wide Strait of Bonifacio from 'pure' *hispaniolensis* in Sardinia, with little evidence of gene-exchange in either direction. According to Steinbacher (1954, 1956) the birds on Sardinia are closer to pure *hispaniolensis* than those on Sicily, not only in plumage characters but also in behaviour. On Sardinia the birds are less associated with man, coming into inhabited areas more in the late summer and autumn, when the island is burnt brown with the long summer drought and food may be short, rather than breeding in this habitat.

Birds that can be referred to *italiae* also occur on Crete (and intermediates between *italiae* and *domesticus* on Rhodes and Karpathos, according to von Wettstein, 1938), and in some of the more remote and isolated Algerian oases. Occasional specimens have been collected in Palestine in March (Meinertzhagen 1921) and Egypt: an undated bird described by Bonaparte as *Passer rufipectus* 1850, as well as a more recent one in January (Meinertzhagen 1921). From its small size, Meise (1934) considered the bird described by Bonaparte to be a hybrid between *P. domesticus niloticus* and a wintering Willow Sparrow that remained to breed, but equally these Egyptian birds could have been casual winter visitors.

Passer hispaniolensis transcaspicus
The birds in Asian Turkey and Cyprus are usually placed in the eastern race, *transcaspicus*, though the boundary between the two races is not clearly marked. For example, Kumerloeve (1970) considered the birds at Ceylânpinar in Turkey to be closer to the nominate race than to *transcaspicus*, and Vaurie (1959) felt the birds in the Near East to be indeterminate as to subspecies.

In the Middle East the Willow Sparrow occurs through Syria, south to Lebanon, where it is a local breeder, and to about the latitude of Jerusalem in Israel and Jordan. Allouse (1953) does not record it as a breeding species in Iraq, though there is a record of attempted breeding in Kuwait (Jennings 1981b). Further east, it occurs sporadically in Azerbaydzhan and Dagestan in the Caucasus, north to the Terek valley, at 44°N on the Caspian coast, round the south of the Caspian Sea into Kazakhstan, where it spreads north up to the valleys of the Syr-Dar'ya and the Chu, eastwards through Semireche as far north as the Karatal river and Panfilov, in the Ili river valley. Since 1940 it has spread as far as Lake Alakol', just north of the Dzungarian (Dzungarskiy) Alatau: not only an extension of range of 300 km, but also the most northerly breeding area for the species (Gavrilov 1963). To the south, what appears to be an isolated population occurs in the western Tarim basin in Chinese Turkestan between the foothills of the Tien Shan and Kunlun ranges.

The southern limit of breeding is in central Afghanistan (Herat and Kandahar), Iran south to Kermān and Fars Provinces, at about the same latitude at Kuwait (30°N) and to Luristan in the west (Witherby 1903, Ticehurst *et al.* 1926, Dementiev *et al.* 1970).

In the more northerly parts, the bird is a summer visitor, the majority of the population migrating south of the breeding range. It occurs in winter in large numbers in northern Egypt, south along the Nile valley as far as Dongola Province in the Sudan. It is common at Elat at the head of the Gulf of Aqaba and is also found in the Sinai peninsula and the northern Hejaz in Saudi Arabia. It is common in Iraq in winter and occurs sporadically along the Persian Gulf as far as Oman. Some birds remain over the winter in the Amur-Dar'ya and Syr-Dar'ya valleys in Turkestan, and the Ili valley in Semireche south of Lake Balkhash, but most move out to winter in southern Iran, southern Afghanistan, south to northern Baluchistan, east across northern Sind, through the Punjab and Haryana, then south through eastern Rajasthan (south to Sambhar Lake and Bharatpur) and east to about Mirzapur in Uttar Pradesh. To the north it extends in winter to the foothills of the Himalayas, with a recent record (*ca* 50 on 16.2.81) in western Nepal (Inskipp 1984), north to Gilgit and Chitral in Kashmir and the North-West Frontier Province of Pakistan. The birds arrive in their winter quarters from early August through to October and depart from early March to mid May. The pattern of migration has been studied by Gistsov and Gavrilov (1984) in the western Tien Shan, where very large numbers pass through in spring and autumn on their way to and from their breeding grounds in Kazakhstan. These observers ringed 278,317 Willow Sparrows in the Chokpakskii pass in the 8-year period 1970–77. Controls in the area showed that at least some birds used the same route in both seasons and, moreover, were remarkably consistent in their timing,

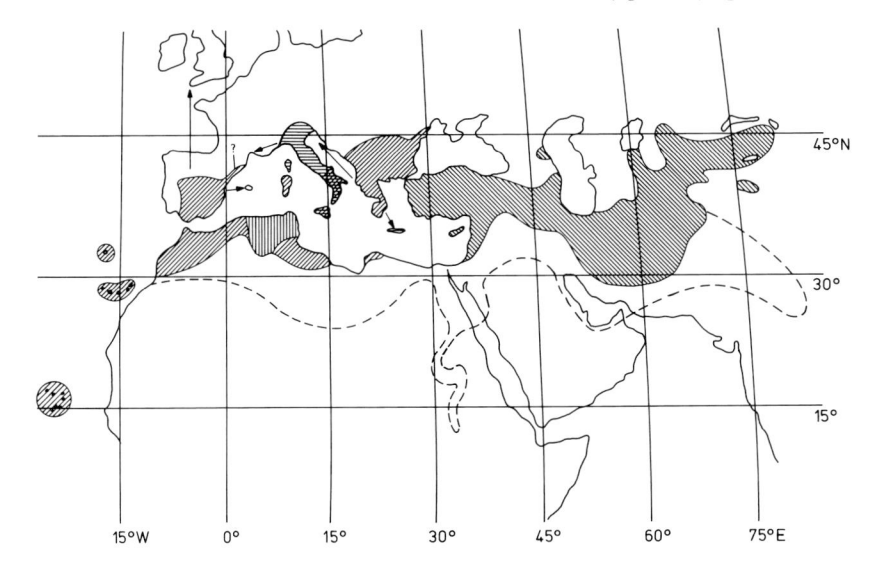

Fig. 63   Distribution of Willow Sparrow

with only a few days separating the dates of spring passage in subsequent years, though there was a suggestion that the birds passed slightly earlier as they grew older. The females lagged somewhat behind the males. The spring migration lasted for over two months from mid April to early July, with an indication of two peaks of passage. Gistsov and Gavrilov hypothesised that these involved two populations, birds from the more easterly population crossing first; they further suggested that the time difference was an adaptation that reduced competition for food along the migration route.

Birds ringed at Bharatpur have been recovered on their breeding grounds in Kirghizia, USSR (Ali 1963). The Willow Sparrow is particularly frequent in the plains of the North-West Frontier Province of Pakistan, the Punjab and Rajasthan in the winter months. Ali (1963) reported a winter roost at Bharatpur of over a million birds that were mainly winter visiting Willow and House Sparrows. Before departing for their breeding areas, these birds cause considerable damage to ripening grain crops.

The breeding range of the Willow Sparrow, together with the main extension to the south outside the breeding season, is shown in Fig. 63. Breeding is restricted to a mere 15° of latitude, lying almost entirely between 30–45°N, with an outlier in the Cape Verdes at 15°N. In the absence of definite information, the boundary between the nominate and eastern races is drawn between Europe and Asia Minor. Because of the nomadic behaviour outside the breeding season, the wintering area is not clearly defined and birds do not necessarily turn up in the whole of the area shown. Members of both races occur in winter in the Nile valley in Egypt and the Sudan. Some sporadic occurrences outside the main range are indicated by arrows.

HABITAT

The Willow Sparrow is a bird of wide distribution that, over much of its range, is sympatric with the House Sparrow. Where the two species occur together, the latter adopts the more arid habitat, the former tends to occur in the more humid areas. It is typically associated with cultivation, provided that trees are available for nesting and roosting, though outside the breeding season bushes and reed beds will serve for the latter and birds may move considerable distances during the day to suitable feeding areas.

During the breeding season it makes use of open, deciduous woodland, plantations, palmeries or rows of roadside trees and is to be found particularly in moist river valleys or lakesides. The association with cultivation is shown by the recent extension of range to Lake Alakol' in Kazakhstan, following the exploitation of the area by man (Gavrilov 1963). In such semi-desert regions it is confined to the damper, cultivated areas and becomes a pest of agriculture.

Where the House Sparrow is absent, as in Italy, Sicily, Malta, Crete and the Atlantic archipelagos (except for the Azores), the Willow Sparrow takes over the role of the former, coming freely into villages and even extensively built-up areas, feeding on scraps just like its urban congener, though only where there are trees. On São Tiago in the Cape Verde archipelago, where it co-exists with the Iago Sparrow, the Willow Sparrow is confined to the larger towns and villages and the richer areas of cultivation with large trees, leaving the poorer cultivation and completely arid areas to the other species.

In its winter quarters it is frequently found in association with the House Sparrow, frequenting open areas – grassy plains, reed beds and the edges of cultivation; it is, however, more nomadic than the House Sparrow and in North Africa it roams more widely, being found on the northern edge of the Sahara, where the more sedentary House Sparrow is absent.

It is normally a bird of low altitudes, but occurs up to 1,370 m in Kazakhstan (Gavrilov 1963), 2,750 m in Afghanistan (Hüe & Etchecopar 1970), 1,200 m in the Zagros mountains, southwest Iran (Witherby 1903) and 1,300 m in the Troödos mountains in Cyprus (Flint & Stewart 1983).

BEHAVIOUR

The Willow Sparrow is the most gregarious of the Palaearctic sparrows, breeding in densely packed colonies and foraging in large flocks outside the breeding season that may contain thousands of birds. In contrast to other members of the genus, it is a relatively shy bird, particularly in those parts of its range that it shares with House Sparrows. The 'open country' birds tend to fly off some distance when disturbed, though the urban Willow Sparrows show the same combination of cheekiness and wariness that the House Sparrow does in a similar situation. This difference was very noticeable on Malta, where the Willow Sparrow appeared to form separate 'town' and 'country' populations, each with its own pattern of behaviour (Summers-Smith 1978).

A predominant characteristic of the Willow Sparrow already referred to is its nomadic behaviour. This has been noted both in North Africa and in the southern USSR, where the post-breeding flocks can move distances of several hundred kilometres in their search for food. This is a dispersal, quite distinct from the regular winter migration southwards undertaken by the same populations, no doubt adopted by the birds because of the irregular nature of their food supply. This pattern of behaviour means that the bird can become a serious pest of cereal cultivation in marginal semi-arid areas. Even within the comparatively small island of Malta the birds make considerable movements: no less than 11 of the 94 recoveries of ringed birds up to the end of 1980 being more than 5 km from the place of ringing, one bird having moved 6 km in less than 24 hours (Sultana & Gauchi 1982).

The winter roosts in thick trees in built-up areas are very conspicuous and noisy, the birds wheeling about above the trees before they finally settle for the night. In other places they roost in scrub, reed beds and in crannies in the walls of wells. The roosts can contain enormous numbers of birds. The flight lines to the autumn roosts have been traced for up to 10 km. Ali (1963) described a mixed winter roost of Willow and House Sparrows in Rajasthan that contained over a million birds.

Several writers have commented on a tendency for the sexes to form separate flocks outside the breeding season. For example, Meinertzhagen (1940) found wintering flocks south of the Atlas in Morocco to contain a preponderance of males; Gavrilov (1962) commented that in 19 post-breeding flocks in Kazakhstan in July, 13 contained 90% males, while six contained 90% females. He concluded that the birds migrated and remained in these sexually separated flocks over the winter, quoting Shulpin (1956) as noting that flocks on autumn passage south of Kazakhstan consisted chiefly of males. In winter they frequently associate with House Sparrows and, in the Sahara, with Desert Sparrows, though in some places the two species are said to remain quite separate, *eg* Iraq (Ticehurst *et al.* 1923).

BREEDING BIOLOGY

The Willow Sparrow is a colonial nester, the breeding colonies in places reaching spectacular proportions. Where it is sympatric with the House Sparrow the nests are predominantly in trees and bushes, both openly among the branches of trees and in the understorey of nests of birds of prey, herons and corvids; the latter is the commonest site in Spain. Deciduous and coniferous trees are used and in the southern parts of its range the crowns of palms are a favourite site. To a lesser extent nests can be placed in reed beds, in holes in cliffs and even on electricity pylons. In the absence of the House Sparrow the bird comes freely into inhabited areas, nesting in holes and creepers on buildings and other man-made structures, such as the walls of wells; an unusual site that I saw in Sardinia was the inside of a hollow, metal, roadside advertisement hoarding that had been punctured by a shot from a shot-gun.

The colony size varies with the local density of the population; for the nominate race colonies range from about 10 to several thousand pairs, but in Kazakhstan Gavrilov (1963) found colonies of 20,000–30,000 nests were common and he even recorded colonies of 100,000–800,000 nests, with an estimated 2.5 million birds nesting in one square kilometre. These very large colonies extend over several kilometres, with an average of 13,000 nests per hectare of forest plantation. The nests are placed close together, so close that they can form a continuous mass; up to 180 nests have been found in one tree in Spain (Alonso 1986) and 120–130 nests in one tree in Kazakhstan (Gavrilov 1963). As many as 20 nests have been recorded in the structure of a single stork's nest. Baumgart (1984) considers that the flank streaking in the Willow Sparrow has evolved to reduce the threatening effect of the bold black throat patch of the House Sparrow and other black-bibbed sparrows, and to allow the close-packed breeding of the species to take place in harmony. He goes on to suggest that the loss of the flank streaking in the sedentary Italian Sparrow populations is a consequence of their evolution to a 'house sparrow' way of life in which the small 'threatening' black bib has a useful function.

The nests are built of green grass, straw, small plants or leafy sprigs from trees; they are rather coarsely woven and form an irregular, roughly spherical mass, 150–300 mm in diameter (weight 170–300 g), firmly attached to tree branches. The nest is domed, with the entrance to the nest cup in the side; the nest cup is lined with finer plant material and only rather occasionally with animal hairs or feathers, the latter not as commonly as with the House Sparrow, and feathers may not be used when they are readily available. The nests can be as low as 0.5 m from the ground in perennial vegetation, to as high as 25–30 m in the tops of poplars.

At the beginning of the breeding season the male may take over an old nest from the previous season or a suitable crotch in the branches of a tree in which it begins to lay the foundations of the nest. From this site he calls and displays to attract a mate. The display is typical of the sparrows: wings held out from the body, slightly drooped and shivered, head held high to expose the black throat, and tail spread and raised, the bird uttering a continuous chirruping that can be heard several hundred metres away. While this mate-seeking display is very similar to that of the House Sparrow, I have, however, never seen in this species a group display (common in the House Sparrow), in which a number of males chase and display in front of a single female, except in the Italian race, doubtless a consequence of its hybrid nature with its infusion of House Sparrow genes. This has been noted in Malta (Sultana & Gauchi 1982), suggesting these birds are more properly allocated to the race *italiae* than to nominate *hispaniolensis*.

Once the male has attracted a mate he completes the nest, mostly by himself, with the female putting in the lining. When the nest is ready, she invites coition by crouching and wing shivering near it, calling a soft *que que que*.

The smaller colonies of 100 or so pairs can be occupied within a few days (Baumgart 1980); the larger colonies take longer to build up as they grow from an initial centre. A nest of a bird of prey may act as a focus for a colony, the birds spreading into the branches and then into neighbouring trees as the first tree is fully occupied (Makatsch 1955). Baumgart (1980) suggests that a breeding colony is only formed when the nomadic flock comes across a sufficiently rich feeding place that the males have the time to take up nesting sites and display to attract females. Much robbing of nest material takes place whenever the owners are absent from their nest or even when they are out of sight inside. This gives rise to frequent disputes in which the owner threatens by raising both the head and tail, with the wings slightly drooped and rotated towards the front.

As with the House Sparrow, the birds in sedentary populations use the same nest for successive broods and in successive years. These birds return to the nest sites in

176 *Willow (Spanish) Sparrow*

Subspecies	Locality	J	F	M	A	M	J	J	A	S	O	N	D	Reference
hispaniolensis	Europe				━|━	━	━|━						Jourdain 1936, Makatsch 1955, Sacarrão & Soares 1975, Baumgart 1980, Flint & Stewart 1983
	North Africa				━	━	━							Makatsch 1957, Heim de Balsac & Mayaud 1962, Bortoli 1973, Mirza 1974
	Canaries			━	━	━								Bannerman 1912, 1963, Ennion & Ennion 1962
	Cape Verdes	━			━					━				Keulemans 1866, Bourne 1955, Bannerman & Bannerman 1968, author
transcaspicus	Turkey					━	━							Makatsch 1955, Kumerloeve 1970
	USSR						━	━						Dementiev *et al.* 1970

Fig. 64   Breeding season data for Willow Sparrow

autumn, when a certain amount of desultory nest-building may take place and, occasionally, even breeding; Sultana and Gauchi (1982), for example, record nests with young in Malta in October and November.

Details of the breeding season are summarised in Fig. 64.

The clutch size ranges from 2–8 eggs. Bortoli (1973) gives an average of 4.3 for Tunisia; 1,079 clutches in Kazakhstan averaged 4.4 eggs (Gavrilov 1963); 57 in Macedonia 5.9 eggs (Makatsch 1955). Incubation begins before the clutch is complete (after the second egg is laid, according to Gavrilov) and both birds take part in incubation with spells of 5–20 minutes on the eggs, the female taking the larger share during the day as well as the overnight stint. The incubation period lasts 11–14 days (Gavrilov), 11–$11\frac{1}{2}$ days (Bachkiroff 1953), 10–14 days (Sultana & Gauchi 1982); during this time the relieving bird often arrives with lining material, which it takes into the nest. The feeding of young in the nest is shared almost equally between the sexes, with feeding rates typically of 10–15 visits per hour. The nestling period is 11–12 days (Heim de Balsac & Mayaud 1962), 11–15 days (Harrison 1975), *ca* 15 days (Bortoli 1973), *ca* 14 days (Sultana & Gauchi 1982). According to Bortoli the young are fed by the parents for 4–5 days after leaving the nest.

Gavrilov (1963) gives the following data for breeding success:

	1959	1960	1961
Average clutch size	4.37	4.51	4.48
% hatched	91.5	97.6	84.8
% fledged	53.5	62.3	53.6

At least some of the pairs may raise two, even three or four broods in one year. In colonies studied by Gavrilov, the proportion of females nesting twice in 1959–61 were respectively 45.4%, 4.6% and 21.0%. This shows a considerable variation between years, but may not, however, be the whole story as there is evidence from Morocco, Tunisia and the Balkans that, after one brood has been raised, the birds may move to a new area for a subsequent brood (Heim de Balsac & Mayaud 1962, Bortoli 1973, Baumgart 1980). Baumgart suggests that in much of its range the species is adapted to take advantage of the short period of favourable conditions for breeding that occur with the Mediterranean-type climate, where there is a quick flush of vegetational growth (and an associated flush of insect life) before the long dry summer; the birds moving north to breed for a second, and even third, time in areas where spring occurs some time later. In this way both the nomadic Willow Sparrow and sedentary House Sparrow, with similar dietary requirements for breeding, are able to co-exist without significant competition when the superabundance of food is more than sufficient for the resident species, whose numbers are controlled by the availability of food at other times of the year. In recognition of this behavioural separation, Baumgart (1984) calls them 'time-differentiated species'.

The use of trees imposes no restriction on nest sites, and the large colonies with close-packed nests, all speedily established, together with the nomadic behaviour are, according to Baumgart (1980), adaptations to exploit this rich but brief flush of food. Although the birds may return to the same area for nesting each year, this is by no means invariably the case.

SURVIVAL

No studies have been made of mortality in the Willow Sparrow. Ali and Ripley (1974), however, record an interesting ringing result in which a first-year bird wintering in India was retrapped nine years later, apparently in good health; and Sultana and Gauchi (1982) record a bird that was at least eight years old when retrapped.

MOULT

Alonso (1984b) found that after the adults had finished breeding they moved to a good feeding area for the moult. In the Tunisian population, moult extends from July to September (Bortoli 1973), in the Spanish birds from late July to late September or early October (Alonso 1984b), lasting 65 days in individual birds. The young have a complete post-juvenile moult that starts about four weeks after they have fledged and is complete by the end of September.

The situation in the migratory populations is rather similar; Gavrilov (1962, 1963) found that adults in his study area in Kazakhstan had begun to moult in August, but no birds were found in full moult and presumably, like the Spanish birds, they moved away to complete the moult elsewhere, before migrating south.

VOICE

I have transcribed the basic call of the male at the nest as a slightly disyllabic *chweeng-chweeng*, *cheela-cheeli*, uttered in pairs. This is a strident, far-carrying call used to indicate nest ownership and to attract a mate, serving the same function as the *chirrup* call of the House Sparrow, but louder and higher pitched. A similar, but rather softer call, is commonly used by birds in the roost, both when assembling in the evening and before departing in the morning; with the large aggregations that attend the roosts the din can be quite considerable.

The *que que que* or *chee chee chee* contact call used between the mated pair at the nest is almost identical with the call used by the House Sparrow in the same situation. The flight call *quer-it* is very similar to the same call of the House Sparrow, as is the threatening *chur-it-it-it*. This last call is used in mild threat, at the nest, by one bird arriving to take over incubation or to feed the young when its mate is inside, or against an intruder of the same sex at the nest.

The similarity of these calls to those of the House Sparrow is very striking and the only significant divergence seems to be the nest ownership proclamation and the mate attracting call.

FOOD

The Willow Sparrow feeds mainly on seeds, both of grasses and cultivated cereals (wheat, millet, barley and oats). This is supplemented by the leaves of young plants (legumes) and fruit (cherry, apricot, date, orange). Just before and during the breeding season the adults start to take some insects (Mirza, 1974, found 10–20% by weight animal food in the stomachs of adults in Libya in March and April). The animal food consists of lepidoptera (caterpillars), hymenoptera (flying ants), orthoptera (grasshoppers and crickets).

Gavrilov (1963) examined the stomachs of 432 birds in Kazakhstan for April–August. The percentages of stomachs in which the different categories of food were present are shown in Table 31.

Table 31: Stomach contents of adult Willow Sparrows (432 birds) in Kazakhstan (Gavrilov 1963)

Food type	% of stomachs in which present
seeds of cereals	75.0
seeds of wild plants	13.6
other vegetable matter	7.0
insects	20.3

The nestlings are fed a much higher proportion of animal food: 75% by weight in Libya according to Mirza (1974), the remainder soft seeds (*ie* grain in 'milk'). Gavrilov's (1963) results, Table 32, show a similar preponderance of animal food in the diet of nestlings and also, to a considerable extent, in fledglings. The ranges in the table show the variation with age of the birds, the figure on the left applying to younger birds, and between different populations, no doubt reflecting local availability of food.

Table 32: Stomach contents of Willow Sparrow nestlings (10,023 birds) and fledglings (233 birds) in Kazakhstan (Gavrilov 1963)

Food type	% of stomachs in which present	
	nestlings	fledglings
insects	100–98	94–73
cereal seeds	4–10	13–68
seeds of wild plants	0	3

Where the bird is common it is considered a serious pest of cereal cultivation. For example, Kashkarov (1926) stated that sparrows (principally Willow Sparrows) reduce wheat yields in Kazakhstan by up to 30%, but the number of birds is so great that control measures against them are ineffective. In some places, however, direct control action is attempted. Matkasch (1955) cites an instance in Asia Minor where a colony was destroyed by local farmers who cut down trees in which the nests were built, and another case in which a whole wood was burnt down to get rid of the nests. In Tunisia, according to Dawson (1969), a law making it mandatory to destroy Willow Sparrows' nests has been in existence since 1892, though how effective this is in limiting numbers is not known. In a recent campaign in Kirghizia, USSR, no less than 1.8 million sparrows were killed by feeding with poisoned grain.

As with most cases of economic ornithology, it is difficult to strike a balance. There is no doubt that the bird does some good by destroying a number of harmful insects.

Moreover, it is doubtful whether such uncoordinated attempts at control can do much to reduce overall numbers in such a successful species, though according to Despott (1917) the population in Malta was so reduced by farmers, sportsmen and netters that the Government issued a notice on 13.10.16 making it unlawful to 'kill or take it by any means between 1st January and 1st July'. How long this was enforced is not clear, but the species there can no longer be considered to be in danger.

11: Dead Sea Sparrow *Passer moabiticus*

NOMENCLATURE
Passer moabiticus Tristram 1864, Proc. Zool. Soc. 1864: 169.
        Palestine [southern end of Dead Sea].
    Synonyms: *Passer yatii* Sharpe 1888.
          Dedadi, [Sistan], western Afghanistan.
    Passer mesopotamicus Zarudny 1904.
          Mochammerah [Khorramshahr, Khuzestān] southwestern Iran
  Subspecies: *Passer moabiticus moabiticus* Tristram 1864.
    Synonym: *P. mesopotamicus*.
    Passer moabiticus yatii (Sharpe) 1888.

The bird receives its scientific name from the Moab region of Jordan. The example from Sistan was named *yatii* by the collector, Dr Atchison, in honour of Captain Yate, a member of the Afghan boundary commission. Alternative trivial names are Moab Sparrow and Scrub Sparrow.

The Dead Sea Sparrow was discovered by R. M. Upcher in 1864 in a few isolated pockets at the southern end of the Dead Sea. Even twenty years later, Tristram (1884) could only write that 'the bird is the most limited in the world in its range and the scarcity in numbers of individuals.' In 1888 Sharpe described as *Passer yatii* a sparrow from the Sistan, lying on the boundary between Iran and Afghanistan. More recent opinion (Hartert 1904–22, Ticehurst 1920–21, Vaurie 1959, Moreau 1962) is that this is a race of *moabiticus*, though this conclusion has been challenged by Boros & Horvath (1953) on the basis of plumage and morphology, together with the physical separation of 2,500 km. Later, Zarudny (1904) described yet a third species, *Passer mesopotamicus*, discovered by him along the lower reaches of the Kārūn river between Ahvāz and Nasrie in southern Iran, that according to him was intermediate in character between the birds of the above two populations. Ticehurst *et al* (1923), however, with larger samples at their disposal, were unable to find any difference between the Mesopotamian and Dead Sea birds.

More recent observations have established further widely scattered populations in Iraq, Syria and Turkey. This again is another of those problems of determining the status of allopatric populations. My own opinion is that, in the absence of clearly established ecological differences, it is best to consider such populations as belonging to one species, recognising separate races where sufficient plumage or morphological differences exist, as they do between the birds from the Dead Sea and those from Sistan.

Tchernov (1961) identified a fossil premaxilla from the Oumm-Qatafa cave near Jerusalem as belonging to *Passer moabiticus* or a precursor of this species. This was recovered from the Middle Acheulean layer, giving a date of *ca* 300,000 years B.P. (Before Present).

DESCRIPTION

The Dead Sea Sparrow is one of the smaller representatives of the genus (length *ca* 120 mm); a typical dimorphic 'black-bibbed' sparrow.

P. m. moabiticus

Male. Crown and nape ashy grey. White superciliary stripe becoming cinnamon behind eye; black stripe through eye; lores same colour as crown. Small black bib separated from lores by white stripe. Yellow spot at sides of throat. Upper back buff with black streaks; lower back and rump ash grey. Wing coverts dark chestnut, remainder of wings streaked black and brown. Underparts pale grey, darker on flanks, becoming yellowish in breeding season with abrasion of feather tips, though brightness of yellow is rather variable; undertail coverts cinnamon brown. Bill black in breeding season, otherwise horn.

Female. Similar to female House Sparrow, but with yellow spots at sides of throat, rather paler than those in male.

Juvenile. Similar to female.

P. m. yatii

Male. Generally paler than nominate *moabiticus* – crown grey-brown, lower back pale grey-brown – underparts grey washed with bright lemon-yellow.

Female. Similarly paler than nominate *moabiticus* and again showing yellow wash on underparts.

BIOMETRICS

Typical body measurements are given for the two subspecies in Table 33.

This is one of the smallest *Passer* species. There is some suggestion that birds from the traditional Dead Sea area are smaller than those from Iraq and Iran, with the Turkish birds closest to the latter, and that the subspecies *yatii* may be larger than the nominate one, though the data are so few that this is uncertain.

Measurements of birds from Sistan (*P. m. yatii*) demonstrate clearly seasonal effects resulting from wear (Boros & Horvath 1954). Wing and tail length show a decrease of 2–3% from winter (after moult) to summer (just prior to moult), Table 34.

Table 33: Biometric data for Dead Sea Sparrow

Feature	Subspecies	Locality	Males		Females	
			Range	Mean (Median)	Range	Mean (Median)
weight	moabiticus		15–20	(17.5)		
	yatii		14–17	(15.5)	14–17	(15.5)
wing	moabiticus	Israel	57.6–64.5	59.8	58.3–62	59.3
		Turkey	63–66	64.8		
		Iran/Iraq	62–66.5	(64.3)	59.3–63.5	(61.4)
	yatii	Sistan	61.9–68	63.9	60.1–65.6	61.5
tail	moabiticus	—	47.5–59.5	50.3	47.5–56.2	49.4
	yatii	—	48.4–55.1	52.4	48–55.5	51.0
tarsus	moabiticus	—	15.5–18	16.8	15.7–17.5	15.9
	yatii	—	16.7–17.6	17.2	16.6–17.5	17.1
culmen	moabiticus	—	8.4–10.5	9.1	8.5–10	8.8
	yatii	—	9.0–10.5	9.5	8.4–10.2	9.1

Table 34: Seasonal change in wing and tail length of Dead Sea Sparrow
(Figures in parenthesis are the number of specimens)

	Winter	Summer	Wear
males: wing	mean 64.7 mm (8)	mean 63.7 mm (8)	1.0 mm (1.5%)
tail	53.0 mm (8)	51.7 mm (8)	1.3 mm (2.5%)
females: wing	62.6 mm (6)	60.6 mm (3)	2.0 mm (3.2%)
tail	51.5 mm (6)	50.1 mm (3)	1.4 mm (2.7%)

In contrast the bill is shorter in winter, through the greater wear caused by the wholly seed diet, than in summer when the diet has a softer invertebrate component (Table 35). Here the difference is about 10%.

Table 35: Seasonal change in bill length of Dead Sea Sparrow
(Figures in parenthesis are the number of specimens)

	Summer	Winter	Difference
males	mean 10.0 mm (8)	mean 9.0 mm (7)	1.0 mm (10%)
females	mean 10.0 mm (3)	mean 8.7 mm (6)	1.3 mm (13%).

DISTRIBUTION

Passer moabiticus moabiticus
The current distribution of the nominate subspecies is a perplexing one. From its apparently very restricted original distribution at the southern end of the Dead Sea where it was discovered in the 19th century, a number of additional widely scattered populations have been found in the present century: along the lower reaches of the Kārūn river in southern Iran, 1904, northwards along the Tigris up to near Bāghdad, 1915; along the Khâboûr (Euphrates) river on the borders of Turkey and Syria near Birecik, 1964; Mosul, Iraq, 1963; Cyprus, 1981. These scattered populations might give the impression of a relict species on the way to extinction, but there is very strong evidence of a recent expansion in range and numbers, so that it is difficult to be

certain how far these are relict populations or possible recent new colonisations, as is certainly the case for the recently discovered breeding population in Cyprus. In order to throw some light on these questions the known history of the bird in the different locations is considered separately.

ISRAEL–JORDAN

Over 50 years after the discovery of the Dead Sea Sparrow at the southern end of the Dead Sea, Carruthers (1910) reported it was still only to be found in three isolated areas in the same region, though outside the breeding season a few individuals had been collected north of the Dead Sea: a male near Jericho in 1893, some in December 1897 near Ain Fescha ('Ain Fashka), and again in November 1899 in the same general area. It was not until 1919, however, that Meinertzhagen (1919) could write that it was now commonly found in the Jordan valley at the north end of the Dead Sea. Since then there has been a remarkable expansion (Mendelssohn pers. comm.; Yom-Tov et al. 1976; Yom-Tov & Ar 1980). This started with a spread up the Jordan valley

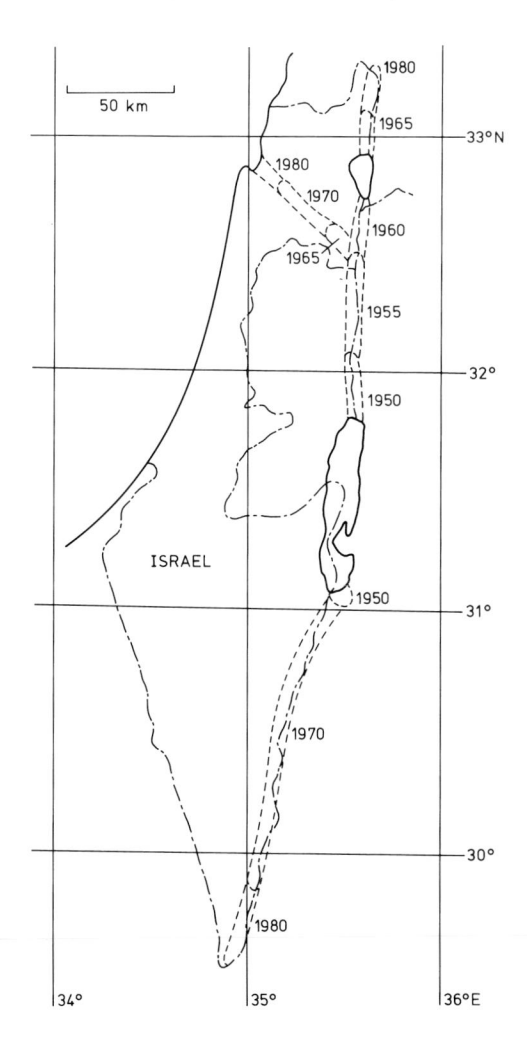

Fig. 65 Spread of Dead Sea Sparrow in Israel

out of the Dead Sea depression, reaching Lake Tiberias (the Sea of Galilee) about 1960, and then further to the north up the Jordan valley to the marshes of 'Emeq Hula in 1965; at the same time there was a spread to the north west along the Esdraelon (River Harod) to the Zevoulum (River Qishon) valleys almost to Haifa on the Mediterranean coast. Between 1965 and 1970 there was also a southward spread along the Rift valley that reached to Elat on the Gulf of Aqaba. Details of the extension of range are shown in Fig. 65. This expansion has also been accompanied by an increase in numbers from the 4,000–5,000 pairs estimated by Mendelssohn in 1950. For example, P. A. D. Hollom (Cramp 1971) considered that the population in the neighbourhood of King Hussein Bridge (formerly Allenby Bridge) to the north of the Dead Sea had increased tenfold between 1955 and 1963.

Mendelssohn attributes this spread to changes in agricultural practice in the modern Jewish state, introducing new settlements with irrigated land and fish ponds that led to increases in grasses along the Jordan, particularly grasses of the genus *Phalaris* (the seeds of which are the principal food of the species), which coincided with the pressure on nesting sites caused by the clearing of dead trees along the Jordan river. An additional factor may have been the reduction in predation following the decrease in wintering Sparrowhawks as the result of secondary insecticide poisoning.

The movements of the Dead Sea Sparrow outside the breeding season are not fully understood. Meinertzhagen (1920) reported the species as plentiful at the north end of the Dead Sea in November and January, but absent in late February and March. Later he wrote that the Dead Sea Sparrow was absent from its breeding quarters (in Israel–Jordan) from November to the end of March, but where it goes is still a mystery (Meinertzhagen 1954). There are records of flocks and of single birds near to the Dead Sea in winter, and Howells (1956) noted birds there in February; again in February, Kumerloeve (1965b) reported numbers in Jordan in the Ain Zarbi valley, near El Rouseifa (Er Ruseifa), 40 km from the nearest breeding colony at the northern end of the Dead Sea. Professor Y. Yom-Tov (*in litt.*) informs me that flocks are often seen in winter in the breeding range (*eg* in the Bet She'an valley). On the other hand, I carried out an extensive search for the bird in its known breeding range in Israel in October 1981, but could only find a single flock of about 60–70 birds in the Hula Reserve ('Emeq Hula), where it breeds.

In contrast, the bird turns up at Elat as a late winter visitor (January) and is a regular passage migrant there between late February and the middle of April: *viz* 26th February to 12th April 1977 (Pihl 1977), March to 12th April 1978 (Argyle & Gel 1978). This suggests a return from winter quarters and even a differential movement between the sexes because in the spring of 1977 only 25% of the birds ringed were males, indicating that the main movement of males had been before the spring opening of the ringing station, or that females were wintering further south than males, though there is complete lack of evidence for Dead Sea Sparrows wintering in those countries lying to the south, *viz* Egypt, Sudan, Ethiopia, Somalia and Saudi Arabia.

The observations suggest that there is some off-season wandering, but the comparatively few autumn and winter reports, considering the present size of the Israeli population and the spring passage at Elat, suggest that part of the population may emigrate, but where to, if this is the case, remains an enigma. As we shall see later, migratory behaviour has been reported for some of the other populations.

CENTRAL IRAQ TO SOUTH-WEST IRAN

Zarudny discovered the Dead Sea Sparrow between Khorramshahr and Ahvāz along the lower reaches of the Kārūn river in southern Iran in 1904, and later to the east at Shellgati and Kulichan on the Gagar (Jarrāhī) river, a tributary of the Kārūn. In 1969, Cramp (1971) found two small colonies on the Mārūn river, 130 km to the east of the Kārūn river. To the north, the bird was found breeding at Jadriya on the

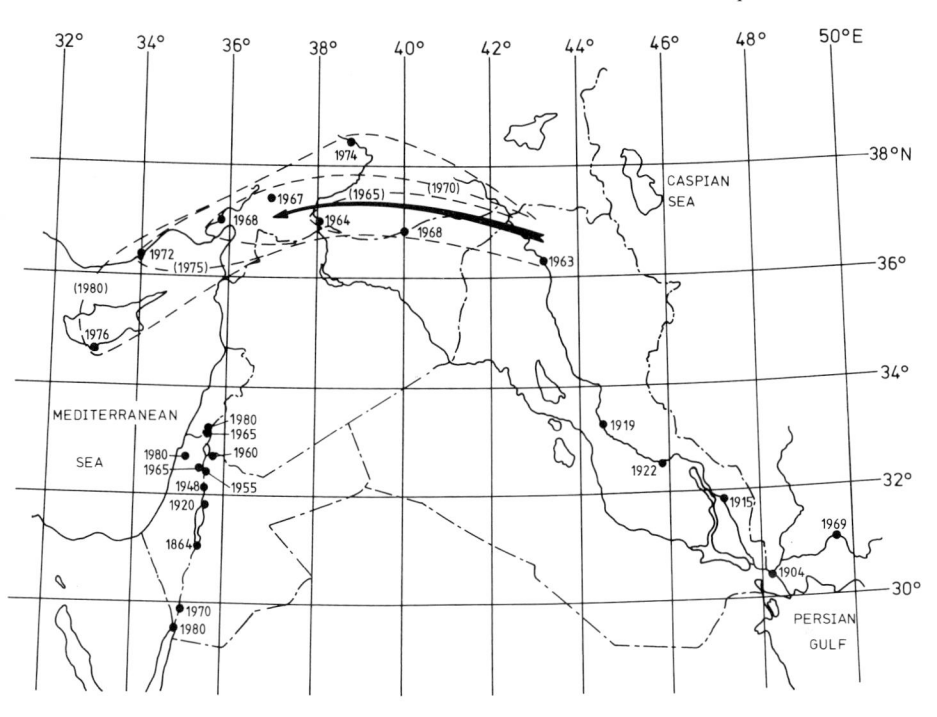

Fig. 66   Distribution of Dead Sea Sparrow in Middle East. Dashed lines give tentative timings for spread through Turkey

Tigris on the southern outskirts of Baghdād in 1919 (Cheesman 1919). Also on the Tigris, there are records from Al 'Amārah in December and January and from near Al Kūt in March and April (Ticehurst *et al* 1923). It would appear that the bird has a more or less continuous distribution along the Tigris from near Baghdād to the extreme southwest of Iran. According to Cheesman, the birds from Baghdād migrate in winter a few hundred kilometres south to the lower reaches of the Tigris.

As with the Israel population there is some evidence of increase in numbers of the southern Iraqi one. Marchant (1963) found the birds at Jadriya much more numerous in 1960–62 than reported by Cheesman in 1919.

NORTHERN IRAQ

Marchant (1963), without further comment, reported that fresh or slightly incubated eggs had been collected in May at Mosul, which, like the previous population, is on the Tigris. Mosul is 360 km north of Baghdād and, in the absence of reports between these two localities, this can be considered as a discrete population. This colony is not referred to by Allouse (1953) in *The Avifauna of Iraq*, but whether this implies that it represents a recent extension of range, or merely the first report of a previously unrecorded population, is not known.

Occurrences in Iraq and southeastern Iran are shown in Fig 66.

SOUTHERN TURKEY – NORTHERN SYRIA

A breeding colony was discovered at Birecik on the River Firat (Euphrates) in southern Turkey in 1964 (Kumerloeve 1965a). In 1967 a second Turkish colony was found at the lake of Gâvur Gölü to the south of Maraş *ca* 110 km to the northwest of Birecik (Vieillard 1968, quoted by Kumerloeve 1969). The following year, 1968, Kumerloeve (1969) found yet another colony at Ceylânpinar on the Khābūr on the Turkish–Syrian border, *ca* 180 km to the east of the Birecik colony with nests also on Syrian territory near Ras el 'Ain, and the same year a further colony was located on the river Ceyhan some 200 km due west of Birecik (Cramp 1971). The range in Turkey was extended even further by the discovery in 1972 of a breeding colony in the Göksu delta near Silifke on the Mediterranean coast, 170 km west of the Ceyhan River colony (Beaman *et al.* 1975) and of another at Kale on the Firat, 170 km north of Birecik in 1974 (Beaman 1978).

The first-recorded dates for the colonies in southern Turkey and northern Iraq are given in Fig. 66. Once again there is insufficient evidence to be able to say whether breeding in Turkey represents a new expansion of range or is merely the result of increased observer coverage. However, the tentative date lines that have been superimposed on Fig. 66 are suggestive of an extension of range of the Iraqi population up the Tigris valley, first into Mosul and then on into Turkey. The dates of the Israeli expansion, which are included in Fig. 66 for comparison, suggest contemporaneous, but separate, expansions of the Israeli birds and Iraqi populations. It is worth noting that, if anything, the Turkish birds appear closer in size to the Iraqi than to the Israeli birds.

Whether the rather discontinuous distribution in Iraq and southern Turkey is genuine, or the result of inadequate observer cover in these areas (for example, it is puzzling that no birds have been reported from the Euphrates in Iraq), or of recent colonisations, is not known; resolution of this question must await further study.

The situation at Birecik is somewhat confused by changes in the location of the nesting colonies, but overall there appears to have been an increase of about 50% over the five-year period 1964–68, again indicative of a recent colonisation. A similar increase was reported at Ceylânpinar between 1968 and 1977 (Kumerloeve 1978).

Nothing appears to be known about the distribution of the Turkish birds outside the breeding season; Beaman *et al.* (1975) reported that there were no winter records for the period 1970–1973 and, in fact, the only non-breeding season record appears to be of some birds near Tarsus, within the breeding range, though some 70 km from the nearest known breeding colony (Vieillard 1968, quoted by Kumerloeve 1969).

CYPRUS

The most recent report of a new breeding area comes from Cyprus, where two nests were found in 1980 and 17 birds, including 8 juveniles were trapped at Akrotiri salt lake (Lobb 1981, Flint & Stewart 1983). Publication of this record triggered the memory of Squadron-Leader R. Foers, who recollected having seen unfamiliar nests in the same area in 1977 and probably in 1976 (P. F. Stewart pers. comm.) Birds were present again in 1981, and 15 nests were located in 1982 (Flint & Stewart 1983). I searched the area in April 1985 and found 35 nests. It was impossible to estimate the number of pairs that this represented as most of the nests were from previous years (the nests are solidly built and remain reasonably intact for several years) and the birds were just establishing themselves in the breeding colony. Pairs were adding nest material at some nests and males display calling at others, but from the birds seen there were probably 20–30 pairs. Judging from Fig. 66 the colonisation of Cyprus would appear to have been a continuation of the westward spread in Turkey.

The Cyprus birds are presumably migratory as they are reported to be absent in the breeding area from mid November until March (Flint & Stewart 1983). What happens to these birds and those of the other western Asian populations remains to be elucidated.

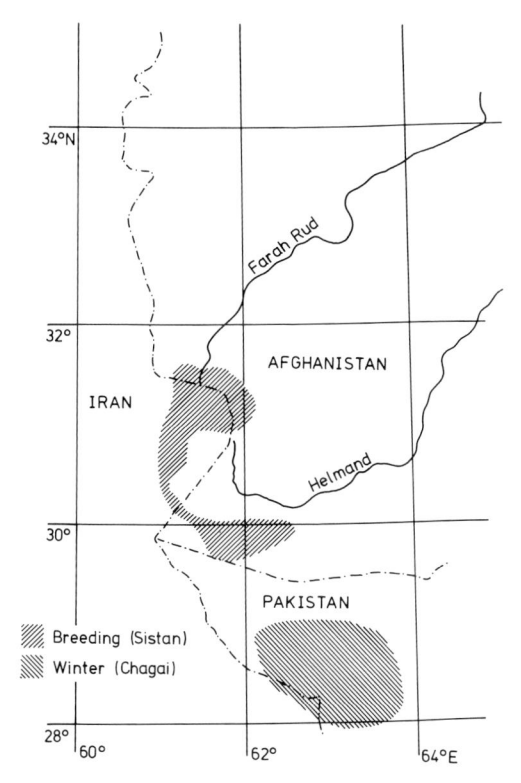

Fig. 67 Range of eastern race of Dead Sea Sparrow

Passer moabiticus yatii

There is a gap of about 1,200 km between the nearest population of *P. m. moabiticus* in southern Iran and *P.m. yatii*; the latter occurs in the Helmand delta in the Sistan region on the borders of Iran and Afghanistan, where according to Hüe and Etchecopar (1970) it is numerous. These authors state that it spreads from Sistan along the Farad Rud, though whether it breeds there or is merely a dispersal from the breeding area is not known. I searched for it without success in May 1976 on the Farah Rud near Farah, Khash Rud and the Helmand delta from Giriskh to Lashkar Gah.

This race appears to have a regular seasonal migration as it has been described by Christison (1941) as a common winter visitor to central Chagai in northern Baluchistan. This is a semi-desert region with an average annual temperature of 37°C and rainfall that is irregular and scanty. Again it would appear that the bird is not tied to water outside the breeding season.

The distribution of *Passer moabiticus yatii* is shown in Fig. 67.

HABITAT

The most detailed investigation of the habitat occupied by the Dead Sea Sparrow has been carried out by Professor H. Mendelssohn of the University of Tel Aviv. Much of what follows is based on his work (H. Mendelssohn pers. comm.). The Dead Sea Sparrow in its original area near the Dead Sea was restricted to regions of high ambient temperature; it occurred predominantly in a riverain or lacustrine habitat surrounded by semi-desert where there was a flush of annual grasses, following the retreat of winter innundation, that provided a good supply of seeds. In this area it appeared to be restricted to places in which there were dead or dying trees that were used as nest sites. This site was apparently selected so that the nest was exposed to full sun, which allowed development of the clutch almost without incubation by the parents, though it required water nearby for thermoregulation when the ambient temperature became excessive. This dependence on water is quite marked. Mendelssohn described, for example, the need for water in the case of a colony that nested in a grove of dead tamarisks at En Gedi on the western shore of the Dead Sea; each year the colony was deserted by the sparrows as the area dried up and the distance to water became too great. Similarly, Yom-Tov and Ar (1980) found a colony of about ten pairs in May in the Rift valley south of the Dead Sea beside some temporary ponds caused by flooding; three weeks later the ponds dried up and the birds deserted even though the nests contained eggs and young. At 'Ein Yahav, just to the south of Hazeva in the Rift valley there is no standing water, but there the birds rely on the field irrigation of the moshav (Yom-Tov pers. comm.).

From about 1950 onwards, however, there has been a move away from this rather restricted habitat following changes in agricultural practices: irrigation of areas that were previously desert has led to increased food in the form of grass and weed seeds, and the creation of fish ponds has provided an additional source of available water. At the same time, the removal of the bare tamarisks, many of which had died through infestation by larvae of the buprestid beetle *Sterapsis squamosa*, markedly reduced the availability of the preferred nest site and forced the birds to move to healthy trees that had grown up in irrigated land and around fish ponds. Professor Y. Yom-Tov (*in litt.*) informs me that, with more efficient fish farming, there has now been some reduction in the number of fish ponds in the northern valleys and with this a decrease in the number of breeding colonies of Dead Sea Sparrows.

The recent extension of range out of the intensely hot Dead Sea depression to the cooler upper Jordan valley and the Esdraelon and Zevoulum valleys has necessitated a change in behaviour. The birds in these areas now nest in leafy trees (though they still require dead twigs for nest building) where they have to incubate the eggs in the normal way, because of the lower temperatures, and are less dependent on open water for thermoregulation. That this is a real change in behaviour is shown by the observation that, in the initial stages of the expansion of range, birds that started nesting in bare eucalyptus trees deserted the nests, even after eggs had been laid, when the growing leaves began to shade the nest, though after a few years such trees were used successfully for breeding. This very neatly shows the adaptability of the species in its ability to change its habitat preference in order to exploit a new food supply.

Much less detail is available about the habitat used by the other populations. Scrub jungle is a typical description used by many authors, but without confirmation, or otherwise, that the birds have a preference for dead trees for nesting, a variety of living trees and shrubs being mentioned. The Turkish population breeds in live trees, but this is no doubt based on immigrants from Iraq that were already adapted to breeding in leafy trees. Marchant (1963) mentions that in the Jadriya colony in Iraq the nests were in dead and living apricots, mulberries, date palms and apple bushes as well as in poplar and tamarisk thickets. In Cyprus the birds are present in a large reed bed and breed in tamarisks growing in a salt lake surrounded by water.

Even less is known about the habitat occupied outside the breeding season, though the records from desert areas round the Dead Sea suggest the bird may be less tied to open water. This would also appear to be true of the wintering population of *P. m. yatii* in the Chagai region of Baluchistan.

BEHAVIOUR

The Dead Sea Sparrow is a social species, nesting in loose colonies and living in flocks outside the breeding season. From the few records for the latter period, it appears that once breeding is over the adults join the bands of juveniles, forming flocks of 50–100 birds. These flocks stay close together in trees or when feeding on the ground, at times in company with House and Willow Sparrows. According to Meinertzhagen (1954) the birds spend more time in trees and less on the ground than other sparrows, though clearly they must come to the ground to feed on the seeds of grasses and small annual herbs that are their staple diet. A flock of 60–70 birds that I watched stayed much closer together than would the individuals in a similar flock of House Sparrows. They were very wary and restless with much bill wiping and tail flicking, flying up to the cover of nearby trees at any real or imagined danger and then trickling back in small groups to the feeding area.

BREEDING BIOLOGY

The Dead Sea Sparrow is a social breeder, forming loose colonies of 10–20 up to 100 nests. The nests are exclusively built in branches of trees, with a distinct preference for dead trees – in Cyprus many nests are in tamarisks that have been killed by fire rather than in nearby living trees – though with the expansion in range and numbers that has occurred in Israel, and probably also in Turkey, there is increasing use of live trees with leaves. There is no particular preference for thorny trees and, in addition to tamarisk, poplar and willow, a large variety of trees has been recorded, including eucalyptus, fruit trees, olive, juniper and shrubs. Nests have been found in large annual plants, such as thistles, several plants being bound together to act as a support for the heavy nest (Lulav 1967). Nesting heights range from as little as a metre to as much as 12 m, depending on the size of tree; in tamarisks they are normally between 3–4 m. Nests at lower elevations are usually in branches overhanging water and those at higher elevations in trees in dry situations. In two study areas in Israel Yom-Tov and Ar (1980) found the mean spacing of nests to be 10.4 and 13.4 m, with occupied nests no closer than 6 m. Occasionally there were two nests in one tree only 1–2 m apart, but these were the successive nests of one pair. In contrast in Iraq, Cheesman (1919) found nests about 100 m apart, with about 100 nests scattered over a distance of 8 km along the river Tigris, though for the same area Marchant (1963) later described the nests as close together.

According to Mendelssohn (1955), males leave the flock at the beginning of the breeding season (March in the Jordan valley) and establish a nest site in a suitable tree. A male may take over a nest from the previous year or build an initial platform from which he calls and displays to attract a mate. The familiar sparrow chirp – transcribed by Mendelssohn for the Dead Sea Sparrow as *tcheep tcheep* – is used as a site ownership call. In the display the male hops about the tree bobbing in an upright posture with the head high, the tail half raised and spread, the wings held out from the body and flicked shallowly, all the time calling loudly, the chirps being extended into a short song. This display reminded me of that of the Golden Sparrow and differs from most *Passer* in which the wings are drooped and shivered. Professor Mendelssohn has told me that if a female, not ready for mating, is approached too closely by the displaying male, she attacks him and this can precipitate a group display similar to that of the House Sparrow; in the Dead Sea Sparrow, however, this involves only the males, that chase each other among the trees without landing on the ground as do the House Sparrows in their group display.

The main structure of the nest is built of dead twigs, quite thick in section, giving an open construction, but well bound together so that the final nest is very secure. The strong construction means that the nest may remain substantially intact for two or even three years and it seems probable that it is the presence of old nests that attracts males in the next breeding season and maintains the constancy of the breeding site. Once the pair is formed the male completes the nest, which is finally lined by both sexes with softer materials, including plant down (reed panicles, tamarisk seed heads), feathers and animal hair. The nest is domed over with the entrance at the top, usually spiralling down to the egg chamber so that the eggs cannot be seen from outside. The final nest is surprisingly large considering the size of the bird, ovoid in shape, 300–400 mm deep, 200–350 mm across the top, and weighs from 200–1,000 g, as much as 60 times that of the bird. The contrast between the size of the bird and its nest is shown in Fig. 68. According to Mendelssohn the nest size depends on the space around it in the tree and the time available before the female starts to lay.

Data on the breeding season are given in Fig. 69. It extends over the northern summer from March to August, depending on the locality; it is probably long enough in Israel/Jordan for up to three clutches to be laid. Clutch size is normally 3–5 eggs. Two study areas in Israel gave the following mean values: 4.16 (n = 177), decreasing from 4.5 at the beginning of the breeding season to 3.78 at the end in one site; 3.90 (n = 140) at the other, where it increased from 3.76 at the beginning of the season to 4.31 at the middle and then dropped to 3.61 at the end. Clutches of 6 have been recorded in Israel (Mendelssohn 1955) and even as many as 7 eggs have been recorded in Iraq (Marchant 1963). These large clutches could have been the result of egg dumping or two females laying in the one nest. Yom-Tov (1980), in fact, has given circumstantial evidence for intraspecific brood parasitism in the Dead Sea Sparrow. Normally eggs are laid in the morning at daily intervals.

The breeding biology of the Dead Sea Sparrow in Israel has been particularly studied by members of the Department of Zoology at the University of Tel Aviv (Mendelssohn 1955 and pers. comm.; Yom-Tov *et al.* 1976, 1978; Yom-Tov & Ar 1980). The structure of the nest is such that in the early and late hours of the day the slanting sun warms the eggs, whereas as it climbs and the temperature rises the thick

Fig. 68   Male Dead Sea Sparrow at nest

open structure of the nest protects the egg chamber at the bottom of the nest from the direct heat of the sun but allows cooling by air passing through. In the Dead Sea depression, with its high ambient temperature, there is no need to incubate the eggs and indeed the female does not develop a brood patch, though she does cover the eggs to reduce the extremes of temperature to which they are exposed. The attentativeness of the female is, in fact, related to air temperature, showing a minimum of 35% about 35°C, the presumed optimum for development of the eggs, and climbing to maxima of 79% and 78% at air temperatures of 20° and 41°C respectively. The female also covers the eggs at night. When the ambient temperature is above the optimum, the female requires ready access to water to compensate for that lost by respiratory cooling.

Measurements have shown that in this way egg temperature was controlled to limits of 25°–39°C (with a mean daily temperature of 33.7°C), when the air temperature ranged from 15°–41°C. During very hot spells (at times the temperature in the Dead Sea depression can rise above 45°C) the female may not be able to control the temperature and as a result there may be considerable mortality of eggs and small young. Support for the hypothesis that eggs are not truly incubated by Dead Sea Sparrows living in the Dead Sea depression is given by the fact that hatching is less synchronous than in most passerines, extending up to two days in most cases. Over this period, and even when the young have hatched, the male does not enter the nest and if he approaches too closely is attacked by his mate. At this time the male may build and defend a second nest a few metres from the one that contains the eggs. The female may later use this second nest for a subsequent brood, but apart from that the same nest is used for successive clutches.

With the spread of the Dead Sea Sparrow from the Dead Sea depression and the destruction of dead trees with changed agricultural practice, the breeding biology has

Subspecies	Locality	Month												Reference
		J	F	M	A	M	J	J	A	S	O	N	D	
moabiticus	Iran (31°N)			──┼──										Allouse 1953
	Iraq (33°N)			──┼──┼──										Cheesman 1919, Ticehurst *et al.* 1923, Allouse 1953, Marchant 1953
	Israel/Jordan (30–33°N)			──┼──┼──┼──										Carruthers 1910b, Hollom 1959, Mountfort 1965, Lulav 1967, Yom-Tov & Ar 1980
	Turkey (36.5–38.5°N)				──┼──┼──									Kumerloeve 1965a, 1970, Beaman *et al.* 1975, Beaman 1978
	Cyprus (34.5°N)			──┼──										Flint & Stewart 1983
yatii	Sistan (31°N)							──┼──						Hartert 1904–22

Fig. 69 Breeding season data for Dead Sea Sparrow. (In 1985 the occupation of the nests in Cyprus was only beginning in the second half of April, with no eggs laid at that time.)

been modified as the bird has adopted leafy trees for building its nest and has moved into areas of lower ambient temperature; in this new habitat the birds have had to adopt a more normal behaviour with the female incubating the eggs in the mornings and the male feeding the young in the nest and even entering it during the incubation period.

The incubation period at the two study colonies in Israel was 12.4 days ($n=82$), and 12.12 days ($n=103$). The nestling period was 14–15 days. The young are fed mainly by the female with feeding visits ranging from 13–22 times per hour for a nest with 3 chicks, 9 days old, to 6–11 times per hour for a nest with one chick. The two colonies in the Rift valley in Israel studied by Yom-Tov and Ar (1980) for three breeding seasons (1974–76) gave very similar values for hatching and fledging success. Approximately 42% of eggs, out of a total of nearly 1,200, hatched and about 68% of these produced fledged young, giving an overall success rate of about 28%. The low hatching rate was attributed mostly to a high desertion rate due to interference by other Dead Sea Sparrows and other species (House Sparrows and Arabian Bulbuls *Pycnonotus capensis*); 5–8% of eggs laid were infertile.

Apparently the Dead Sea Sparrow is able to raise three broods in its original breeding place in the Dead Sea depression, but this is reduced to two, in both the north and south, because of the greater demands on the female during the incubation period for thermoregulation. The overall effect is that a pair could raise about 7 young in a season in the original breeding area, compared with only 4.5 in the more recently occupied cooler and hotter zones.

Over much of its range the Dead Sea Sparrow is sympatric with the House Sparrow and in the northern part of its range also with the Willow Sparrow. At times the nests of the different species can be quite close together and the House Sparrow has been recorded evicting a Dead Sea Sparrow from its nest. Mendelssohn has told me of a mixed pairing between a male Dead Sea Sparrow and a female House Sparrow in captivity at Tel Aviv University. The nest was a composite, with the male attempting to build a *moabiticus*-type nest with the entrance at the top, and the female a *domesticus*-type nest with the entrance at the side. The clutch was infertile. Despite their nesting in close proximity there has been no report of mixed pairings in the wild.

MOULT

In Israel adults and young birds both begin the annual moult at the beginning of August (Mendelssohn 1955).

VOICE

The bird is very noisy at the nesting site, but otherwise rather quiet. I have transcribed the call of the male as *chip-chip-chip* or *chip-chip-chizz*; other authors represent it by *tcheep tcheep* (Mendelssohn 1955), a clear, Chaffinch-like *dli-dli-dli* (Kumerloeve 1965b), a liquid *dli-dli-dli* (Cramp 1969) and a bunting-like *jew-ee* repeated three or four times (Mountfort 1965). This is transformed to an excited, far-carrying *chillung-chillung-chillung* or *tweeng-tweeng-tweeng*, with wings raised and flicked, by the male that is trying to attract a female (display calling). Both sexes use a similar churring call to that of the House Sparrow—*chur-it-it-it, chit-it-it-it, chittup chittup*, that is associated with tail flicking and is used in mild alarm, *eg* as a reaction by a bird at the nest to an approaching human. The female uses a soft *tweeng tweeng tweeng* as an invitation to coition.

Dead Sea Sparrows, like the House Sparrow, also indulge in social singing, in which the birds collect in trees. This occurs on emerging from the roost, before going to roost and also during resting periods between feeding bouts during the day.

FOOD

Like all the sparrows the Dead Sea Sparrow is predominantly a seed eater, specialising on smaller seeds such as those of grasses (particularly *Phalaris* sp. according to Mendelssohn), rushes and sedges (*eg.* bulrush, papyrus), shrubs (*Sueda*) and trees (*Tamarix*). The nestlings are fed both on invertebrates and seeds (Yom-Tov & Ar 1980).

12: Sind Jungle Sparrow *Passer pyrrhonotus*

NOMENCLATURE
Passer pyrrhonotus Blyth, J. Asiat Soc. Bengal 1844 13: 946.
Bahawalpur, Sind.

Following the discovery of this species in 1844, 36 years passed before it was reported again, despite searches by that indefatigable student of Indian ornithology, A. O. Hume, in the intervening period (Hume 1873). The rediscovery of the species was made in 1880 by Doig (1880). Hartert (1904–22) considered that *pyrrhonotus* was a small race of *Passer domesticus,* though this is clearly not so, as Doig found the two species breeding separately in the East Narra (Eastern Nara) district of Sind, and later Ticehurst (1922) was able to confirm this by finding nests of both species within 100 yards of each other. Meise (1936) suggested that *pyrrhonotus* was originally a race of *domesticus* that, having lived for a long time in India (Pakistan), had become specifically distinct from a later group of House Sparrows arriving in the area. He points out that this is an area that is a centre for small bird forms. Baker (1921), who considered *pyrrhonotus* as a race of *domesticus,* named it the Rufous-backed Sparrow, but, as this could have caused confusion with the already named Rufous Sparrows of Africa, Ticehurst proposed the name Sind Jungle Sparrow and this is now the accepted name in the English language.

DESCRIPTION
The Sind Jungle Sparrow closely resembles a small House Sparrow. No doubt this was why it was overlooked for so many years, though it is considerably smaller with a length typically 130 mm, compared with 150 mm for the Indian race of the House Sparrow that is sympatric with it.

Male. The most striking difference between male Sind Jungle Sparrow and House Sparrow is that black bib of former is much narrower, parallel-sided and does not extend on to breast. Further distinguishing characters are that cheeks are ashy grey, quite distinct from Indian race of House Sparrow, in which white cheeks stand out in contrast to grey crown; black streaking on back is more distinct; rump reddish brown and underparts paler, becoming almost pure white on belly. Bill is black in breeding season, upper mandible becoming brown, lower mandible yellow-brown, in non-breeding season.

Female. Closely resembles a small female House Sparrow, though in the field I thought it had a sleeker appearance.

BIOMETRICS

Typical body measurements are given in Table 36.

Table 36: Biometric data for Sind Jungle Sparrow

Feature	Males		Females	
	Range	Mean (Median)	Range	Mean (Median)
wing	62–70	67.4	62–67	63.6
tail	47–57	54.1	48–54	52.5
tarsus	15.5–19	(17.3)	15.5–17	(16.3)
culmen	11–12.5	(11.8)	11–12.5	(11.8)

DISTRIBUTION

The Sind Jungle Sparrow is a species of restricted distribution, mainly confined to the Indus valley and the lower reaches of its main tributaries in the Punjab. It extends from the Indus delta, being recorded from Maradani (south of Tatta) in the south, north to about Nowshera on the Kabul river near its confluence with the Indus, up the Jhelum river to the Nurpur Escape, the Chenab to Waziribad, the Ravi to Lahore, into the Indian Punjab on the Beas river near Gurdaspur, and along the Sutlej to Ladhowai (10 km north of Ludhiana). It is common, but locally distributed in many places near these major waterways. Whistler (1913–14) reported it as abundant at Ferozpore (Firozpur) and I found it regular along the Sutlej from Harike, east to the bridge on the main road between Ludhiana and Jullundar, but not further upstream at Rupar. Away from the Indus valley it is a local breeder in Baluchistan, extending west into Baluchestan (Iran), where it has twice been collected in March (Zarudny & Härms 1902, quoted by Vaurie 1949). The distribution is shown in Fig. 70.

HABITAT

The Sind Jungle Sparrow is predominantly a riverain species, found in flooded tamarisk and acacia (kikur) jungle and tall grass by the Indus and its tributaries; it spreads into the jungle along canals and other waterways, but is never far from water. The area near the river Sutlej, where I found the bird, was intensively cultivated with corn and rice paddy, with large reed beds between the bund walls and the river itself. The Sind Jungle Sparrows were in kikur and eucalyptus trees planted in the bund walls to stabilise them, as well as in small plantations of eucalyptus and sesame *Sesamum indicum* planted close to the river and sometimes standing in water. This

196 Sind Jungle Sparrow

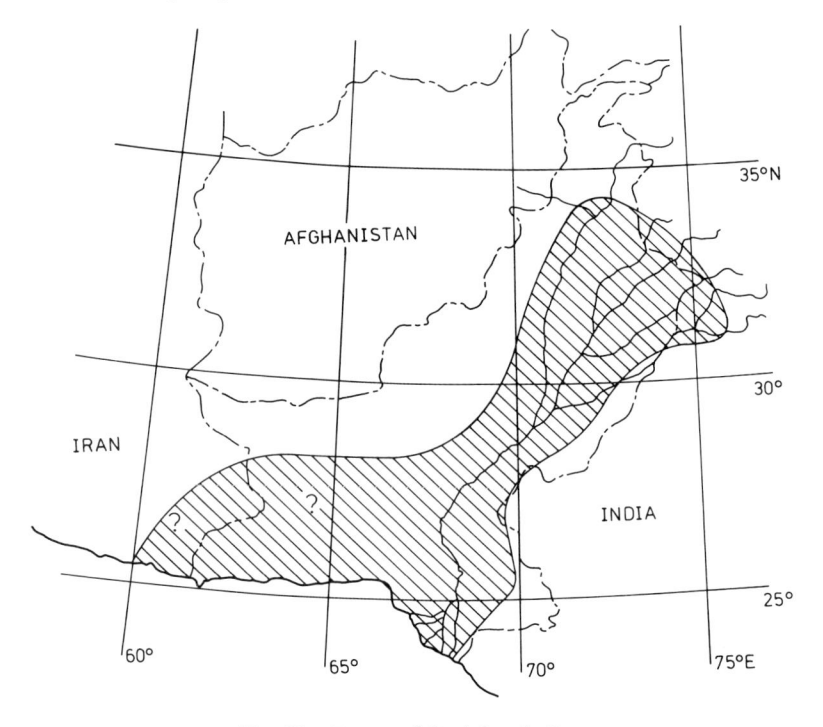

Fig. 70   Range of Sind Jungle Sparrow

area has only been intensively cultivated in the last 10–20 years and is probably not part of the bird's habitat requirement. Outside the breeding season the bird moves into semi-desert jungle with *Capparia* and *Salvadora* bushes, but only as long as there is water.

The Sind Jungle Sparrow co-exists with House Sparrows where there are habitations in riverain forest and tamarisk jungle; it may occur round habitations, where these are in deep cover, but this sparrow is not specifically associated with man.

BEHAVIOUR

This species is only moderately gregarious, usually occurring in small bands of 5–6 birds, though larger flocks of perhaps up to 30 birds can be seen outside the breeding season in places where it is common. In the winter months they may join flocks of other species, such as House Sparrows, Avadavats *Estrilda amadava* and Rock Buntings *Emberiza cia*. Normally the birds spend much time in the cover of the trees, or in the tall grasses on which they feed, and are quite difficult to see. Most observers have reported the bird as shy, though Whistler (1910–11) on the contrary described it as tame. I did not find the birds particularly shy and they allowed reasonably close approach, both when they were at their nests in the trees and when collecting food on the ground. They roost socially in acacias or tamarisk bushes, with a preference for those standing in water.

Latitude	Month												Reference
	J	F	M	A	M	J	J	A	S	O	N	D	
24°45N (Maradani)			----										Holmes & Wright 1969
25°N (Sadnani)					────								Ticehurst 1922
25°30N (Eastern Narra)			----										Doig 1880
26°30N (Manchhar Lake)			----										Ticehurst 1922
27°50N (Sukkur)													Holmes & Wright 1969
31°N (Jhang District, Harike)				----									Whistler 1922, author
32°N (Punjab)						────							Jones 1912, Currie 1915–16

Fig. 71 Breeding season data for Sind Jungle Sparrow (nest building ---- breeding ────)

BREEDING BIOLOGY

The Sind Jungle Sparrow breeds mainly in free-standing nests in branches of trees and bushes. Acacia and tamarisks have been reported and I found them in a plantation of sesame trees; trees standing in water are selected in preference. The nests are usually in loosely connected colonies; they are typical of those built by tree-nesting sparrows: untidy masses, globular in shape, domed over with a flatish top and an entrance in the side or on the top, very similar to a tree nest of a House Sparrow, but smaller (90–180 mm in overall diameter). The nests can be constructed of tamarisk twigs (Ticehurst 1922), grass (Whistler 1922) or both (Baker 1926), or according to my observations a mixture of coarse twigs, roots and dried reeds may be used; in all cases the nest is lined with finer plant materials and feathers. Currie (1915–16) describes a nest in a hole in an iron gate, and breeding in the old nests of Baya Weavers *Ploceus*

philippinus has also been recorded (Jones 1912). Both sexes take part in building. The height above ground, or water, depends on the size of the tree or bush and can range from 1.5–10 m. More than one nest may be found in some of the larger trees, but they are usually not close together.

Data on the breeding season are summarised in Fig. 71.

Breeding has been recorded from April until September, though probably not as long as this in any one year, being dependent on the rains. There is also some suggestion from the published data in Fig. 71 that breeding is correlated with latitude, starting earlier in the south and finishing later in the north; Whistler (1922) stated that, in the Jhang District, breeding did not start until July, and at the same latitude along the river Sutlej in the Indian Punjab I did not see any evidence of active breeding in May, though the birds were nest building. According to Baker (1926) two clutches are raised. Both sexes take part in incubation (Whistler 1922). Clutch size is normally 3–5 eggs, with 4 most frequent, but 2 has also been reported.

MOULT

The post-juvenile moult, as in all *Passer* species, is complete. The timing of moult has not been recorded.

VOICE

I transcribed the call of the male as a soft *chup*, much softer and less strident than that of the House Sparrow. This confirms Doig's observations (Hume 1880), in contrast to Ticehurst (1922), who described the bird as noisy with the call rather wagtail-like. Currie (1909–10) gives the 'song' as consisting of the usual sparrow chirrups interspersed with a note, often repeated, resembling the call-note of the Pied Wagtail *Motacilla alba*.

FOOD

No detailed study has been made of food taken by this sparrow. Like all the sparrows it is basically a seed-eater, feeding on the seeds of grasses and small weeds (*Polygonum plabaya* being particularly mentioned), taken from the plants and collected on the ground. Insects are also taken – I have watched birds searching for caterpillars among the branches of trees – and no doubt insects are fed to the nestlings.

13: Pegu Sparrow *Passer flaveolus*

NOMENCLATURE
Passer flaveolus Blyth, J. Asiat. Soc. Bengal 1844 13: 946–947.
    Arakan, Burma.
Synonym: *Passer assimilis* Walden 1870.

Passer assimilis was described from the same area as *flaveolus,* apparently the immature male being mistaken as a different species. The English name comes from Pegu, a town and district lying to the southeast of Arakan, from which the bird was originally described. The shorter name, Pegu Sparrow, is preferred to Pegu House Sparrow and to the more recent, but rather prosaic, Plain-backed Sparrow.

DESCRIPTION
Medium-sized, black-bibbed sparrow, length about 140 mm; sexes dimorphic.

Male. Greenish yellow with typical 'House Sparrow' pattern. Crown and nape olive green, becoming yellowish on forehead. Mantle and back plain chestnut (not streaked as in many black-bibbed sparrows). Rump and upper tail coverts greyish yellow. A broad chestnut band from behind eye, encircling back of cheek, which is greyish yellow. Black around eye continued in narrow band to base of bill. Tips of lesser wing coverts yellow, giving yellow wing bar, remainder of wing blackish. Black bib from throat to breast. Underparts yellowish grey, becoming more yellow on belly. Bill black in breeding season, otherwise yellowish horn.

Female. Similar to female House Sparrow, but paler and less streaky above, giving very 'clean' appearance. Upper parts grey-brown with creamy stripe from above eye to nape. Broad white wing bar. Underparts pale greyish yellow. Prominent black eye.

Juvenile. Similar to female.

BIOMETRICS

Typical body measurements are given in Table 37.

Table 37: Biometric data for Pegu Sparrow

Feature	Males		Females	
	Range	Mean (Median)	Range	Mean (Median)
weight	**17–22**	**19.4**	17–23	19.3
wing	68–75	72.0	64–71	69.2
tail	53–64	61.3	48–60	57.7
tarsus	18–20	18.9	18–20	18.9
culmen	11–12	(11.5)	11–12	(11.5)

DISTRIBUTION

The Pegu Sparrow was first described from Arakan in Burma. It occurs in west Burma from Arakan on the coast, where it is local, through the Chin hills to the upper Chindwin about 23°N. The northern limit extends through the Shan State in central Burma, central Laos to the Annam region of Vietnam at about 16°N on the coast. Southwards it extends to the Pegu Division of Burma, where it is found sparingly as far south as Rangoon, through Thailand north of the peninsula, Campuchea and the Cochinchina region of Vietnam (Deignan 1945, Delacour & Jabouille 1931, King *et al.* 1975, Lekagul & Cronin 1974, Riley 1938, Smythies 1953, Wildash 1968).

In the last 60 years it has spread south of the Isthmus of Kra into Malaysia, being first recorded at about 10°N in 1919; the first reported occurrence in Malaysia was at Alor Star (Alor Setar) in 1938 and it now extends south to northern Pahang State on the east coast (1976) and central Selangor (Kuala Selangor) in the west (1969). It has also reached the islands of Langkawi and Penang (Medway & Wells 1976, Wells pers. comm.). Ward (1968) has suggested that massive deforestation in the peninsula, creating continuous corridors of open land, has opened the way for the spread of the Pegu Sparrow and other open country birds; he predicted that it would not be long before the Pegu Sparrow reached Singapore. However, Dr David Wells has pointed out to me that the species is by no means successful in peninsular Thailand and Malaysia, being restricted to small pockets in low lying coastal areas, so that there may be ecological or physiological factors that limit its widespread colonisation, other than that previously provided by the forest barrier, possibly the transition from an area with a clearly defined dry season to one without. The most southerly record up to 1980 was of two birds on the outskirts of Kuala Lumpur in January 1980 (D. R. Wells pers. comm.).

The current distribution is shown in Fig. 72.

HABITAT

As might be expected from one of its vernacular names, the Pegu House Sparrow, this species occurs around habitations. Over most of its range it is sympatric with the Tree Sparrow and in Burma with the House Sparrow as well. These other two sparrows are the ones most associated with man and, although the Pegu Sparrow does come into villages and the outskirts of towns, it is much more of a rural species, being found most commonly in cultivated areas, where it lives in the trees (particularly sugar palms and bamboos) bordering the fields, though it also occurs in semi-wooded

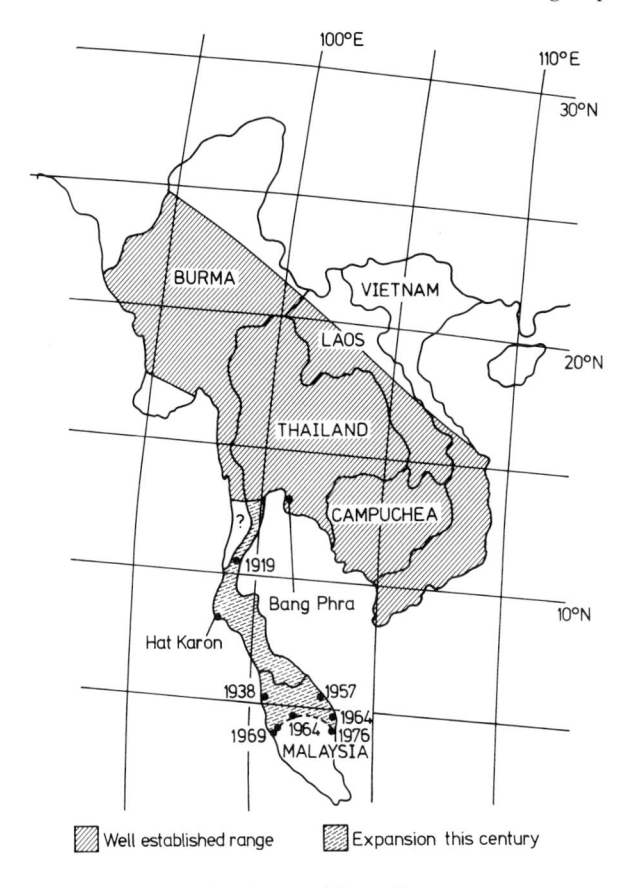

Fig. 72   Range of Pegu Sparrow

country, such as coconut plantations and stands of casuarinas along the coast, and even in small copses, the edges of woodlands and forest clearings.

BEHAVIOUR

The Pegu Sparrow is a social species, forming small colonies of usually 5–10 pairs, but sometimes up to 25–30 pairs. Outside the breeding season it may be seen in flocks of up to 100 or more birds that gather to feed in ripening rice fields. These flocks may consist of birds that have come a few kilometres from the breeding area, where they roost at night. This was the case with the birds studied by Pantuwatana *et al.* (1969) at the Bang Phra Horse Serum Farm southeast of Bangkok, that remained throughout the year, staying in the immediate breeding area from February to October, but for the remainder of the year flying 2–4 km during the day to feed on ripening rice in the paddy fields. In other parts, however, as at Wat Phai Lom for instance, the birds were present only from January to July (McLure & Kwanyuen 1973). Further support for some nomadic behaviour was given by weekly roadside counts in a farming area northwest of Bangkok, where H. E. McLure (*in litt.*) only saw birds from May to December. It is perhaps also significant that numbers of Pegu

Sparrows are for sale in the Bangkok bird market only from March to November. No doubt these differences in behaviour reflect changes in the availability of food. The Bang Phra Horse Serum Farm, with its irrigated pastures, possibly provides a rich supply of food throughout the year, whereas in other parts this may not be the case and the birds are forced to wander further afield.

As the breeding season approaches, the birds gradually forsake the communal roost, the female to roost at the nest and her mate in nearby trees. At the Bang Phra Nature Reserve, when I was present in March, the birds were just taking up their nest sites and most birds of the breeding colony of 50–60 birds were still roosting together at night in some thick bushes about 300 m from the main breeding area. In the same month at Hat Karon on Phuket Island, where breeding had already taken place, the main flock of 60–70 birds dispersed from the roost in a group of casuarina trees, *Casuarina* sp., about 15 minutes after sunrise and did not return until about 25 minutes before sunset, leaving behind only the few birds still breeding (Summers-Smith 1981).

At this time of year, at least, pairs are very faithful, staying close together in the roost and spending much time in each other's company when feeding, nest building and indulging in other activities.

In the threat display used by the male against rivals or intruders of other species near the nest, the wings are held out slightly from the body and drooped, with the head thrust forward, not raised to display the black bib as it is in the invitation display to the female.

BREEDING BIOLOGY

As already mentioned, the Pegu Sparrow is a social breeder; the breeding colonies are often quite small with only about 5–10 pairs, though sometimes up to 30 pairs.

In the precopulatory invitation behaviour, the male displays in front of the female by bowing and flicking the tail, with the wings held out and slightly opened. I have seen this display both near the nesting site and also in the roost. If the female is not ready for mating she flies at the male and displaces him. This display is very similar to that of the House Sparrow, but with the latter species the bowing display in front of the female attracts neighbouring males and results in the well-known group display, in which several males posture in front of the female; such a group display does not take place with the Pegu Sparrow, though nearby males hop around excitedly. I have also seen a male at a nest fly up to a neighbouring female to present her with nest material; he then hovered over the female as if to mount, though in fact he did not do so. Presentation of material to the female was seen by me on several occasions when pairs were beginning to take up nest sites, but she did nothing with it and dropped it after a short interval.

When the female is ready to mate she solicits the male by crouching and wing-shivering; this occurs in trees and on the ground close to the nesting tree.

Breeding is mainly in free-standing nests in the branches of trees and bushes, though quite commonly, in addition, a number of more enclosed sites may be used; these range from the crowns of palms, to holes in trees, in dead bamboos, under the eaves of buildings and in electricity junction boxes (a favourite site with *Passer* species); at times the bird will also use the old nests of other species, eg Spotted Munia *Lonchura punctulata*.

The open-site nest is the usual large, untidy, globular structure (250–300 mm in diameter) typical of sparrows, built of dry grass, small twiglets and rootlets, domed over with an entrance at the side and lined with feathers; those I have seen in mango trees were well hidden and built amongst the leaves. These open nests in trees are very like those of the House Sparrow built in a similar situation. Nest heights range from 3–20 m. Both sexes take part in nest building, usually arriving at and departing from the nest together, though at times only one bird will actually collect material, being accompanied to and from the nest by its mate.

Pegu Sparrows quite often form mixed colonies with Tree and House Sparrows. Data on the breeding season are summarised in Fig. 73.

Locality	Month												Reference
	J	F	M	A	M	J	J	A	S	O	N	D	
Malaya													Medway & Wells 1976
Thailand													Deignan 1945, Pantuwatana *et al.* 1969, H. E. McLure, author
Burma													Oates 1882, Harington 1914

Fig. 73   Breeding season data for Pegu Sparrow

The presence of fledged young in the Hat Karon colony on Phuket Island in March, suggests that breeding must have started in late January; published information indicates that breeding continues in Thailand at least into June (Pantuwatana *et al.* 1969). No information is available from the more easterly parts of the range.

The adults I watched that were feeding young in the nests were collecting caterpillars and aphids by searching acrobatically in the small, outer branches of trees; larger insects were caught on the ground. Feeding of the young, both in the nest and until they have become independent, is shared between the sexes but, as with nest-building, quite often is by only one bird, the other accompanying it to and from the nest on its food-gathering sorties. Harington (1914) stated that in Burma the clutch size was normally 3, though he had once found a nest with 5 eggs. One clutch from Malaya had 3 eggs (Medway & Wells 1976).

MOULT
Nothing has been recorded about the moult.

VOICE

The main call is a disyllabic *chirrup* (a loud, clear *chirip,* Medway & Wells 1976), with the second syllable rather slurred; this call is used by both sexes at the nest and in the roost. It is very similar to the familiar call of the House Sparrow, somewhat harsher than the monosyllabic *chip* of the Tree Sparrow.

There is also a mild threat call very similar to that used by the House Sparrow and a number of other sparrow species – *chit-chit-chit* or *chi-chi-chip-chip*. This is given by the male at the nest when the female is present and is regularly heard from the roost. A deeper *chu chu* or *chu chu weet* is probably a mild alarm call, similar to the *quer quer* alarm of the House Sparrow; this call was heard frequently from the roost as the birds were settling for the night.

FOOD

The Pegu Sparrow is predominantly a grain eater, in this case rice from ripening crops and from spillage along the roadsides. In addition I have seen the birds feeding on casuarina seeds, the seeds of small ground plants and berries (unidentified) in trees. The young in the nest receive mostly animal food. I watched adults that were feeding young in the nest collecting Homoptera (aphids), Lepidoptera (caterpillars) and Heteroptera (bugs). At the Bang Phra Horse Serum Farm, where Pegu and Tree Sparrows were both nesting, the two species were collecting food for their young from different places and did not seem to be in competition with each other.

14: Cinnamon Sparrow *Passer rutilans*

NOMENCLATURE
Fringilla rutilans Temminck, Planches Color, Oiseaux 1835: 99.
    Japan.
Passer rutilans (Temminck) 1835.
    Synonyms: *Pyrgita cinnamomea* Gould, Proc. Zool. Soc., London 1836: 185.
        'apud montes Himalayenses'; [restricted to NW Himalayas by Ticehurst,
        J. Bombay Nat. Hist. Soc. 1927 32: 347.].
        Passer russatus Temminck & Schlegel, Fauna Japon; Aves. 1850: 90.
            Japan.
    Subspecies: *Passer rutilans rutilans* (Temminck) 1835.
        Synonyms: *Passer rutilans kikuchii* Kuroda, Bull. Br. Orn. Cl. 1924 45: 16.
            Horisha, Nanto [Nan-t'ou] District, central Formosa [Taiwan].
        Passer rutilans parvirostris Momiyama, Annot. Orient 1927 1: 127.
            Quelpart [Cheju] Island.
        Passer rutilans ignoratus Deignan, Proc. Biol. Soc., Washington 1948 61: 3.
            Mount Omei, Szechwan [Emei Shan, Sichuan].
        Passer rutilans batanguensis Zheng & Tan, Bull. Zool. 1963 15: 307.
            Batang, Sichuan.
    Passer rutilans cinnamomeus (Gould) 1836.
        Synonyms: *Passer rutilans debilis* Hartert, Vög. pal. Fauna 1904 1: 162.
            Sind-Tal in Kaschmir [Sind valley in Kashmir].
        Passer rutilans schaeferi Stresemann, Orn. Monatsb. 1939 47: 176.
            Süd-Tibet, Shigatse.
    Passer rutilans intensior Rothschild, Bull. Br. Orn. Cl. 1922 43: 11.
        Mekong Valley, [Yunnan].
        Synonyms: *Passer rutilans yunnanensis* La Touche, Bull. Br. Orn. Cl.
            1923 43: 120.
            Lotukow, SE Yunnan.

Passer rutilans lisarum Stresemann, Mitt. Zool. Mus., Berlin 1940 24: 172.
Mount Victoria, [Chin Hills, southern Burma].
Passer rutilans annectans Koelz, J. Zool. Soc., India 1952 4: 154.
Mawryngkneng, Khasi Hills.

The original description of *cinnamomeus* by Gould was from the western Himalayas. Hartert (1904), in error, attributed this to Bhutan and named the birds from the western Himalayas, *P. rutillans debilis*. Stresemann (1939) proposed the name *P. rutilans schaeferi* for the birds from southern Tibet on the grounds of larger size. It is now recognised that only one race occurs along the length of the Himalayas, for which Gould's name *cinnamomeus* has priority, though there is a cline in size, with the smallest birds occurring in Kashmir, the size increasing to the north in Tibet and to the east in Bhutan (see Table 38).

The English name Cinnamon Sparrow has wider usage than the alternative Russet Sparrow and the somewhat unnecessary Cinnamon Tree Sparrow.

DESCRIPTION

The Cinnamon Sparrow is one of the black-bibbed type. The sexes are dimorphic, the male having the typical House Sparrow type plumage and the female resembling the female House Sparrow. It is medium sized, length about 140 mm. There are three recognisable races.

Passer rutilans rutilans
Male. Upper parts from head to rump bright cinnamon red with heavy streaking on upper back. Cheeks and sides of neck whitish. Black around eye, extending to bill, separated from crown by narrow white streak. Small black bib extending from chin to upper breast, but not spreading sideways onto breast; remainder of underparts pale grey. Lesser wing coverts cinnamon red, double white wing bar, upper one broad; remainder of wing streaked black and grey. Tail streaked dark grey and black. Bill black in breeding season, otherwise horn.
Female. Similar to female House Sparrow, but more brightly marked, with upper parts tinged a dull cinnamon. Very conspicuous creamy supercilium extending back from eye and almost meeting at back of head. Bill blackish brown in breeding season, otherwise horn.
Juvenile. Similar to female, but duller.

Passer rutilans intensior
Male. Differs from nominate race in that it is generally darker, with underparts washed yellow; black bib spreads a little onto breast.
Female. Similar to nominate *rutilans*, but darker above.

Passer rutilans cinnamomeus
Male. Intermediate in tone between nominate *rutilans* and *intensior*, but with yellow patches on sides of throat. Birds from Tibet, Sikkim and Bhutan were separated by Stresemann (1939) as *schaeferi* on the grounds that they are larger and paler; Vaurie (1949), however, pointed out that plumage tends to become paler with wear of feather tips and he did not consider differences sufficient to justify separation of these northern birds as a distinct race.

BIOMETRICS
Typical body measurements are given in Table 38.

Table 38: Biometric data for Cinnamon Sparrow

Feature	Subspecies	Locality	Males		Females	
			Range	Mean (Median)	Range	Mean (Median)
weight	rutilans	—	18–22.5*	(20.3)		
	intensior	—	19.3–19.5*	(19,4)		17.3*
	cinnamomeus	—		21*		21*
wing	rutilans	—	70–74	71.3	65–71	68.0
	intensior	—	68–75.5	(71.8)	67–72.5	(69.8)
	cinnamomeus	W Himalayas	68–76	(72)		
		C Himalayas	73–82	78.2	68–77	74.8
		E Himalayas	72–82	75.5	69–78	73.3
tail	rutilans	—	43–49	(46)		ca 44
	intensior	—	46–51	(48.5)	46–50	(48)
	cinnamomeus	—	46–54	(50)	46–53	(49.5)
tarsus	rutilans	—	16.5–17	(16.8)	16.5–17.5	(17)
	intensior	—	16–18	(17)		(16.5)*
	cinnamomeus	—	16.5–18	(17.3)	16.5–18	(17.3)
culmen	rutilans	—	11–13	(12)	11–13	(12)
	cinnamomeus	—	12–14	(13)	12–14	(13)

* very small samples

The race *cinnamomeus* that breeds at the highest altitudes is generally larger than the other two races; within the range of *cinnamomeus* the largest birds occur at the higher altitudes (central and east Himalayas).

DISTRIBUTION

The distribution of the three recognised races is not well documented in detail. There is some suggestion that *P. rutilans rutilans* may intergrade with *P. rutilans intensior* and with *P. rutilans cinnamomeus,* but the latter two races appear to be allopatric and the neighbouring populations differ considerably in size (Table 38). The Cinnamon Sparrow is predominantly a montane species and thus its range is largely restricted to hilly and even mountainous country.

Passer rutilans rutilans

The nominate race occurs over eastern, central and southern China from Kwangtung (Guangdong) and Fukien (Fujian) Provinces in the southeast, through Kwangsi (Guangxi) and eastern Szechwan (Sichuan) to extreme eastern Tsinghai (Quinghai). In the north it reaches southern Kansu (Gansu), Shensi (Shaanxi), Shansi (Shanxi) and Shantung (Shandong) and there are records from Hopeh (Hebei) as far north as Peking (Beijing), though whether of stragglers or breeding birds is not clear (Vaurie 1959, Zheng 1976).

In the northern part of this range, in Shensi, Shansi and Shantung Provinces, the species is a summer visitor and it is probably only a winter visitor in the south, with one record of a straggler to Hong Kong (Webster & Phillipps 1976). It is a resident in Taiwan, breeding at 500–1,800 m (Severinghaus & Blackshaw 1976). In Korea, it is by no means common and is mainly a summer visitor, withdrawing to the southern coast in the winter (Vaurie 1959, Gore & Won 1971). There are also records for the offshore islands of Ullung (Dagelet) and Cheju (Quelpart), for which the subspecific name *parvirostris* has been proposed (Momiyama 1927). In far eastern USSR, it is a summer visitor to southern Sakhalin (up to 49°N) and the westernmost Kuril Islands (Kunashir, Iturup and possibly Shikotan); there is also a record of a straggler on Askol'd Island, SE of Vladivostok (Vaurie 1959, Dementiev et al. 1970).

In Japan, it is a summer visitor to the northern islands of Hokkaido and central and northern Honshu (south to Nagano, Gumma, Tochigi and Niigata Prefectures). It is a numerous passage migrant on the northern and northeast coasts of Honshu from late August to November, no doubt involving birds from Sakhalin and Hokkaido; further south it is a rare migrant at Tokyo and the Seven Islands of Izu (Izushotō). In winter it moves to lower ground and is found in western Honshu, Shikoku and Kyushu, and south to Amami Island in the Ryukyu group (Vaurie 1959, Yamashina 1961, Dementiev et al. 1970).

The distribution of *P. r. rutilans* is shown in Fig. 74. The solid line encloses all the records; the status within this area is less certain. The bird withdraws from the more northerly parts of the range after the breeding season and there is also some movement into southeast China, though according to Zheng (1976) part of the population of Szechwan winters in the west of Yunnan Province (where the race *intensior* occurs). The areas in which these movements occur are not well understood; the dashed lines, based on my interpretation of the best available evidence, are sketched in to show where the bird is a summer visitor, on the one hand, and present only in the winter on the other.

The situation in west central Szechwan is unclear. Zheng and Tan (1963) consider the birds from Batang as belonging to a separate race, *P. r. batanguensis,* and Zheng (1976) places the birds from Erlang Shan and the Emei Shan as intermediate between nominate *rutilans* and *intensior*.

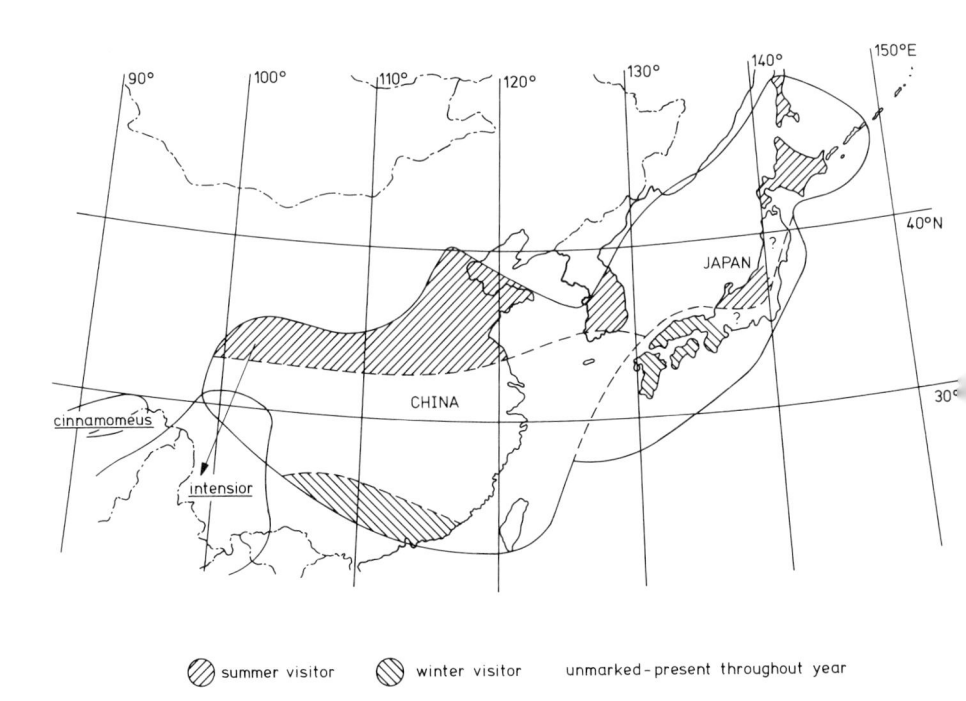

Fig. 74 Range of Cinnamon Sparrow P. r. rutilans

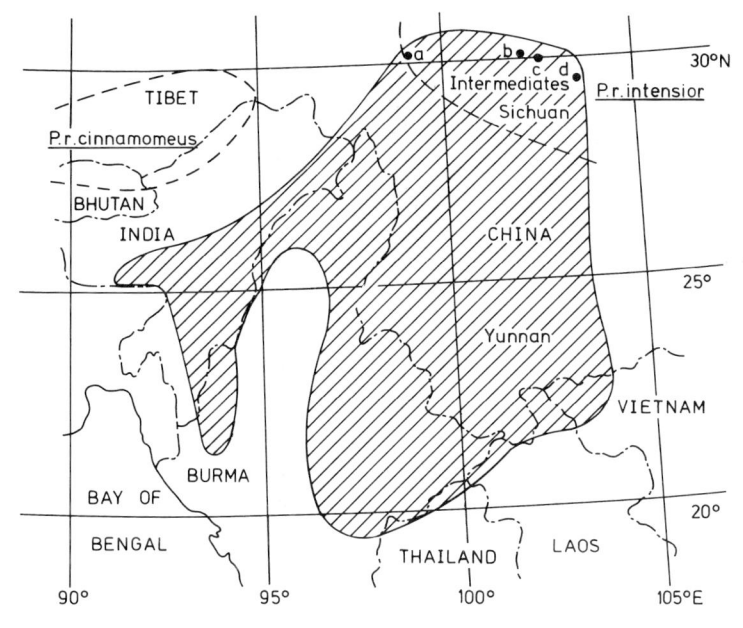

Fig. 75 Range of Cinnamon Sparrow P. r. intensior

Passer rutilans intensior
The race *intensior* abuts the nominate race at the extreme southwest of its range where, as already mentioned, there appears to be a zone of intergradation.

In China, *P. r. intensior* occurs over southwest Szechwan and most of Yunnan to the west of about 103°30′E (La Touche 1923, Zheng 1976). To the south it extends into extreme northwest Vietnam at Cha Pa (Delacour 1930), extreme northern Laos, and as a straggler into northern Thailand (Lekagul & Cronin 1974). In Burma, it is found from the Karen hills (at *ca* 1,500 m), north to the Adung valley (1,500–2,000 m) and also in the east in the Chin hills. These two areas are linked by a continuous population in eastern India that extends from the Naga hills to Manipur (1,400 m and upwards) and Mizoram. It also extends westwards in India to the Khasi hills in Assam (Smythies 1953, Ali & Ripley 1974, Ripley 1982). The distribution is shown in Fig. 75.

Birds of this race appear to be largely sedentary, though there may be some local movement to lower ground in the winter months; for example, Stanford and Ticehurst (1935) found it in the Triangle at 750 m in March, and, as already discussed, the population in western Yunnan may increase in winter through immigration of the race *P. r. rutilans* from northern Szechwan and Kansu.

Passer rutilans cinnamomeus
The race *cinnamomeus* appears to be allopatric with both the other subspecies. It occurs as a breeding species in a narrow band, about 200 km wide, along the length of the Himalayas from Arunchal Pradesh (at 1,800–4,000 m) and southeastern Tibet, east to about 95°30′E (at 2,100–4,500 m), through northern Bhutan (at 1,800–3,000 m), probably in Sikkim, though there are no definite breeding records, northwestern Nepal (at 1,800–4,270 m, occasionally down to 370 m), the Garhwal Himalayas in Uttar Pradesh, India (at 1,900–2,600 m, sporadically down to 1,200 m), Himanchal

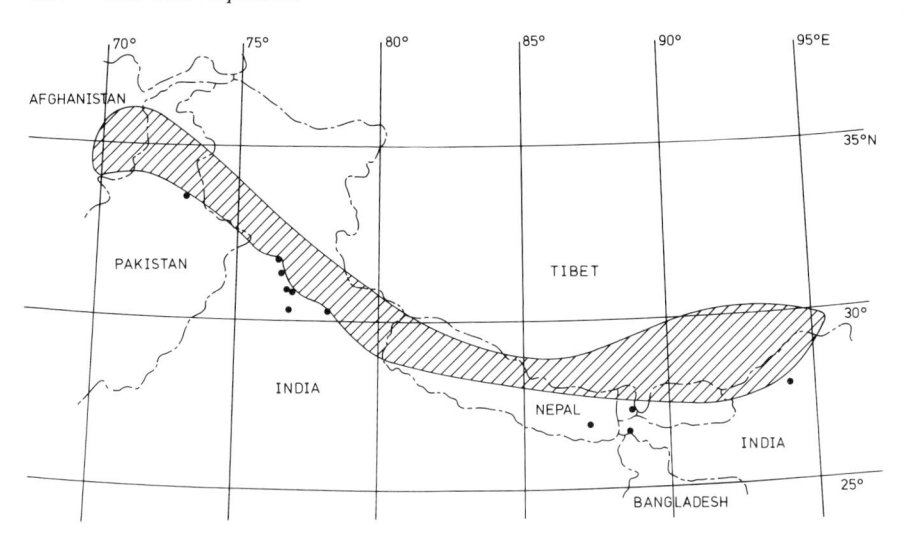

Fig. 76   Range of Cinnamon Sparrow P. r. cinnamomeus. *Non-breeding records (winter)* •

Pradesh (mainly 1,700–2,100 m, but occasionally down to 1,200 m and up to 2,750 m), Kashmir (up to 2,700 m), the Chitral region of Pakistan (1,800–2,400 m) and Kafiristan in extreme eastern Afghanistan. (Records mainly from Ali 1977; Fleming *et al.* 1979; Kinnear 1922; Ludlow 1937, 1944, 1951; Ali & Ripley 1974; Vaurie 1972; Whistler 1926, 1930, 1949). No doubt the restricted north-south distribution is determined by its altitudinal tolerance at this latitude.

In winter it moves south, often only a few kilometres, to lower altitudes, mainly 500–1,500 m, returning to the breeding quarters about March. For example, I found it breeding round Mussooree, Uttar Pradesh, India, at 1,900 m in May, while in December the birds were less than 15 km away near Dehra Dun at 640 m. Some birds, however, winter at higher altitudes, *eg* at 2,592 m at Tukucha, Nepal.

The distribution of *P. r. cinnamomeus* is shown in Fig. 76.

HABITAT

Over much of its range the Cinnamon Sparrow is a montane species, though this is influenced by latitude as can be seen from Fig. 77, where the breeding altitude is plotted against latitude. It breeds at sea level at the northern limit of its range between 49° and 38°N. At the extreme south of its range at 20°N, the minimum breeding altitude is about 2,500 m. The maximum reached is 4,500 m at 28°N, but south of this the height is governed by the availability of high country within its breeding range rather than by altitudinal tolerance. The impression given by Fig. 77 is that the race *cinnamomeus* is less tolerant of low altitudes than the other two races, irrespective of latitude. In the Garhwal Himalayas near Mussooree, the birds occupied a very limited altitudinal range, occurring from about 1,900 m in Mussoorie up to 2,600 m at the small hamlet of Buraskanda, but absent from Thathur, lying in a deep valley just to the north at 1,200 m, where House Sparrows were present in strength.

The Cinnamon Sparrow is a bird of light woodland, particularly on the edges of terrace cultivation, but also penetrating into towns and villages. Where House or Tree Sparrows are present, the Cinnamon Sparrow keeps to the rather less built-up areas of the towns where there are open gardens with the congener taking over the

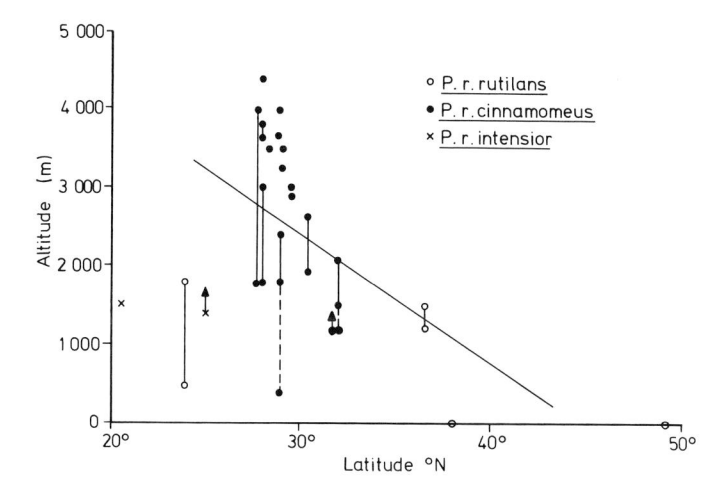

Fig. 77   Altitudinal breeding limits of Cinnamon Sparrow

more urban role. In such places the Cinnamon Sparrow freely perches on trees and houses, while in those villages in which it is the only sparrow, it occurs among the houses, even where there are no trees, behaving as a complete 'house sparrow', coming down into the streets to feed on spilled grain and scraps just like its more familiar cousin.

In the northern part of its range in Sakhalin, where it occurs at sea level, it is a bird of open woodlands and forest edges, particularly in river flood plains, coming into inhabited areas only during migration (Dementiev *et al.* 1970).

In winter it occurs in lower lying ground in mountain foothills and near the coasts, where it is to be found in open cultivated land and riverain grassland, but always where there are nearby trees and bushes in which it can take refuge at any disturbance.

BEHAVIOUR

The Cinnamon Sparrow forms loose colonies in the breeding season, collecting in flocks during the rest of the year. These flocks can be large, reaching spectacular proportions during autumn migration on the northern and northwestern coasts of Honshu, Japan, from late August to November (Yamashina 1961), before dispersing to form small bands further south. The winter flocks tend to keep away from

roller-feeding

212   *Cinnamon Sparrow*

human habitations. They feed in a close group on the ground, flying together into neighbouring trees when disturbed and then trickling back to the feeding area after the disturbance has passed in a similar way to many other sparrow species. On the ground the birds remain close together, the flock gradually moving forward as individuals from the rear overfly the leading birds in a continuous progression, searching for fallen grain or seeds (roller feeding).

Many writers describe the species as shy and retiring, but it frequently perches on exposed branches of trees or on telegraph poles and I found it very approachable, both at the nest and in its winter flocks: a pair continued feeding their young in a nest while I photographed them, without cover, from less than three metres, and I was able to approach within 20 m of a winter flock feeding on the ground without the birds flying away. Cinnamon Sparrows do not appear to associate with their congeners nor with other seed eaters.

During the breeding season the female roosts in the nest with her mate in some cover nearby. In the winter the birds roost communally in trees or thick bushes (Smythies 1953).

BREEDING BIOLOGY

Although the Cinnamon Sparrow is gregarious outside the breeding season, it cannot be described as a highly social breeder. Even when the bird occurs at high densities, the nests are fairly uniformly dispersed rather than clustered. It breeds in isolated trees and light woodland, and also in built-up areas. Fig. 78 shows the

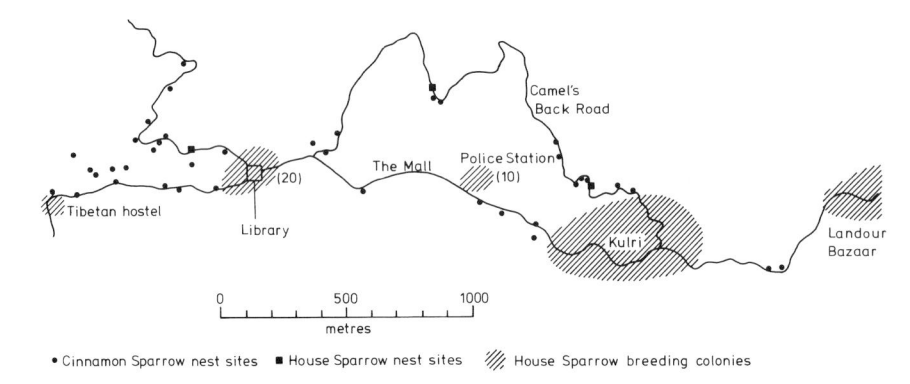

Fig. 78   Distribution of Cinnamon and House Sparrow nests in Mussooree

distribution of nests in Mussooree, in May 1981, where there were about 50 pairs in an area of approximately 150 ha; the pairs were fairly scattered, compared with the dense colonies of House Sparrows in the built-up bazaar areas. The species nests most frequently in holes, mainly holes in trees, though also in a variety of other situations, such as embankments, stone walls, in buildings (in thatched roofs, under eaves), in the old nests of Red-rumped Swallows *Hirundo daurica* and even that typical urban sparrow site, an electricity junction box. Less commonly, though this has been reported from Sakhalin, free-standing nests are built in shrubs (Dementiev *et al.* 1970).

Over most of its range, where it is exclusively a hole nester, the breeding density must be determined by the availability of sites, though, even with a superfluity of these, the birds still remain dispersed; in the small village of Dhanaulti, 24 km from Mussooree, for example, where there were about 30 stone-built houses providing ample nesting opportunities, the six pairs of Cinnamon Sparrows were dispersed throughout the village, rather than close together in a colony; the same dispersal occurs in Sakhalin where the bird nests in bushes and there are no constraints on the availability of sites (Dementiev *et al.* 1970). Only in a few cases did I find nests close together under house eaves; here, once ownership of the sites had been established, the birds appeared to be quite tolerant of each other. I have watched a pair of Cinnamon Sparrows trying to oust a pair of Black-crested Tits *Parus melanocephalus* that were feeding young in a nest under the eaves of a stone house; they were unsuccessful, as the following day the tits were still in possession, and the sparrows were nowhere to be seen. The height of the nest varies widely, ranging from less than a metre to over 10 m. The nest is an untidy, loose bundle of dried grass, filling the nesting cavity and warmly lined with feathers; both sexes take part in building.

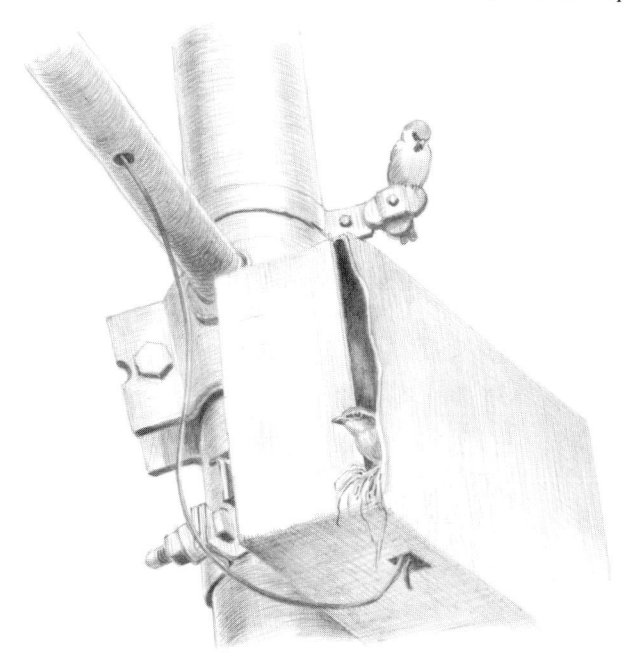

The male chooses the nest site, spending much time calling nearby. He displays with the head up, chest pushed forward and tail raised; the wings are held slightly

Subspecies	Month												Reference
	J	F	M	A	M	J	J	A	S	O	N	D	
rutilans													Kinnear 1929, Dementiev *et al.* 1970
intensior			-?-										Stanford 1941
cinnamomeus													Whistler 1930, 1949, Ali 1949, Ali & Ripley 1974, author

Fig. 79 Breeding season data for Cinnamon Sparrow

out from the body and in this stiff posture the bird bows up and down in front of the female. The unreceptive female lunges at the male and flies off, but there is no development of the group display, such as occurs in the House and Somali Sparrows and to a lesser extent with the Tree Sparrow, as might be expected with the greater dispersion of the Cinnamon Sparrow nests.

Published data on the breeding season are summarised in Fig. 79. The season is fairly short, doubtless a reflection of the late spring and early autumn in its montane breeding localities. Clutch size is 4–6, most frequently 4. Two clutches are laid. Both sexes take part in incubating and feeding the young; though in the nests that I watched the male took the principal role in the latter activity.

MOULT

According to Dementiev *et al.* (1970), moult in the Russian population probably takes place from August to September; that is, immediately after breeding and prior to migration.

VOICE

The main call, used by both sexes, is a monosyllabic *cheep* or *chilp,* softer and more musical than the similar call of the House Sparrow. This call is given frequently by the male at nest sites, where he uses exposed perches, such as the bare branches of trees and telephone wires. I have transcribed variations of the male's basic call at the nest as *chweep,* and a trilled *cheeep,* though never as obviously disyllabic as the calls of the House Sparrow.

On occasions a series of notes can be strung together – *cheep chirrup cheeweep* – to give a short musical song. Ali and Ripley (1974) describe the song as *chwe cha cha* and Ali (1977) also mentions a thin *swee swee* call, but does not describe the circumstances in which it is given.

In disputes among males at the nests, a rapid *chit-chit-chit* call is uttered frequently; by analogy with the House Sparrow, I believe this to be an assertive or threat call, no doubt given by the nest owner.

All observers agree that the calls are among the sweetest and most musical of those given by any of the sparrows.

FOOD

The Cinnamon Sparrow is basically a seed eater, taking grain (rice and barley) and the seeds of small herbs. To a lesser extent berries are taken; for example, I have watched them feeding on the berries of the Kingoor, a thorny shrub with elongated fruits. The young in the nest are fed almost exclusively on animal food: caterpillars and grubs obtained by searching the leaves of trees, and flying insects taken from leaves or caught in the air.

GENERAL COMMENTS

In the Himalayas, the race *cinnamomeus* appears to be associated solely with terrace cultivation, villages and towns. As terrace cultivation in this area is a fairly recent development (probably not earlier than about 3,000–4,000 years B.P. (A. K. Ghosh, *in litt.*), the arrival in the Himalayas and the evolution of the race *cinnamomeus* must have been comparatively recent and we have to look further east for the area in which the species evolved, probably China or Japan.

Eastwards the degree of association with man appears to be less common. This association has not been recorded for *intensior* in Burma, it is uncommon in China and Japan and has not been reported from the Kuril Islands and Sakhalin, where it is predominantly a bird of light woodland.

15: Tree Sparrow *Passer montanus*

NOMENCLATURE
Fringilla montana Linnaeus, Syst. Nat. 1758 edn. 10: 183.
    Northern Italy; restricted to Bagnacavallo, Ravenna, by Clancey, Bull. Br. Orn. Cl. 1948 68: 135.
Passer montanus (Linnaeus) 1758.
  Subspecies: *Passer montanus montanus* (Linnaeus) 1758.
    Passer montanus malaccensis Dubois, Fauna Ill. Vert. Belg. Ois. 1: 572.
      Malacca [Melaka].
      Synonym: *Passer montanus hepaticus* Ripley, Proc. Biol. Soc., Washington 1948 61: 106.
      Tezu, Mishmi Hills, NW Assam.
    Passer montanus transcaucasicus Buturlin, Ibis 1906: 423.
      Akhalzykh [Akhaltsikhe], Transcaucasia.
    Passer montanus saturatus Stejneger, Proc. U.S. Nat. Mus. 1835 8: 19.
      Ryu Kyus [Ryukyu Retto]; Okinawa according to Phillips, Auk 1947 64: 126.
    Passer montanus dilutus Richmond, Proc. U.S. Nat. Mus. 1895 18: 575.
      Kashgar, Eastern Turkestan.
      Synonyms: *Passer montanus pallidus* Zarudnyi, Ptitsy Vostochnoi Persii 1904: 262.
      Eastern Iran [eastern Khorasan].
    Passer montanus iubalaeus Reichenow, J. Ornith. 1907 55: 470.
      Caucasus to Tsingtao [Qingdao, Shandong]; the type locality is Tsingtao according to Hartert & Steinbacher, Vög. pal. Fauna, Ergänzungsband 1932: 80.

Passer montanus pallidissimus Stachanow, Ois. Rev. Franç. Orn. 1933: 789.
    'Harma Bouroung', eastern Zaidan, northern Tsinghai. [Qaidam, Qinghai].
Passer montanus zaissanensis Poliakov, Mess. Orn. 1911; 150.
    Kara Irtysh in Zaisan Nor region [Lake Zaysan, Kazakhstan].
Passer montanus tibetanus Bates, Bull. Br. Orn. Cl. 1925 45: 92.
    Khumbajong, Tibet.
    Synonym Passer montanus kansuensis Stresemann, Orn. Monatsb. 1932 40: 55.
    Heitsuitse, northern Kansu; [=above Sining on the Sining river, northeastern Tsinghai according to Vaurie, *Birds of the Palearctic Fauna* 1959: 578.].

A large number of races has been described from the extensive range of the Tree Sparrow, but, with growing availability of specimens, there has been increasing rationalisation. Vaurie (1949, 1956, 1959) and Moreau (1962) have made critical evaluations that show that much of the variation is clinal rather than the result of disjunct differences. In his most recent contribution, Vaurie recognises seven subspecies; I have amalgamated *zaissanensis* with *dilutus* as it appears to be an intergrade between *dilutus* and *montanus,* insufficiently distinct from either to justify separation. This leaves the six races shown above.

This species is sometimes given the names European, Old World or Eurasian Tree Sparrow by some American authors to distinguish it from the American Tree Sparrow *Spizella arborea,* and to avoid confusion in North America, where *Passer montanus* has been introduced. Tree Sparrow is, however, the common usage for *Passer montanus*; it is not particularly appropriate, though no less so than the German 'Feld-Sperling' (=Field Sparrow) or the latin name *montanus* (=montane), but despite these qualifications, I have retained the traditional name Tree Sparrow for this species.

DESCRIPTION

The Tree Sparrow is a typical 'black-bibbed' sparrow, but is unique in the genus in that the female has also adopted the black-bibbed 'male' plumage, so that the sexes are indistinguishable. It is a medium to large sized sparrow, length about 150 mm, smaller than the House Sparrow.

Passer montanus montanus
Adult. Head from bill to nape chocolate, becoming redder in worn plumage. Cheeks white with prominent black patch, white of cheeks extending to sides of nape. Back and rump yellowish brown, becoming more rufous in worn plumage, upper back boldly streaked with black. Double white wing bar. Small black bib. (According to Szlivka, 1983, bib is generally smaller in female; his results, however, show that there is so much overlap that it is not possible to use this feature to separate the sexes in the field.) Underparts greyish white. Bill of both sexes black in breeding season, otherwise blackish horn.

The white collar is a prominent field character, making easy differentiation from somewhat similar House Sparrow, even at long range.

Juvenile. Similar to adult, but duller.

Passer montanus transcaucasicus
Similar to nominate *montanus,* but duller above and whiter below. Bill slightly smaller. Intermediate between *montanus* and *dilutus.*

Tree Sparrow

Passer montanus dilutus
Much paler than nominate race.

Passer montanus saturatus
Rather darker and more richly coloured than nominate race. Bill longer than in any of other races.

Passer montanus tibetanus
Similar to nominate race, but distinctly larger (see Table 37).

Passer montanus malaccensis
Similar to *saturatus*, but smaller.

BIOMETRICS

Typical body measurements are given in Table 39. The analysis of the biometric data is complicated by the monomorphism of the sexes, with a large proportion of the published measurements referring to unsexed birds. It is possible, however, to detect a few features.

There is a cline of increasing size (wing length) in the nominate race from the United Kingdom to central Europe (regression equation $y = 0.114x + 69.92$, $r = 0.79$); measurements are lacking from Siberia to show whether this continues to the east. As can be seen from the following data on average wing length (both sexes combined), the smallest birds occur in the extreme south of the range, viz

P. m. montanus		
	Malta (36°N)	66.9 mm
	Central Europe (50–55°N)	69.6 mm
P. m. malaccensis		
	S. Malaysia and Sumatra (5°S–5°N)	66.4 mm
	Thailand, Indo-China and Burma (20–25°N)	69.9 mm

However, any real south–north trend is obscured by the fact that the largest birds occur at the highest altitudes, which lie at mid latitudes; this is shown by the following average male wing lengths:

P. m. montanus	Switzerland	72.3 mm
P. m. tibetanus	Bhutan, Tibet	*ca* 78 mm

Measurements on birds from the western Palaearctic show that the bill is more than 10% longer in summer compared with winter (Clancey 1948). This again shows the effect of the seasonal change in diet from a mixture of seeds and softer invertebrate food in summer to one exclusively of hard seeds in winter, as has already been described for House and Dead Sea Sparrows.

DISTRIBUTION

The Tree Sparrow has an extensive range in the Palaearctic and Oriental regions. In addition, the natural range has been increased by introductions to parts of Indonesia, the Philippines, Micronesia, North America and Australia. The different races and their distributions have been examined in detail by Vaurie (1949, 1956, 1959), and valuable contributions have been made by Dementiev *et al.* (1970), Ali and Ripley (1974) and Zheng (1976).

The ranges of the six races that I recognise, based on my interpretation of the above contributions, are approximately as follows.

Table 39: Biometric data for Tree Sparrow

Feature	Subspecies	Locality	Males Range	Males Mean (Median)	Females Range	Females Mean (Median)
weight	montanus	UK		22.0*		
		European Continent	**20.2–30**	23.9	**21–27.4**	23.2
	transcaucasicus	—		22.3*		
	dilutus	—	24–25	(24.5)	20–25	23.2
	malaccensis	—	17.5–21.4	(19.5)	16.8–24	(20.4)
	saturatus	—	22.8–28.7	25.2		
wing	montanus	UK	65–73	69.3	65–70	67.8
		European Continent	**66–76**	71.1	65–74	69.0
	transcaucasicus	—	64–72	68.1	64–68	66.1
	dilutus	—	**65–76**	71.1	**65–74**	**69.6**
	tibetanus	—	72–82	(77)		
	malaccensis	—	64–76	(70)	66–74	(70)
	saturatus	—	65–71	68.7	61–69	67.2
tail	montanus	UK	50–56	53.2	49–55.5	52.5
		European Continent	**52–58**	53.6	**49.5–59**	52.4
	transcaucasicus	—	47.0–56.0*	51.1*		
	dilutus	—	50–60*	54.3*		
	tibetanus	—	61–63*	(62)*		
	malaccensis	—	52–57	(54.5)	50–56	(53)
tarsus	montanus	—	**16–18**	17.4	**15.9–18**	17.0
	malaccensis	—	15–19*	(17)*		
culmen	montanus	—	13–14	(13.5)	12.5–14	(13.3)
	transcaucasicus	—	10.3–12*	11.1*		
	dilutus	—	11.5–13.5*	12.3*		
	malaccensis	—	12.5–14*	13.1*		
	saturatus	—	**12.5–16.0***	14.1*		

* not sexed

Passer montanus montanus

The nominate race has an extensive distribution from the British Isles, east to the Pacific coast of the USSR, south to the Mediterranean coast, the line of the Himalayas and most of northeast Asia; in the north it extends to approximately 70°N in Norway and eastwards across northern Siberia.

In the south, it occurs from the Iberian peninsula, excluding south Portugal, through mainland France and Italy. Wardlaw Ramsey (1923) stated that it was a winter visitor to Corsica and Sardinia. According to Bezzel (1957), however, the Tree Sparrow was introduced to Cagliari at the end of the last century. Initially, it appears to have spread little as Moltoni (1923) stated that it was present only in the city and its immediate surroundings; in 1951 Steinbacher (1952b) was unable to find it even there, though Bezzel (1957) reported it not only present in Cagliari in 1955, but also in two isolated locations on the west coast. I visited the island in 1979 and found the species well established in the towns and villages in a narrow belt, about 10–20 km wide, on the east coast from Cagliari in the south to Olbia in the north (Fig. 80), displacing the Willow Sparrow from the built-up environment that it occupies elsewhere on the island (Summers-Smith 1979). This colonisation appears to have

220 *Tree Sparrow*

Fig. 80 Range of Tree Sparrow in Sardinia

● Urban areas occupied by Passer montanus
▨ Range of P. montanus in Sardinia

taken place in the last 30–40 years. The Tree Sparrow is the urban sparrow in Naples on the adjacent Italian mainland and it seems to me that the breeding stock on Sardinia could well have been derived from ship-borne immigrants from Naples.

It is present on Sicily; it has bred on Malta and Gozo in small numbers since 1959 (Sultana & Gauchi 1982) and is said to have bred in 1940 on Zembra Island off the coast of Tunisia (Thomsen & Jacobsen 1979).

It is absent from northern Yugoslavia and does not breed in the area bordering the eastern Adriatic, but it occurs sparingly as a breeding species in Bosnia (Bosna i Hercegovina) – I saw one near Jajce in June 1955 – and eastern Bulgaria. It is a regular breeding species in Macedonia through to northern Greece, but is absent from south-central Greece and the Peloponnese. It breeds in small numbers in Asia Minor – on the Black Sea coast and the central plateau of Turkey. East of the Black Sea, it extends from the northern Caucasus, along the north shore of the Caspian Sea and across southern USSR, north of the chain of the Pamirs, Tien Shan and Altay mountains, through northern Kazakhstan, Transbaicalia and northern Mongolia, and extreme north China in Manchuria to the Pacific coast in central Korea.

It is widespread in Great Britain, though absent from the extreme southwest and is sporadic in south Wales. In Scotland, it is mainly confined to the lowlands and the east coast, though it also breeds on many of the western islands, Orkney and Shetland. Numbers in the British Isles are subject to extreme fluctuations; this is particularly so in Ireland, where it became extinct as a breeding species in 1959–60, though it has since built up again to give a small, rather scattered population round the coast.

It formerly bred on the Faeroes (on Skuø from 1866 until the early 1960s and in Thorshavn, Strømø in 1888, Salomonson 1935) but is now extinct. In Scandinavia, it occurs in southern Norway and Sweden, with extensions up the Norwegian coast to about 70°N and up the Baltic coast of Sweden to about 65°N; it formerly occurred in Finnmark, the northernmost province of Norway, but according to Blair (1936) it is no longer present. It is scarce in Finland, and in 1958 Merikallio (1958) considered it had decreased as a breeding species, estimating the total population as only 100 pairs. In Russia it spreads north to the southern coast of the Kola (Kol'skiy) peninsula, just north of the Arctic circle, eastwards across Russia and Siberia at about 65°N, with northward extensions along the river valleys – the Pechora to its mouth about 68°N, the Yenesey to 70°N (occasionally to 72°N) – reaching the Sea of Okhotsk at the mouth of the River Amur at about 60°N.

In the north of its range it is a summer visitor, but elsewhere it is both a resident and a partial migrant, though this does not appear to follow a consistent pattern and the picture is complicated by dispersal of the young after the breeding season as will be described later.

Birds of the nominate race were introduced (reportedly 20 individuals, Widmann 1889) to St Louis, Missouri, USA in 1870. These birds initially bred in this area, where at that time House Sparrows were not present. Within 10 years, however, House Sparrows moved in and began to dislodge the Tree Sparrows, which then moved to the north along the Mississippi into Illinois, extending to *ca* 160 km north of St Louis and now occupying an area of about 22,000 km^2 in extreme east-central Missouri and west-central Illinois (Barlow 1973), with stragglers being reported from Modoc and Galesburg in Illinois, Oconomowoc in Wisconsin, Lone Oak, and possibly

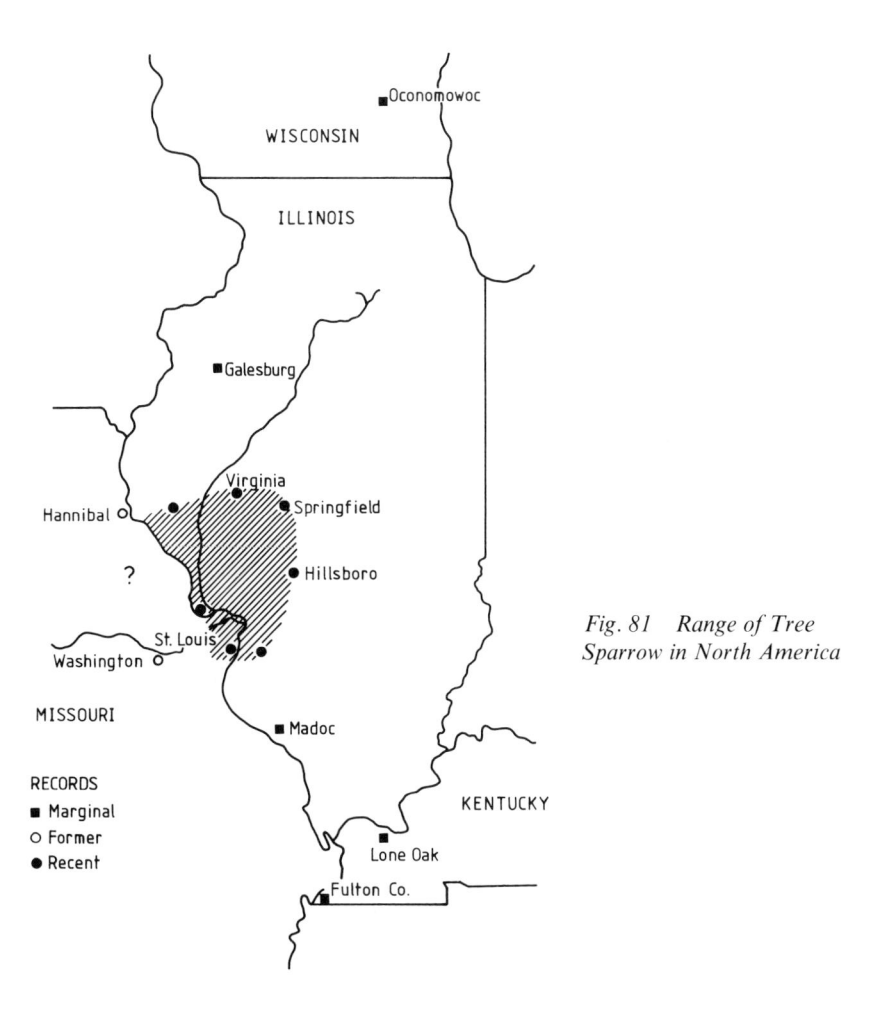

Fig. 81 Range of Tree Sparrow in North America

Fulton County, in Kentucky (Barlow 1973, Lever 1987). The present distribution is shown in Fig. 81. Barlow estimates the current American population at about 15,000 birds.

A similar introduction was made to New Zealand in 1868, but apparently the birds died out, perhaps through competition with the introduced House Sparrow (Williams 1953). An introduction to Bermuda in the second half of the 19th century also did not prosper, possibly for the same reason (Bourne 1957a).

Passer montanus transcaucasicus
This race has a restricted distribution to the west and south of the Caspian Sea, from Transcaucasia in the USSR to northern Iran. The northern limit runs through Georgia and Azerbaijan (Azerbaydzhan), from Sukhumi on the Black Sea coast through Tibilsi to Baku on the west Caspian coast. To the south it occurs through Armenia (Armenya) and eastwards in Mazandaran Province in Iran, between the Elburz mountains and the Caspian Sea as far as Gorgān.

This race is mainly sedentary, though there is a possible winter record from near Baghdād in Iraq (Ticehurst *et al.* 1923).

Passer montanus dilutus
The well-differentiated race, *dilutus,* occurs south of the nominate race: from Transcaspia south to extreme northeast Iran, much of Afghanistan, northern Baluchistan and western Pakistan, north to Gilgit in Kashmir and Chamba in Himanchal Pradesh, eastwards through Uzbekistan, Tadzhikistan, Sinkiang (Xinjiang), the Lake Balkhash area to northern Mongolia, Manchuria, and from there south through mainland China, east of Kansu (Gansu) and Szechwan (Sichuan), south to Hong Kong. Some authorities (*eg* Zheng 1976) have considered the Chinese birds to belong to the race *saturatus,* that occurs offshore to the east, others have separated them as a distinct race *iubalaeus.* As *iubalaeus* forms a cline with *dilutus* in northern China, and the mainland China birds tend to be intermediate between *dilutus* and *saturatus,* I think they are best retained in *dilutus.*

The birds occurring in extreme eastern Kazakhstan and northeast Mongolia, from the Russian Altay and Targabatay mountains to the Gobian Altay (Govĭaltay) and Hangayan mountains, have also been separated as a distinct race, *zaissanicus* (originally described from Lake Zaysan). These birds, however, appear to be intermediate between *dilutus* and nominate *montanus* and, in my opinion, are best retained in *dilutus.*

The majority of the birds of this race are presumed to be sedentary, though they range in winter to the Makran coast in Baluchistan. Movement has been recorded off the east coast of China. Styan (1891) found 'thousands of birds in October' on Gutzleff Island in Hang-chou bay and la Touche (1912) reported the Tree Sparrow as a passage migrant on Shaweishan Island, 30 miles east of the mouth of the Yangtze Kiang (Chang Jiang) from the end of March until the middle of April. It is not clear, however, whether these movements are an annual occurrence or of an irregular nature, like that off the east coast of England (see later).

Birds from mainland China were introduced, about 1728, to the Pescadores Islands (Penghu Liedao), lying to the west of Taiwan.

Passer montanus tibetanus
The range of the race *tibetanus* lies to the south of *dilutus,* through Tibet, Sikkim and Bhutan, east to Szechwan (Sichuan) and north to Tsinghai (Qinghai) and Kansu (Gansu). The birds in the eastern part of this range have been separated as *kansuensis,* but they are essentially intergrades between *tibetanus* and *dilutus* and are best retained in the latter.

Passer montanus malaccensis

This is a distinct race that occurs in the Himalayas, mainly below 3,000 m to the south of *tibetanus*. In Nepal, the two races are separated by forest, but further east there is some intergrading about 2,000–2,500 m. To the south it ranges through Assam, Burma, east to Yunnan and south through Thailand, the Indo-China countries, to Hai-Nan and south down the Malay peninsula to western Indonesia. In Burma, where it is sympatric with the House Sparrow, it is said to be partially migratory (Harington 1909–10), but its status elsewhere has not been described.

It is considered that the spread down the Malay peninsula only occurred in the 16th and 17th centuries, via the coastal trading stations to Singapore (Medway & Wells 1976), and from there on to Sumatra, Java and Bali. Ward and Poh (1968) suggest that Singapore itself was only colonised in the 19th century. Some authorities have suggested that the spread to the Indonesian islands has been the result of introductions (*eg* Decker 1968, Voous 1960), but the continuous distribution is quite consistent with a natural spread. More recently the bird has been recorded on the Philippines and more of the Indonesian islands. The arrival dates have not been clearly documented. The earliest appears to be the Philippines, where they were first reported in 1893 from Cebu City on Cebu island (Whitehead 1899); they were described as common there in the 1940s (Delacour & Mayr 1946). They are now also present on the neighbouring Negros Island (du Pont 1971). The next report is from Sulawesi (Celebes), where in the 1930s they were confined to the southern peninsula (Stresemann 1936), but have since become widespread (Escott & Holmes 1980). Borneo is next on the list, with the first report at Sandakan, Sabah, in 1964 (Gore 1964), followed by Kuching, Sarawak, in 1965 (Harrisson 1974), Labuan Island 1966 (Gore 1968), Bandar Seri Begawan, Brunei, 1969 (Harrisson 1970). By 1973 the Tree Sparrow was present all over the coastal plain of Brunei and to the north on the west coast of Sabah at Kota Kinabalu and Kota Belud (Harrisson 1974).

It is most probable that the original colonists were ship-borne immigrants from Singapore or even Hong Kong. The former would of course have involved birds of the race *malaccensis*, the latter *saturatus* (see below); no taxonomic study has been made, however, to enable the point of origin to be resolved.

Finally, Moreau (1962) mentions the presence of birds of the race *malaccensis* on Lombok, Lesser Sunda Islands (possibly a natural spread from Bali) and Ambon in the Moluccas.

The discontinuous nature of these occurrences, both in space and in time, suggests that they are the result of introductions rather than a natural extension of range, probably involuntarily as ship-borne immigrants, as noted for Sandakan and suggested for Brunei (Medway & Wells 1976). The absence of migration or nomadism in this subspecies is suggested by the fact that Boswall and Kanwanich (1978) considered it worth recording the appearance of a single individual in November on Phi Phi Le Island, Krabi, 25 km off the west coast of peninsular Thailand, quoting Dr D. R. Wells as stating that there are very few offshore records of this species for Thailand and Malaysia.

There was also an introduction to the Andaman Islands from Moulmein in Burma, but this does not appear to have been successful (Beavan 1866, Abdulali 1964).

A small isolated population of Tree Sparrows has recently been reported (1972) from Lammasinghi in the Visakhapatnam section of the Eastern Ghats, in Andra Pradesh, India (Raju & Price 1973, Price 1979). Nothing is known about the origin of this population, but it could well have been an introduction and the birds probably belong to the race *malaccensis*. Price (1979) estimated the population as less than 500 birds and thought it was decreasing to extinction.

The birds from the Himalayas, Assam and Burma have been placed in a separate race, *hepaticus*, but this seems merely to be a cline within *malaccensis*.

Passer montanus saturatus
The birds on the islands lying off the continental coast and those in southern Korea belong to another race, *saturatus*. This race occurs from Sakhalin, where it is absent from the north and not common elsewhere, and is possibly a recent arrival as it was not reported in 1889 (Dementiev *et al.* 1970), south through the Kuril Islands and the Japanese islands to Taiwan. These birds are largely resident, though most move out of Sakhalin in the winter months.

Birds of this race occur on Luzon in the northern Philippines, having been present in Manila (where they were introduced) at least since the middle of the 19th century (Blyth 1867) and are now common in settled parts of Luzon (Delacour & Mayr 1946).

Tree Sparrows are now also present in Micronesia, being reported from Guam in 1962 (King 1962); Saipan, Rota and Guam in the Marianas in 1979 (Ralph & Sakai 1979) and on Kwajalein in the Marshall Islands in 1978, possibly 1977, (Temme 1985). Nothing is known of these introductions; the fact that Guam and Kwajalein are American occupied territories suggests it is most probable that the original colonists were ship-borne immigrants. It seems likely that the birds belong to the race *saturatus*.

OTHER INTRODUCTIONS

Tree Sparrows were introduced to Australia in the 19th century. Sage (1956) considered that the Australian birds matched those from southern China (*saturatus* according to Sage, but presumably *malaccensis* according to my distribution of the subspecies) and traced them back to 'Chinese Sparrows' introduced, probably to Melbourne, in 1864, 1866, 1867 and 1872. In contrast, Keve (1976b) placed the Australian Tree Sparrows as a mixture of *saturatus* (presumably *malaccensis*) and the nominate race, the latter possibly inadvertently, introduced with House Sparrows from Europe in 1863 and 1872. Just as in North America, the Tree Sparrow has been much less successful in Australia than the House Sparrow and is today confined to the southeast of Victoria and New South Wales, north to about Sydney (Slater 1975).

Introductions were made to New Zealand in 1868 and 1871, but the bird failed to become established.

The distribution of the Tree Sparrow in the Palaearctic and Oriental regions is given in Fig. 82. The ranges of the subspecies are indicated by different hatching, but, as there are differences of opinion between the sources of information, and in any case there tends to be some intergradation between them, it has to be appreciated that the boundaries must be somewhat approximate. Those populations that I have considered to be synonymous with the six subspecies described above, but which further research may show to be worthy of separation, are indicated by a-f in the figure (see caption). The first recorded dates for the recently established introduced populations are also shown in the figure. Winter and passage records of birds outside the breeding season are indicated by arrows.

HABITAT

There is some variation in the habitat occupied by the Tree Sparrow over its extensive range. The factor that influences this more than any other appears to be the presence or absence of the House Sparrow. As we shall see later, I consider the Tree Sparrow to have evolved in southeast Asia, where the House Sparrow is not present; the presumably preferred habitat that is occupied in this part of the bird's range will be described first.

The birds belonging to the subspecies *saturatus, malaccensis* (excluding the '*hepa-*

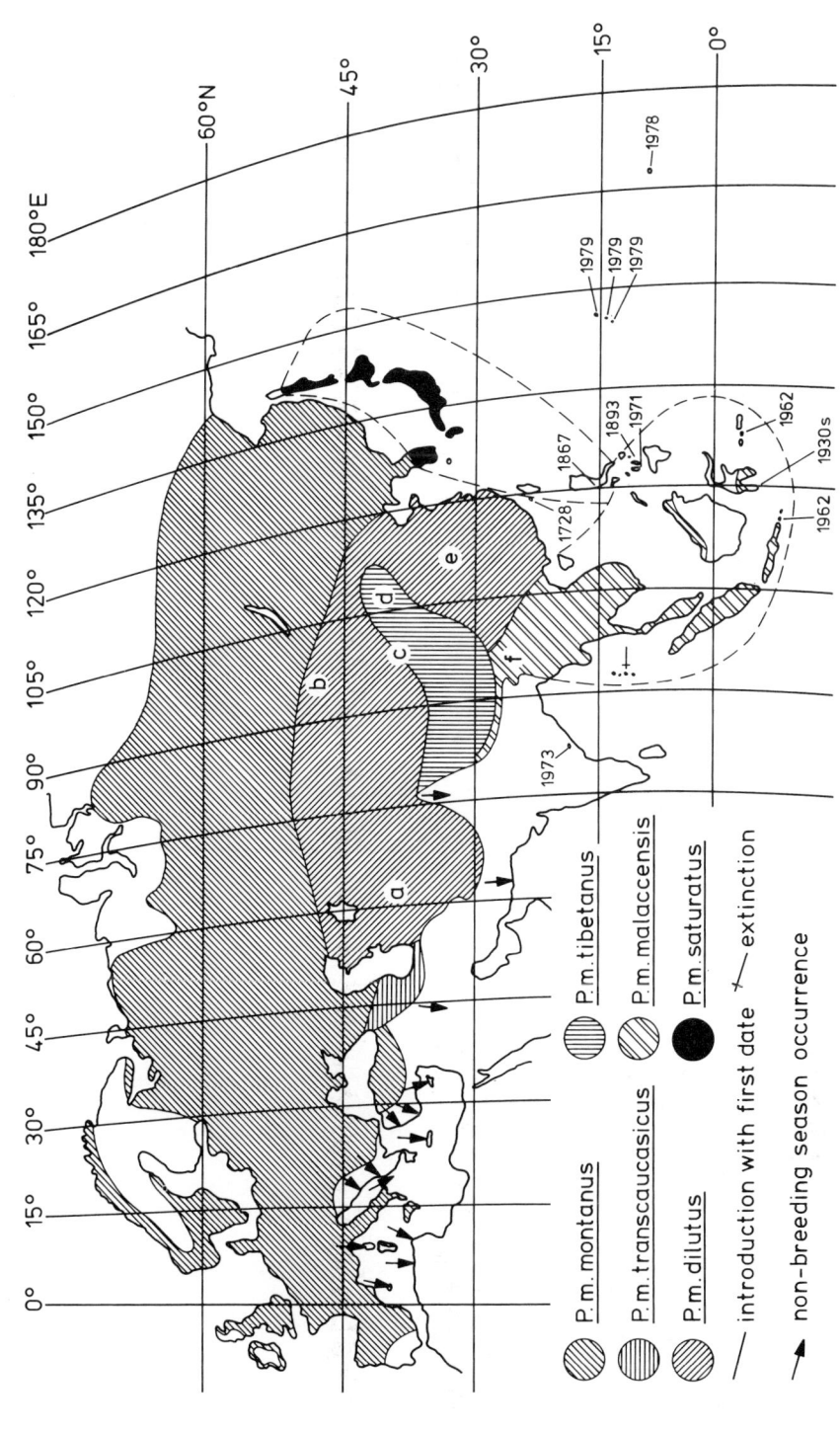

Fig. 82 *Distribution of Tree Sparrow in Palaearctic and Oriental regions Suppressed races:* (*a*) pallidus; (*b*) zaissanensis; (*c*) pallidissimus; (*d*) kansuensis; (*e*) iubalaeus; (*f*) hepaticus

ticus' group), the '*iubalaeus*' group of *dilutus* and *tibetanus*, that do not overlap with the House Sparrow, are predominantly associated with man, living in man-dominated areas, ranging from towns, where they penetrate to the completely built-up centres as well as occupying the more open suburbs, to villages and the surrounding cultivated land; the birds move out to arable fields at harvest time but still remain basically attached to man's buildings, which they use for roosting as well as breeding.

In the East, the Tree Sparrow is largely a bird of lowland areas, though in places it ascends, as a breeding bird, to about 2,800 m, but only as long as there are villages and cultivation.

In this part of its range, the Tree Sparrow can be described as an obligate commensal of man and, despite its English name, it occurs even in parts where there are no trees (*eg* in Mongolia).

The same situation largely obtains in Burma, where although the House Sparrow is also present in the north, the latter is more of a rural species and the Tree Sparrow is still the sparrow of the houses in the majority of villages and towns. The species also overlap in Assam; here I found them breeding side by side in the houses with, if anything, the Tree Sparrow in greater numbers. The House Sparrow does not penetrate to the high Himalayas in Bhutan and the extreme east of India and here again the Tree Sparrow is the sparrow of the villages up to altitudes of 2,500 m. Further west along the Himalayas the Tree Sparrow extends through Sikkim to Nepal above about 1,500 m; here the House Sparrow, subspecies *parkini*, is mainly a summer visitor, arriving after the Tree Sparrow has taken up its nesting sites in the villages. The ecological separation of the two species is, however, unclear, both being confined to inhabited areas and the surrounding cultivation. In Nepal, the Tree Sparrow lives up to its name of *montanus* and occurs up to 3,500 m. One was even obtained at over 4,000 m during a Mount Everest expedition (Stevens 1925). Further west in Baluchistan, Afghanistan, Turkmenistan and northern Iran, the western end of the range of *P. m. dilutus*, the Tree Sparrow is once more the bird of the towns and villages, with the House Sparrow (subspecies *bactrianus*) a summer visitor that does not arrive until May and lives in open country, largely away from man and his cultivation, nesting in trees, earth banks and under bridges. Fig. 83 shows an area of northeast Iran and Afghanistan that I traversed in May–June 1976: the dots indicate inhabited places where only Tree Sparrows were present; the hatching where House Sparrows were nesting in open country away from habitations. The separation appeared to be almost complete, though both species were present at the more isolated habitations, *eg* petrol stations away from the towns. The same situation apparently obtains in southern Afghanistan and Baluchistan (Christison 1941) and in the North-West Frontier Province of Pakistan (Whitehead 1909). This was also the case in Russian Turkestan when Carruthers (1910a, 1949) visited the area in 1907–08; the Tree Sparrow was the dominant urban bird, though the situation was changing as the sedentary nominate race of the House Sparrow was spreading from the north into the area. The current status of the two species in this area has not been described.

The situation in the rest of its range in Siberia and through Europe is much less clear cut. Over most of Siberia the Tree Sparrow is still a bird of human habitations, tending to replace the House Sparrow in the east, where the latter is a comparatively recent arrival. Here the Tree Sparrow occupies the larger villages and the centres of towns, with the House Sparrow on the edges or in the smaller villages where Tree Sparrows are lacking. As one moves west, however, the situation tends to reverse and the Tree Sparrow inhabits the more rural areas for breeding, withdrawing to inhabited areas in winter. Over Siberia the Tree Sparrow is most frequent in the lower-lying areas, penetrating even into the inhabited areas of deserts, though it is much less of an arid-loving species than the House Sparrow; it also occurs, however, in places, in the mountains up to elevations of 3,500 m. In Europe and the Caucasus, the House

Fig. 83   Habitat utilisation by House and Tree Sparrows in Afghanistan

Sparrow tends to become the dominant urban species, with the Tree Sparrow less closely associated with man, occurring in light woodland as well as the cultivated areas (particularly market gardens and orchards), though this is not invariably the case. In Bessarabia, for example, Haviland (1918) found that, in addition to breeding freely in open country, the Tree Sparrow also bred side by side with its congener in villages. In Switzerland (Haller 1936) and at higher latitudes in Europe (northern Germany, Scandinavia) some villages, or parts of towns, are inhabited solely by one species, others solely by the other. For example, in a scattered settlement at Motzener See, GDR, where Deckert (1972) carried out her observations, House Sparrows were absent and Tree Sparrows bred in buildings; and Stephan (1965) reported on observations on a small colony of Tree Sparrows breeding in a built-up district of Berlin, apparently in close proximity to House Sparrows. My impression is that the Tree Sparrow occupies the smaller, more dispersed, villages and the more open parts of towns, much like the separation of the Cinnamon Sparrow and the House Sparrow in the Himalayas. This may, however, be an oversimplification and the ecological factors determining which species is present are by no means clear. Some of these places may have been occupied only comparatively recently (the fact that this vast region is occupied by only a single race of each species is consistent with the area having been recently colonised) and an equilibrium may have yet to be established. This is particularly true of Siberia, where development by man to create the species' preferred habitat is very recent.

In Great Britain, the Tree Sparrow is typically a bird of arable farmland with hedgerow trees, parkland and even light woodland, but is scarcely ever found in villages and towns, and then only in the larger gardens on the outskirts. On the other hand, Tree Sparrows, where they occur in Ireland, tend once again to be birds of houses, living in villages rather than in open country, though even the House Sparrow tends to be rather local so that competition between the two species is minimised.

Broadly speaking, in Europe the Tree Sparrow is a bird of the cultivated areas surrounding inhabited regions, with some penetration, on the one hand, into the smaller villages and more open suburbs, where for some not very obvious reason the House Sparrow is lacking, and, on the other, into light woodland or the edges of forests. Pinowski (1967a) describes the optimum biotope in Poland as village areas surrounded by farms, with forest edge a suboptimum habitat occupied in years of high population density. In a study of the occupancy of nest boxes put up over a wide variety of habitat in Poland, Graczyk (1970) found that, of 938 pairs of Tree Sparrows, 72% were in field and roadside hedgerows and orchards, 16% in villages and the remainder in young or light woodland, with negligible penetration of older woodland and forest. In a study of the distribution of the Tree Sparrow in the countryside surrounding London, Sage (1962) showed a distinct concentration in damper areas, such as river valleys, with a noticeable preference for sewage farms.

The introduced birds in North America and Australia, where they are sympatric with the House Sparrow (introduced at about the same time), occupy a rather similar habitat to that in Europe. Although the Tree Sparrows liberated in St Louis, Missouri, USA, in 1879, initially remained in the city, once the House Sparrow began to increase and spread, so the Tree Sparrow moved to suburban and surrounding rural areas, eventually abandoning the city, and is now restricted to gently rolling countryside with small tracts of deciduous woodland lying to the north and east (Barlow 1973).

Haviland (1926) pointed out the difficulty of isolating the environmental factors that determine habitat selection by these two sparrows, where one species appears to replace the other in a random, even capricious, way. In the 60 years that have passed since she wrote these words, we are not very much further forward in this regard.

BEHAVIOUR

The Tree Sparrow has been the subject of a number of detailed studies covering its ecology and ethology. The most important of these are: Balat (1970, 1971, 1972, 1974a, 1976), Berck (1961, 1962), Bethune (1961), Boyd (1932, 1933, 1934, 1935, 1949), Creutz (1949), Deckert (1962, 1968), Eisenhut and Lutz (1936), Gauhl (1984), Pinowski and co-workers (Pielowski & Pinowski 1962, Pinowski 1965a, 1965b, 1966, 1967a, 1967b, 1968, 1971, Pinowski & Wieloch 1973), Scherner (1972a, 1972b), Seel (1964, 1968a, 1968b, 1970). The following sections draw heavily on these papers, supplemented in part by my own observations.

The Tree Sparrow is a social species, occurring almost invariably in flocks outside the breeding season. The young join up in flocks as soon as they become independent of their parents; for about 2–6 weeks these birds remain in the breeding area, but soon they combine with other similar flocks in suitable feeding places within 1–2 km, such as ripening grain fields, and later the flocks are augmented by adults that have finished their breeding activities. In areas where the bird is common, these autumn flocks may amount to several hundred, even thousand, birds that collect at grain fields, vegetable gardens or orchards. The movement involved is probably quite small, though more pronounced than with the House Sparrow (see Rademacher 1951); in Poland, for example, Pinowski (1965b) found that 81% of birds of the year, two to three months after fledging, were still within 1.5 km of their birthplaces. Bibby (1975) suggested that there could be a differential dispersal of the young, with those from the first broods remaining closer to the natal area than those from subsequent broods. In the late autumn the flocks, at least in those parts of the range in which the bird is largely sedentary, begin to break up, the adults returning to their breeding areas, the young dispersing randomly and more widely.

In addition to this dispersal of the young, a proportion of European Tree Sparrows, just as already described for the House Sparrow, undergoes a more directed type of migratory movement from late September until early November. These movements tend to be somewhat erratic. For example, there were significant movements off the east coast of Britain in the second half of the 19th century, continuing up to 1907, and then again in the period 1957–1962, when the Tree Sparrow population of the British Isles showed a dramatic build up, with apparently little movement in the intervening period. Movement also occurs in southern Europe; this is shown by: passage through the high Swiss alpine passes (Jenni & Schaffner 1984), its occurrence as a winter visitor at la Tour du Valat in the Camargue in southern France (Hoffmann 1955), its appearance as an irregular, at times common, autumn visitor on Malta, where it breeds only in small numbers (Sultana & Gauchi 1982), and sporadic occurrences south of its breeding range. I have seen Tree Sparrows in October in Majorca, Corfu and Crete, and similarly there are winter records from Albania, Tunisia and Pantelleria.

The picture further east is rather similar: Korzyukov (1979) observed a southward movement of Tree Sparrows over the northwest corner of the Black Sea in October 1975, but not in the following years. There is an increase in the population in Turkey in winter, where it is found on the Mediterranean coast as well as in those parts in which it breeds; it is an autumn passage migrant or a scarce, though perhaps regular, visitor to Cyprus in winter months (Flint & Stewart 1983) and has recently been reported as a vagrant to Israel (Keve 1976a). Autumn movements have also been recorded at Gorki in Belorussia, Kamyshin and Astrakhan, north of the Caspian Sea (Dementiev et al. 1970).

In general, only a small proportion of the European birds is involved in migratory movement, though Verheyen (1957) estimated that up to 25% of the Belgian population wintered away from their breeding areas, and the distances as shown by ringing recoveries are small, with only a few records as far as 600 km. The direction of the autumn movement is to the southwest or south, with a return in spring. Again, young birds are mostly involved: Jenni and Schaffner (1984) found that 96.5% of birds passing south over the Col de Bretolet in Switzerland were immature and Rademacher

(1951), in an analysis of the German ringing data, showed that only 3.2% of adults were recovered more than 10 km from the place of ringing, compared with 23% for immatures (total recoveries: adults 128, immatures 130). There is some indication of a greater tendency for migration in Japanese populations, with 19% of ringing recoveries between 100 and 300 km, and 12% over 300 km (Kuroda 1966).

This pattern of behaviour, with dispersal of birds of the year and migration by a small proportion of the population, serves to prevent inbreeding. For example, Creutz (1949) estimated from ringing results that a breeding 'colony' consisted of approximately 45% of adults from the previous year, 5% of young birds reared in the colony and 50% of young immigrant birds from other areas. In addition, the movements allow the colonisation of new areas that become favourable through changing land use or the provision of nest boxes in areas with a shortage of suitable natural hole (Boyd 1932, Creutz 1949, Dornbusch 1973, Gauhl 1984). Further, irruptions in years of successful breeding could be responsible for the fluctuations in the numbers that have been a characteristic of the British population.

The pattern of movement described above takes place where there is a continuous supply of food. This is not the case in the north of the range, where all birds move south; while, in the more arid parts of the range in the east, the bird adopts a nomadic way of life, still apparently remaining in social flocks containing adults as well as young, though less is known in detail of the behaviour of these birds. In parts of central Asia, where the birds breed in the country, they move into settlements during winter. Birds living in large towns spend most of their time there, but, like House Sparrows, move out to feed in ripening grain fields in autumn.

On the break up of autumn flocks in those parts of the range in which the bird is essentially non-migratory, adults return to their nesting colony area, where they roost and spend much of the morning hours. There is some recrudescence of sexual activity at that time as shown by the way that pairs take over nest holes, add material to the nest and even, at times, copulate. Some observers have suggested that this nesting activity is merely the building of roost nests as a protection against the cold nights (Niethammer 1937, Creutz 1949), but most see this as an indication of autumn sexual recrudescence (Berck 1961, 1962; Deckert 1962; Pielowski & Pinowski 1962; Pinowski 1967a). This latter view has been well confirmed by Löhrl (1978), who showed that nest building is confined to late autumn and, if nests are removed before winter sets in, no attempt is made to rebuild them as would be expected if their purpose was merely protection against the cold. Moreover, many of the 'nests' built at this time are no more than a covering on the floor that would give little thermal insulation, though as the nests are used for roosting by one or both members of the pair those that are more complete do provide some advantage in heat retention as a secondary effect.

Some birds of the year will take over unoccupied holes and may even attract a mate; these new pairs may also indulge in some nest building. Bibby (1975) has suggested that these pairs are likely to be formed by first brood birds, as later-fledging young would still be moulting at this time. With the addition of young birds, the population in the breeding colony area will be at a maximum and will probably exceed the available number of holes or the carrying capacity of the area. Pinowski (1965a) considers it is this overcrowding in the breeding areas that is responsible for the dispersal of young birds that are unsuccessful in obtaining a hole. The young birds are, however, clearly less sexually motivated at this time and, just like birds in the nomadic flocks, roost communally in trees, thick hedgerows, creepers on houses and even in reed beds.

When the leaves fall, and these more exposed sites cease to provide adequate shelter, the birds separate and increasingly look for holes in which to roost; in the case of non-migratory and non-nomadic birds these can later be used for breeding.

The use of holes for roosting has been most thoroughly researched by Creutz (1949) in his nest-box study from their first adoption in October up to the following breeding season. Usually the nest holes are occupied individually or by two birds (94% of the cases found by Creutz), though up to seven birds have occasionally been found in the same hole. Creutz found the maximum occupation of his nest boxes for roosting occurred in January and February, and from then on pairs that subsequently bred were increasingly found together. Strangely, in his study, the box was abandoned for roosting some weeks before egg-laying began, though other studies (*eg* Stephan 1965) suggest that this is not invariably the case. Once egg-laying has begun, only one bird, presumably the female, roosts in the nest overnight.

Somewhat similar behaviour can also occur with migratory birds; for example, I watched a pair of Tree Sparrows nest-building in October–November on Majorca, where they do not breed.

Even when communal roosting is abandoned, roosting is still very much a social activity, with birds collecting in trees with conversational chirping and skirmishing before they separate to roost in individual holes. During the day, most activities away from the breeding area are carried out socially; when feeding, a flock on the ground faces in one direction and birds at the rear overfly those in front (roller feeding). At any disturbance, real or imagined, the birds fly to the protection of trees or a nearby hedge and then gradually drift back once the danger is past. The feeding bouts last for 10–30 minutes, before the birds fly to a hedge or tree to rest and preen. Particularly in the afternoon, such flocks may indulge in social singing from their resting place. The significance of this behaviour is not known, but very similar behaviour also occurs with House Sparrows. The winter flocks may also join up with other small seed-eating species, including other sparrows, finches and buntings.

Just as there are habitat changes over the extensive range of the Tree Sparrow, so there are differences in behaviour. In Britain, for example, the country-living birds tend to be shy and even secretive in the breeding season, slipping quietly off the nest at the approach of an observer, though becoming bolder when the young have hatched and are nearly ready to fledge. In the East the Tree Sparrow is a much more confident and conspicuous bird, like the House Sparrow in Europe, regularly coming into houses. It is well described by Creutz as cunning, suspicious, cautious, hard to catch and quick to learn, characteristics that guarantee it the possibility of living successfully in close contact with man.

The Tree Sparrow shows aggression towards rivals of its own kind and to food competitors by adopting a threat posture with head raised and thrust forward, the tail stuck up and slightly fanned, the closed wings held out and slightly lowered.

threat

tail flicking

Stephan (1965) described how, in a small colony of five pairs, the birds apparently recognised and tolerated each other, reserving this aggressive behaviour for strange Tree Sparrows that came close to the breeding holes, particularly towards apparently hole-seeking individuals that appeared at the beginning of the breeding season (April); in the case of the most persistent intruders this could result in actual fights. Anxiety is shown by tail flicking and bill wiping; actions that are common to most sparrows. The Tree Sparrow regularly indulges in both water and dust bathing; it also occasionally sunbathes.

BREEDING BIOLOGY AND POPULATION DYNAMICS

Most detailed studies on breeding have been carried out in the European part of the range, where the bird is largely sedentary.

The mechanism of pair formation has not been clearly established, but it seems most likely that an unpaired male takes over a nesting hole and calls to attract a female. This occurs in March or April with birds of the year (possibly also occasionally in autumn), though replacements of lost mates of either sex also occur throughout the breeding season and again in autumn, but here possibly only of a lost female. Bethune (1961), from the result of studies on ringed birds, considered that pairs once formed remained faithful for life (up to four years in the case of some ringed birds), attachment to a chosen nest site being the controlling factor, individuals showing a lifetime attachment to a nest site unless disturbed by human or other interference. Creutz (1949) and Deckert (1962) similarly confirmed pair and nest site faithfulness for life with ringed birds. This was also considered to be the case in England where the birds remain in flocks in winter away from the nesting area and appear to be much less attached to nest sites than is the case with other populations (N. Moore pers. comm.).

In sedentary populations, adults that have previously bred return to the nest in September and October and, if one of the pair fails to return, a new mate can be attracted from birds of the year. It is unlikely that many pairs involving two first-year birds are formed at this time; as already described the young birds visit unoccupied holes and increasingly use them for roosting at night, though not necessarily in pairs as from 1–7 have been recorded occupying the same hole. Pair formation involving two young birds more probably occurs at the nest site in early spring (March onwards according to Creutz), rather than in the flock where the absence of sexual dimorphism and obvious differences in the behaviour of the sexes make pair formation less likely. Polygamy is rare among these sedentary populations (Dyer *et al.* 1977). Whether pair faithfulness also holds with nomadic and migratory populations of Tree Sparrows has not been established, but could well be the case as the result of nest faithfulness. Once the pair bond has been established the pair remains increasingly together, not only when feeding or bathing, but also in resting near the nest, where the two birds often sit pressed close together.

The Tree Sparrow is predominantly a hole nester. In eastern parts of its range, where the House Sparrow is absent, it takes the place of the latter in the built-up environment and most of the nests are in holes in buildings, under eaves, in holes in the ends of hollow bamboos and in nest boxes. To a lesser extent holes in trees, earth banks, haystacks, fissures in rocks are used and, more rarely, free-standing nests are built openly in branches of trees (see Vaughan & Jones 1913, Spennemann 1934). Where the range of the Tree Sparrow overlaps that of the House Sparrow, the former breeds more frequently away from houses, though in many places man-made structures are still used, particularly old stone walls, but also houses, sometimes even side by side with House Sparrows. Tree holes are also a common site in Europe (natural cavities and old woodpecker holes), together with holes in earth banks and the understorey of nests of corvids, herons and birds of prey.

Two studies involving large numbers of nests have been made in the Balkans on the different types of nest site used. The authors use slightly different categories but

Table 40: Situation of Tree Sparrow nests in the Balkans

	Yugoslavia (Szlivka 1983)		Bulgaria (Nankinov 1984)		Total	
	No.	%	No.	%	No.	%
a) Holes in houses and other man-made structures:						
houses	594	47.4	683	38.0	1,277	41.7
pylons, bridges, street lights	0	0	534	29.0	534	17.4
wells	8	0.6	32	2.0	40	1.3
nest boxes, dovecotes	179	14.2	35	2.0	214	6.9
sub-total	781	62.2	1,284	71.0	2,065	67.3
b) Other holes:						
cliffs, earth banks	215	17.2	247	13.5	462	15.1
trees	134	10.7	171	9.5	305	9.9
haystacks	59	4.7	10	0.5	69	2.2
sub-total	408	32.6	428	23.5	836	27.2
c) Other sites:						
understory of nests of large birds	64	5.1	42	2.0	106	3.5
open nests on trees, poles	0	0	61	3.5	61	2.0
sub-total	64	5.1	103	5.5	167	5.5
Grand total	1,253		1,815		3,068	

I have combined them in Table 40 to provide a general picture. This shows the strong preference for man-made structures, even where the species is sympatric with the House Sparrow, with the two species regularly nesting side by side. Nankinov remarks that up to the beginning of the present century the Tree Sparrow was largely a forest species in Bulgaria and that the use of man-made structures for nesting has coincided with an expansion in numbers. In some of the more rural sites, such as in cliff holes and the understorey of the nests of large birds, the Tree Sparrow shares the nest site with Willow Sparrows.

Free-standing nests can be built in quite open sites, such as the branches of trees and on telegraph poles, but they are most often in conifers, palm crowns, ivy-covered walls and thick hedges, where they are well concealed, though poplars, sycamores, limes and similar trees are also used (see Kalitsch 1927, Grote 1935, Ruthke 1955). According to Seel (1964), 33 of 1,073 British Trust for Ornithology Nest Record Cards, *viz* 3%, were from tree nests, and in a separate study in Great Britain, Sage (1962) reported 2% of nests being in trees. Free-standing nests are probably built when there is a shortage of suitable holes. For example, Bethune (1961) found a pair, that was displaced in March from a nest box it had occupied over the winter, by a pair of Starlings *Sturnus vulgaris* that enlarged the entrance hole. This pair did not leave the nesting area to seek another hole, but instead built a free-standing nest in a holly tree about 3 m away.

The free-standing nest is similar to that built by the House Sparrow: globular in shape, domed over and sometimes with an entrance funnel, but generally rather smaller (typically 150–250 mm in external dimensions, compared with 200–300 mm in the House Sparrow). It consists usually of an outer structure of dried grass, straw, rootlets and leaves, though in towns a variety of rubbish, including rags, string and paper, may be incorporated as well; the nest cup is lined with feathers and, at times, with animal fur. The nest in a hole is normally also domed and, unless very large, the cavity is completely filled with nesting material; in the case of small holes only a nest cup may be built and the roof omitted.

Cavities with small entrance holes (about 30 mm diameter) are preferred. In a study with ample nest boxes, with both 32 mm and 34 mm diameter entrance holes, 45 pairs of Tree Sparrows nested in the boxes with the smaller holes, compared with only 24 in those with larger ones (Löhrl 1978). No doubt, holes with the smaller entrances are chosen as they give good protection against larger hole-nesting competitors, such as House Sparrows and Starlings. Larger holes (up to 60 mm diameter) and cracks are, however, readily adopted and the entrance hole is reduced to the preferred size as the nest is built up. The Tree Sparrow is able to defend the nest against rival hole-nesting species, such as tits (*Parus* sp.) and Pied Flycatchers *Ficedula hypoleuca*, and will successfully usurp holes from these species, even when they contain eggs and young (Löhrl 1978). Like the House Sparrow, it will take over the nests of other species, particularly those of hirundines: the House Martin in the west, the Red-rumped (Striated) Swallow *Hirundo daurica* (*striolata*) in the east, as well as the nests of Sand Martins *Riparia riparia* in earth banks.

The adoption of holes in autumn for roosting, together with nest building, gives the Tree Sparrow an advantage over these potential small-hole rivals. Not only does this mean that Tree Sparrows are in possession of a hole long before these other species start looking, but also the rivals do not appear to be interested in holes that already contain a substantial Tree Sparrow nest.

The preference for holes places a restriction on social breeding; although the birds will nest side by side when there are suitable opportunities. There is, however, little real evidence of true colonial breeding and in England, at least, many pairs nest solitarily in holes in hedgerow trees.

Nest building, of the breeding nest as distinct from that built in autumn, may begin a month before the first egg is laid. Both sexes take part and, if re-using a hole from a previous season, the birds will clear out the old nesting material before starting to build. Addition of lining material continues until the young hatch.

In courtship, the male displays by hopping around the female with head and tail stuck up, the wings drooped and shivered, all the time uttering a rhythmic series of calls. If he approaches too closely, the female bites him or flies off if she is not ready; the male pursues and continues to display where she has landed. This may attract other neighbouring males, who then parade in front of the female in a similar way. Group displays of this sort are much less frequent and less vehement than those of the House Sparrow. Their function appears to be sexual stimulation of a female not ready for mating. When the female is ready for copulation, she crouches in front of the male, shivers the wings and utters a soft call.

Moore (1962) also describes another display of the mated pair that occurs prior to breeding, which he calls 'copy preening'. In this the birds perch close together, often touching, and go through a series of preening movements, the actions of one bird being exactly copied by the other. Presumably this has the function of strengthening the pair bond. It is odd that it has not been reported by either Berck (1961, 1962) or Deckert (1962, 1968) in their very extensive studies of this species in Germany, nor for that matter by any other workers on this well-studied bird. The way that a pair of Tree Sparrows will snuggle up close together when perched near the nest, either quietly sitting side by side or at times indulging in mutual preening, is a feature that, as far as I know, is not shared by any other member of the genus. It is tempting to associate it with the monomorphic plumage of the sexes.

Published data on the breeding season are summarised in Fig. 84. The beginning of breeding is positively correlated with latitude (regression equation $y = 1.47x + 41.2$, where $y =$ day of year and $x =$ latitude; correlation coefficient $r = 0.86$), beginning early in the tropics of southeast Asia – December or January in Malaysia (Ward & Poh 1968, Medway & Wells 1976) – but not until the end of April or beginning of May over most of Europe. The duration of the breeding season, the time from the laying of the first egg to the fledging of the last young, decreases with latitude (regression equation $y = -1.23x + 158.7$, correlation coefficient $r = -0.76$).

Results from the most significant studies on breeding, mainly in Europe, are summarised in Table 41. The clutch size ranges typically from 2–7 eggs, with records of up to 10 eggs, though some of these larger clutches are probably the product of two females; it probably increases with latitude, with the mean values ranging from about 4 eggs near the equator to about 5 eggs at 55°N, though the correlation is weak ($y = 0.024x + 3.85$, $r = 0.34$, $n = 21$). The clutch size varies seasonally, reaching a maximum about the end of May or beginning of June, annually and from place to place, no doubt reflecting in some way the favourability of conditions for breeding, though the relevant factors have not been isolated. Most of the studies have not been on ringed populations, but assuming that the same nest site is used by the same pair throughout the breeding season, and the few ringing studies support this, and that

Subspecies	Locality	Month
		J F M A M J J A S O N D
montanus	36–55°30′N	
transcaucasicus	36–43°N	
dilutus	29–51°N	
tibetanus	27°30′–42°N	
saturatus	31–53°N	
malaccensis	N of 15°N	
	S of 15°N	

Fig. 84 Breeding season data for Tree Sparrow

observers have correctly differentiated genuine new clutches from replacements after the loss of clutch or brood, the results suggest that in central Europe approximately 65% of pairs have two clutches (range in 13 studies 8–97%) and 25% have three clutches (range 4–66%), with one study in Belgium finding 3.5% of the pairs laying four clutches; Szlivka (1983) also reports a pair in Yugoslavia with four clutches. Again, there is considerable annual and local variation, as can be seen from the ranges given in parenthesis. The average number of clutches per pair per year apparently decreases with latitude, though there is only a weak correlation ($r = -0.24$, $n = 16$).

Incubation begins before the last egg is laid and is by both sexes, with the female taking the major share (about 70%) and the overnight stint. Incubation ranges from 11–14 days. The young are brooded intermittently until they are about 8 days old; both sexes take an almost equal share in feeding the nestlings. Deckert (1962) found a feeding rate of about 12 visits per hour for the first five days of life, increasing to 20 visits per hour with the older young; Szlivka (1983) recorded 10 visits per hour at day 5 and 32 per hour at day 10, with the male bringing the major share. Different observers give the fledging period as from 12–20 days; it seems probable that the shorter periods are the result of observer disturbance and that under more natural conditions the fledging period is 15–20 days. Nankinov (1984) remarks that, where nests are on the inside walls of wells, nestlings remain up to five days longer in the nest before attempting their first, demanding, vertical flight. It is likely that the length of the fledging period is influenced by ambient temperature; for example, according to a study in Japan by Abé (1969), time spent by young in the nest decreased from 17.5 ± 0.3 days in spring (mean temperature 13.5°C) to 14.0 ± 0.2 days in summer (mean temperature 20.5°C). Myrcha et al. (1973), from measurements of three broods, found that homoiothermy was not achieved by the nestlings until they were 13–14 days old.

As with the House Sparrow, young can be successfully reared by the female if she loses her mate. After fledging, the young continue to be fed by their parents for 10–12 days.

The breeding success figures given in Table 41 are to be used with some caution as they are obtained by different observers in different ways, some being expressed on the basis of total eggs laid, others on the success or failure of complete clutches. The results indicate an overall success ranging from 50–65%; once again there are considerable annual and local variations. Calculations based on these figures give an annual production per pair of 3–10 young birds, with an overall mean of 5.8 and an apparent tendency for decrease with latitude, though the correlation is rather weak ($r = -0.29$, $n = 14$).

238　*Tree Sparrow*

Table 41: *Tree Sparrow breeding statistics*

Locality	Clutch size Range	Mean	% of 1st clutches 2nd	3rd	4th	Av. No. of clutches	Incubation period	Fledgling period	Breeding success % Eggs hatched	Young fledged	Overall	No. of young/pair/year	Reference
Russia: Kursk 51.5°N		5.5											Eliseeva 1961
Poland:													
Dziekanów Leśny 52.3°N:													
1960–66	3–8	4.92	90	65	0	2.55			76.7	87.6	69.0	8.7	Pinowski 1968
1969–79		4.92	69	42	0	2.11			67.4	79.7	53.7	5.6	Pinowski & Wieloch 1973
Gdańsk 54.6°N													
1969–70		4.63	59	0	0	1.59			73.0	63.9	46.6	3.4	ditto
Kraków 50.1°N													
1969–70		5.11	37.5	8	0	1.46			80.1	94.7	75.8	5.7	ditto
Nowy Targ 49.5°N													
1969–70		4.91	8	0	0	1.08			86.5	74.8	64.7	3.4	ditto
Czechoslovakia:													
Bzenec 49.0°N													
1968–69	2–7	4.65	77.5	35	0	2.12							Balat 1971
Sokolnice 49.3°N													
1968–69			81.5	56.5	0	2.38			85.1	75.4	64.1	6.8	Balat & Touskava 1972
Bzenec 1970–71	2–7	5.00				2.65			76.9	80.9	68.4	8.8	
Sokolnice 1970–71	2–7	4.86				2.47			87.5	75.5	66.0	7.8	ditto
Břeclav 48.8°N	2–7	4.77											
Zidlochovice 49.0°N													
1971–74	3.7	4.80					12–19	13–16	92.6	71.2	65.9		Bauer 1875
Germany:													
Radolfzell 47.7°N 1965	5–7	5.51	97	35	0	2.12	11–13						Eisenhut & Lutz 1936
Dresden 51.5°N 1936–43	1–8	4.65	83	17	0	2.00	13–14	13–18	60.1	73.7	44.3	4.1	Creutz 1949
Motzener See 52.2°N							11–12	16–17					Deckert 1962
1965–69													
Wolfsburg, Lower Saxony (Niedersachsen) 52.4°N													
1968–69	3–7	5.56	48.8	4	0	1.53	11–14	12–17	61.9	86.2	53.4	4.5*	Scherner 1972a, b
Halle, GDR, 51.5°N	3–8	5.26							93.5	93	87		Schönfeld & Brauer 1972

Location												Reference
Rottenau 52.0°N 1969–71		4.96	87.2	50.9	0	2.38*				81.7	9.6	Kaatz & Oldburg 1975
Brenkhausen 48.4°N 1975	3–7	5.1	28	9	0	1.37	10–15	13–17	58.3	28.6	16.7	1.25	Gauhl 1984
Belgium: Marke 51.3°N	2–7	4.95	64.9	31.6	3.5	2.00	11–13	12–17.5	57.5	87.8	50.5	5.0	Bethune 1961
Great Britain: Oxford 51.8°N 1961, 1963–4	2–7	5.05	77.1	20.5	0	1.6			82.9	58.5	48.5	3.9	Seel 1968a, b, 1970
Yugoslavia: Vojvodina		4.3					10–16		79.2	(70–75)	(55–59)		Szlivka 1983
Spain: Guadaljara "El Seranillo" 40.8°N 1979–80		4.71				} 1.76							Sanchez-Aguado 1984
"La Bomba" 40.5°N 1979–80	2–7	4.95											
USA: Portage des Sioux 38.9°N 1968–73	2–7	5.02				2.67			72.8	71.8	52.2	7.0	Anderson 1975
Sikkim	4–6	4.7											Ali 1962
Tibet	3–5												Ludlow 1928
China: Beijing 39.9°N		5.6				2.0	10–12		85				Chia et al. 1963
Japan: Sapporo 43.3°N												9.1	Abé 1969 (quoted by Dyer et al. 1977), Abé 1970
Burma							14–18						Smythies 1953
Singapore 1.3°N	2–6	3.7											Ward & Poh 1968
Malaysia	3–6												Medway & Wells 1976

*recalculated from data in paper

SURVIVAL

The high breeding productivity shown in Table 41, implies that mortality, too, must be high if the population is to remain stable. Much of this mortality occurs with the young birds. Dyer et al. (1977) estimated that only 15–20% of young would survive to their first breeding season and only 12% for their first 12 months of life, the expectation of life increasing from 6 months at fledging to about a year for those birds surviving until their first December. Life expectation is not high, with only 3–4% of birds breeding for two years and 1–2% for three years.

Ringing shows that a number of individuals have survived for at least four years, though Creutz (1949) considered that Tree Sparrows have only exceptionally survived in the wild for more than four years. Deckert (1972) had one ringed bird in her study population that lived at least seven years and Bethune (1961) had one 10 years old. In captivity they may live for 10–12 years.

All the above mortality data for the Tree Sparrow have been obtained from western European and North American populations, where it is sympatric with the House Sparrow and where it occupies a habitat less well protected against severe weather conditions. The high turnover shown for the Tree Sparrow from these studies is thus not necessarily typical of the situation in southeast Asia.

BREEDING DENSITY AND NUMBERS

Pinowski and Kendeigh (1977) have summarised over 250 estimates of breeding density for the nominate race of the Tree Sparrow. These are arranged in Table 42 in order of decreasing population density for the same habitat types used in Table 28 for the House Sparrow. This shows that the Tree Sparrow, in its range of overlap with the House Sparrow, well merits its vernacular name of 'Tree' Sparrow, with maximum densities occurring in light woodland, orchards, parkland and open country with trees. The factors controlling breeding density have not been established, though

Table 42: *Tree Sparrow breeding densities in different habitat types.* (*Based on Kendeigh & Pinowski 1977, Appendix 3.2*)

Biotype	No. of census results	Breeding density (birds/ha)	
		Mean	Range
Deciduous forests over 50 years old with nest boxes	43	13.7	1.2–35
Old parks in villages and open fields with trees	46	5.0	0.2–22
Old orchards	14	4.5	0.5–11
Young orchards with nest boxes	4	3.0	0.5–7.5
Pine forests 1–50 years old with nest boxes	7	2.8	0.03–9.0
Riparian areas and cemeteries	25	2.3	0.1–7.2
Deciduous forests over 100 years old	11	2.1	0.1–4.3
Villages	14	1.9	0.2–6.9
Old parks in larger towns	30	1.4	0.2–3.0
Towns: small allotments and gardens	8	0.76	0.1–4.0
Deciduous forests 50–100 years old	8	0.58	0.1–1.2
Pine forests 50–100 years old	3	0.56	0.06–1.0
Deciduous forests 1–50 years old	20	0.44	0.05–1.6
Towns: suburban areas with one-family houses	9	0.39	0.1–0.9
Pine forests 1–50 years old	3	0.24	0.1–0.3
Towns: residential areas with apartment complexes	5	0.09	0.02–0.2
Towns: commercial and shopping areas	1	0	—

suitable holes for nesting may well be a limiting factor. I have already pointed out that the provision of nest boxes can result in the build-up of breeding colonies; for example, Boyd (1932) in England was able to increase the number of pairs nesting around a seven-acre field from a 'few pairs' to 24–25 pairs; Creutz (1949) from a 'few pairs' to 85 pairs in 50 ha in two years; Dornbusch (1973), in a 40 ha study area in pine plantations in Halle, GDR, built up a colony of 75 pairs, where none had previously bred, and Gauhl (1984) describes how a colony of 36 pairs was established in a bird reserve at Brenkhausen, FRD, where they had not bred before the nest boxes were put up.

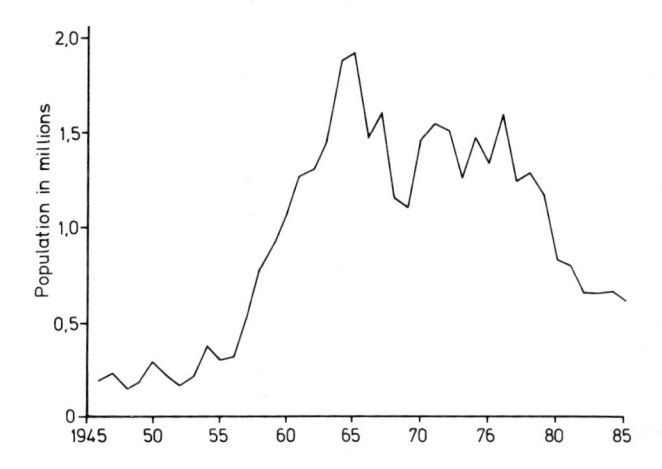

Fig. 85   Changes in Tree Sparrow population in British Isles, 1945–1985

As already mentioned, the numbers of Tree Sparrows in the British Isles have been subject to extreme fluctuations. I give my estimate for the changes in breeding population in Fig. 85 for the period 1945–1985, based on the British Trust for Ornithology ringing, nest record and farmland census data, supplemented by data from County bird reports and my own observations. I estimate the winter population at about 250,000 in the late 1940s; it began to increase dramatically in 1957, reaching a new plateau of about 1.5 million from 1961 to 1979, and then began to fall to its present level of about 750,000 birds. There is also some evidence of a decrease in northern Germany during the same period, with a decline in breeding numbers in Lower Saxony (Niedersachsen) and a reduction in passage birds on Heligoland (Helgoland) from an annual average of 58.4 birds for 1959–69 to 30 birds for 1970–80 (Moritz 1981).

MOULT

As with all members of the genus, the Tree Sparrow undergoes a complete post-juvenile moult. According to Stresemann and Stresemann (1966) this moult begins for the young from the earliest broods in Germany approximately eight weeks after they have fledged, reducing to five weeks for birds from later clutches. Bibby (1975) found a similar interval in England between fledging and the beginning of moult for birds from first broods, but with one third brood it was estimated that moult could have started as little as three days after they had left the nest. Bibby found the average duration of moult in juveniles to be 77 days (probably slightly less in adults), but

Month	J	F	M	A	M	J	J	A	S	O	N	D
Natural food availability (Ward & Poh 1968):												
breeding												
primary moult												
Superabundant food availability (Wong 1982):												
breeding												
contour feather moult												
birds in moult, %	8	6	3	7	0	11	14	50	33	29	27	8
primary moult												
birds in moult, %	0	9	0	27	17	0	14	25	33	21	0	0
Rainfall, mm	283	111	106	175	104	195	58	214	170	371	287	247

Fig. 86 Comparison of the effect of food availability on the timing of moult in two populations of Tree Sparrows in Singapore (more than 10% in moult ——— less than 10% in moult _ _ _ _ _)

with considerable variation between individuals. He found that the juveniles increased in weight 2.0–2.5 g during moult and suggested that this weight increase and moult might be physiologically related; thus a bird unable to obtain enough food to put on weight might also lack the energy for a fast moult. The duration of moult decreases in young from the later broods: those from first broods are fully moulted by the time they are 4 months old, those from the second broods by the time they are $3\frac{1}{2}$ months old and those from the last brood at age $2\frac{1}{2}$ months.

Adults begin to moult as soon as they have completed breeding activities. For European birds this means that the first individuals begin in late June or early July, the last in early September (Bibby 1975, Ginn & Melville 1983). Near Berlin, in north Germany, Deckert (1962) found that all birds in her study population had completed the moult by the end of October.

As soon as moulting is complete, in September–October, adults enter into a period of sexual recrudescence, indulging in defence of the nest and even sporadic nest-building, before winter sets in.

Less attention has been paid to moult in tropical populations, though, as in temperate regions, Ward and Poh (1968) found in Singapore that it followed immediately after breeding (May) and was completed by late August before conditions for survival appeared to deteriorate, suggesting that limitations in food availability kept these two energetically critical activities separate. However, in rather special circumstances at the Petamian University in Singapore, where a plentiful supply of food was available at a poultry unit throughout the year, Wong (1983) found that the breeding season was extended some 2–3 months and, although the peak of moult followed the end of breeding (August), some also occurred when the birds were involved with breeding. Data from the two studies are compared in Fig. 86. Wong considered that a factor controlling the timing of the breeding and moult cycles could have been the need to complete the moult before the period of heavy rains from October to January.

As soon as moult starts, the black bill fades to blackish horn and remains this colour until the onset of breeding in spring.

VOICE

The basic call of the Tree Sparrow is *chip*, higher pitched and quite distinct from the similar call of the House Sparrow; it is clearly monosyllabic. The call is used by the male to indicate ownership of a nest site and in pair formation to attract a mate. In these situations it may be elaborated to form a rudimentary song: *chip chip chicki, chip chi chit, chippa chippa chip.*

Variants of the basic call are used in contact situations, and again in a conversational type of singing in social gatherings in hedgerows and in pre-roosting activity.

A variety of other calls is used in different situations. These have been analysed in detail by Berck (1961, 1962) and Deckert (1962, 1968).

FOOD

The Tree Sparrow is essentially a granivore, specialising on the seeds of grasses, cultivated cereals and small annual herbs but, just before and during the breeding season, adults take a proportion of animal food and the young are raised largely on this diet. The animal food consists of insects and spiders. Initially, nestlings are fed on small prey, such as aphids, but, as they grow older, so larger prey, such as caterpillars, grasshoppers and beetles, particularly weevils, are given, the specific animals depending on local availability at the time.

Szlivka (1983) has published results of his analysis of the stomach contents of 541 birds (455 adults, 86 nestlings) taken in Yugoslavia. These are summarised in Table 43. This shows that adults feed more on weeds than on cultivated plants. Some animal matter is taken during spring and summer, but plays only a minor role in overall diet. In contrast, animal matter forms the major part of the diet of nestlings. Other workers have obtained similar results for nestlings. Abé (1970), in Japan, found that the proportion of animal food varied with the time of year (*ca* 67% by volume with spring nestlings, increasing to 81% in summer) and with the age of the young (89% for the first five days, decreasing to 69% for the second five and finally to 60%, the overall average being *ca* 72%). Chia *et al.* (1963) found that insects made up 91% of the food of the nestlings. Clearly these figures are at best only illustrative; much will depend on the availability of different foods. When fewer insects are available a greater proportion of vegetable food is given to nestlings, and with later broods cereal grains in 'milk' are included in food for the young. While adults are feeding their young on insects, they also take some animal food for themselves, as Table 43 shows, though not in the same proportion as they feed to their young.

The adults obtain much of the animal food for their nestlings by scrambling acrobatically among the leaves of trees, to a lesser extent on the ground and by catching flying insects in the air.

In parts of its range, particularly the Far East, the Tree Sparrow is considered a pest of rice and millet cultivation, though in cereal growing areas it takes a much

Table 43: *Stomach analyses of Tree Sparrows in Yugoslavia (Szlivka 1983)*

Sample	No.	Vegetable matter				Animal matter	
		Cultivated plants		Weeds			
		No. of stomachs in which present	No. of items	No. of stomachs in which present	No. of items	No. of stomachs in which present	No. of items
Adults:							
Jan–Mar	197	90 (52%)	192 (21%)	155 (79%)	727 (79%)	—	—
Apr–Jun	100	34 (34%)	48 (14%)	73 (73%)	142 (42%)	75 (75%)	150 (44%)
Jul–Sep	98	60 (61%)	232 (35%)	92 (94%)	386 (59%)	23 (23%)	37 (6%)
Oct–Dec	60	50 (83%)	157 (50%)	55 (92%)	143 (46%)	9 (15%)	14 (4%)
Total	455	234 (51%)	629 (28%)	375 (82%)	1,398 (63%)	107 (23%)	201 (9%)
Nestlings:							
May	86	—	—	20 (23%)	23 (13%)	77 (90%)	148 (87%)

smaller proportion of cereal grains than do the larger House and Willow Sparrows, specialising much more on seeds of grasses (especially *Panicum* sp.) and small annuals, such as *Chenopodium* sp. (*eg* goosefoot), *Polygonum* sp. (*eg* knotgrass), *Stellaria* sp. (*eg* chickweed); Grun (1975) actually lists 169 plant species that have been taken. Basecke (1949) gives a record of Tree Sparrows feeding on seeds of reeds (*Phragmites* sp.), a food also taken by some other sparrow species, including the Dead Sea Sparrow. Tree Sparrows, unlike House Sparrows, do not appear to eat fruits. Birds living in inhabited areas readily take bread and other scraps left by humans.

The economic importance of the Tree Sparrow as far as man is concerned is by no means clear. In China, it was declared a pest by Chairman Mao and in April 1958 three million citizens were mobilised for three days to wage a war against the sparrows by shooting, trapping, poisoning and by keeping the birds in the air until they fell exhausted. The writer Han Suyin (1959), who was visiting China at the time, gives a dramatic account of the campaign in Beijing, in which an estimated 800,000 birds were killed in the city and its immediate surroundings. The net effect was that the Tree Sparrow was almost eradicated in China. This was followed by serious losses of grain through insect pests and it was realised that the campaign against the sparrows had been an oversimplification of a complex biological balance. Subsequently, the sparrow was exonerated and even encouraged; it has since recovered and is once again a common bird throughout China.

Deckert (1962) reports how a population of Tree Sparrows, built up in orchards at Steckby, Halle, GDR, by putting up nest boxes, kept fruit trees free of pests, so that it was not necessary to spray them. An infestation of asparagus fields by the asparagus beetle (*Crioceria aspergi*) in the same area was similarly controlled.

16: Saxaul Sparrow *Passer ammodendri*

NOMENCLATURE
Passer ammodendri Gould, Birds of Asia, Vol. 5, 1862.
    'Turkestan'; the type locality is 'Djulek above Kyzl Orda on the Syr Darya' according to Hartert, Vög. pal. Fauna 1904: 158.
  Synonyms: *Passer stolickzae* Hume, Stray Feathers 1874 2: 516.
    About 4 miles east of Kashgar [Kashi].
  Passer timidus Sharpe, Cat. Birds Brit. Mus. 1888 12: 339.
    Gobi Desert.
Subspecies:
  Passer ammodendri ammodendri Gould 1872.
  Passer ammodendri stolickzae Hume 1974.
  Passer ammodendri timidus Sharpe 1888.
  Passer ammodendri korejewi Zarudny & Härms, Orn. Monatsb. 1902 10: 53.
    Eastern part of Transcaspia between the foothills of the Paropamisus and the Amu Darya.
  Passer ammodendri nigricans Stepanyan 1961.
    northern Sinkiang [Xinjiang].

There is currently little information on this last subspecies; it may be a synonym of *P. a. ammodendri*.

This bird occurs in remote parts of Asia and less has been published on it than on any other *Passer* species. (It is also the only one of the 20 species of *Passer* that I have not been able to study in the field.)

DESCRIPTION
The Saxaul Sparrow belongs to the 'black-billed' group. It is one of the larger sparrows, length about 160 mm. The sexes are sesquimorphic.

Passer ammodendri ammodendri
Male. Broad black stripe from bill through centre of crown to nape, flanked by broad band that is white in front of eye, becoming tawny behind eye and fanning out

on sides of nape. Black stripe from bill through eye, circling back of cheek. Cheeks greyish. Upper parts sandy grey, with black streaks extending to rump and upper tail. Black bib, spreading sideways on breast; remainder of underparts dirty white. Wings streaked brown and black with broad white bar. Bill black.

In freshly moulted plumage, black of nape and crown is obscured by ashy margins of feathers.

Female. Basic pattern similar to that of male, but colours are more muted and less contrasting. Crown streaked with grey, bib ash brown, eyebrow paler.

Passer ammodendri stolickzae
Male. Back buff instead of grey, rump and upper tail coverts unstreaked.

Passer ammodendri timidus
Male. Similar to *stolickzae* in general colouring, but stripe on sides of head paler, streaks on back wider and blacker; bill thicker and longer.

Passer ammodendri korejewi
Male. Similar to *ammodendri*, but rump and upper tail coverts unstreaked.
Female. Dark bib very faint, almost absent.

BIOMETRICS
Typical body measurements are given in Table 44.

Table 44: Biometric data for Saxaul Sparrow

Feature	Subspecies	Males		Females	
		Range	Mean (Median)	Range	Mean (Median)
weight	ammodendri		(28.1)*		
	korejewi	25.2–27.0	26.1*		
wing	ammodendri	72–81	77.9	71–77	75.5
	stolickzae	78–82	80	75–76*	75.5*
	timidus		ca 77.5		ca 77.5
	korijewi	70–77	73.9	70–75	72.4
tail	ammodendri	63–69.5	(66.3)		
tarsus	ammodendri	19–20	(19.5)	19–20	(19.5)
	timidus		ca 19		ca 18
culmen	ammodendri	10–13	(11.5)		
	timidus		(13)		(13)

DISTRIBUTION
Largely because of the remoteness of regions in which this species occurs, it has been relatively infrequently collected and records of its occurrence are somewhat scattered, both in space and time. The range of each subspecies is discussed separately and distribution maps have been drawn on the basis of such published records as are available. These may well neither reflect the complete nor the present distribution, particularly in areas, such as northern Sinkiang (Xinjiang), where developments in agriculture have recently given rise to significant changes in land use.

Passer ammodendri ammodendri
The nominate race is found in two widely separated populations. The western

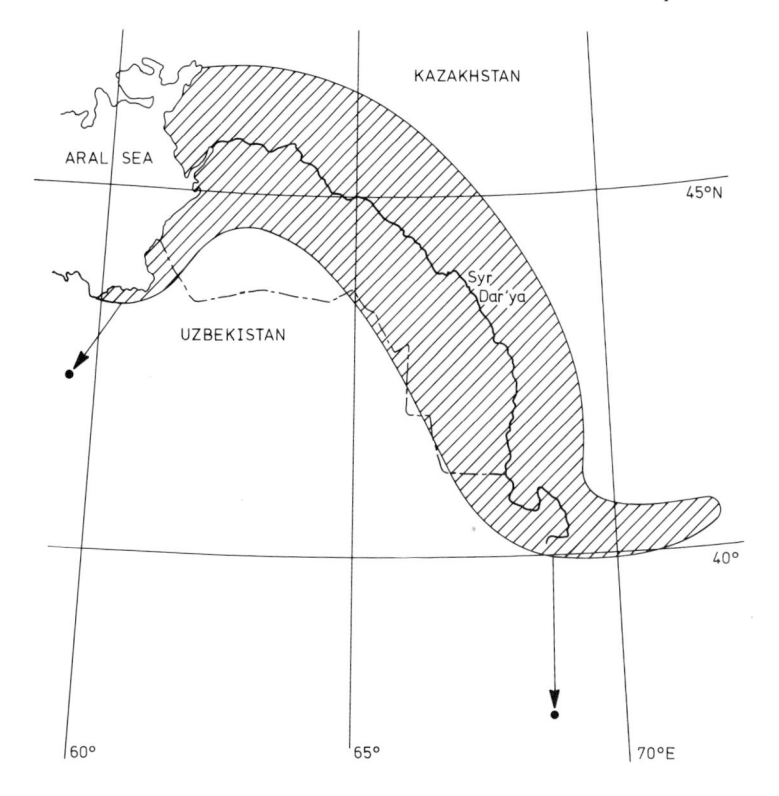

Fig. 87 Range of Saxaul Sparrow. P. a. ammodendri *(western population)*
Extra limital records •

population occurs in the northern Kyzylkum in Kazakhstan, from the eastern and southern shores of the Aral Sea along the course of the Syr Dar'ya as far as Ferghana and the foothills of the Alayskiy Khrebet. The precise distribution is unclear as, according to Dementiev *et al.* (1970), it no longer occurs at Dzhulek (= Djulek), the type locality. The presumed range is shown in Fig. 87. Occasional stragglers have been reported away from the breeding range: a female at Kungrad-Kul, in the Amu-Dar'ya delta, and a small flock 40 km from Dzhilikul' on the lower reaches of the Vakhsh river (Dementiev *et al.* 1970).

The eastern population occurs in eastern Khazakhstan, from Lake Balkhash, along the Ili river to the Issyk-Kul' district, eastwards to the Ketmen' mountains and just into China at I-Ning (Yining) in western Sinkiang (Xinjiang) Province; in the northeast it reaches to the Dzhungarskiy Alatau mountains. It is reported as common in the Ili valley, but elsewhere it is not numerous (Dementiev *et al.* 1970). This range is shown in Fig. 88.

Passer ammodendri nigricans
Birds in the Manas river valley in extreme northern Sinkiang (Xinjiang) to the northeast of the eastern population of the nominate race have been separated as *nigricans* (Fig. 89). The boundary between these birds and *P. a. ammodendri* is not clearly defined, but they may be separated by mountain ranges.

248 *Saxaul Sparrow*

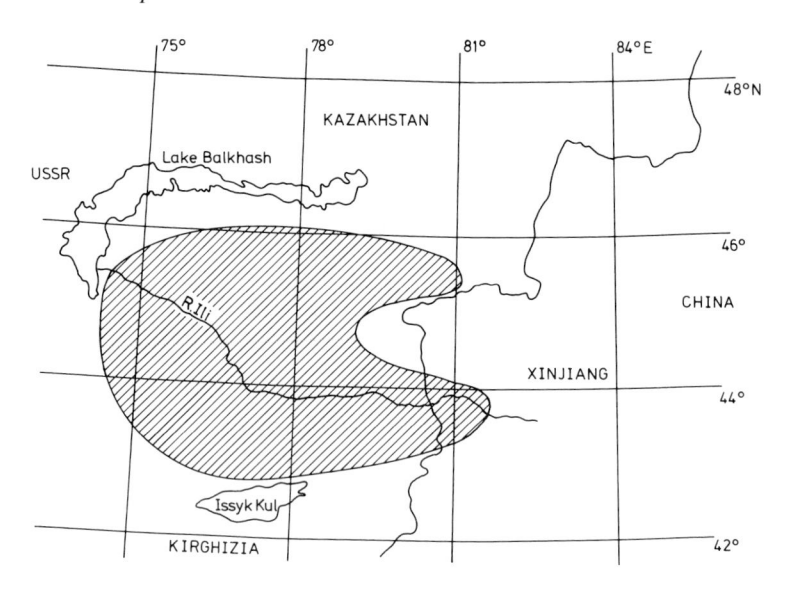

Fig. 88   Range of Saxaul Sparrow P. a. ammodendri *(eastern population)*

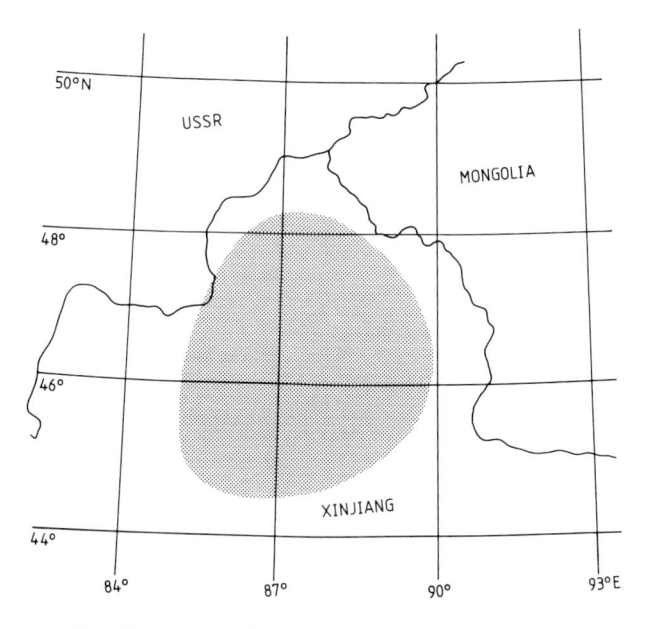

Fig. 89   Range of Saxaul Sparrow P. a. nigricans

Passer ammodendri korejewi

The race *korejewi* is separated from the western population of nominate *ammodendri*; it lies to the south of the Karakum, between the Amu Dar'ya in the north and, according to Hüe and Etchecopar (1970), just extending into the extreme northeast of Iran and the extreme northwest of Afghanistan in the foothills of the Paropamisus, though possibly only as a winter visitor. The range is shown in Fig. 90. It is said to be common at Repetek, but is only sporadically distributed and total numbers are small (Dementiev *et al.* 1970).

Passer ammodendri stolickzae
P. a. stolickzae has the most extensive and best-documented distribution of all the races of the Saxaul Sparrow. It occurs in Sinkiang (Xinjiang) from Kashgar (Kashi) in the west, along the Tarim river basin and the foothills of the Tien Shan (Tian Shan) in the north, and along the foothills of the Kunlun Shan in the south, but probably absent from the Takla Makan (Taklimakan Shamo), the desert region lying between the Tien Shan and the Kunlun ranges. From there it occurs eastwards along

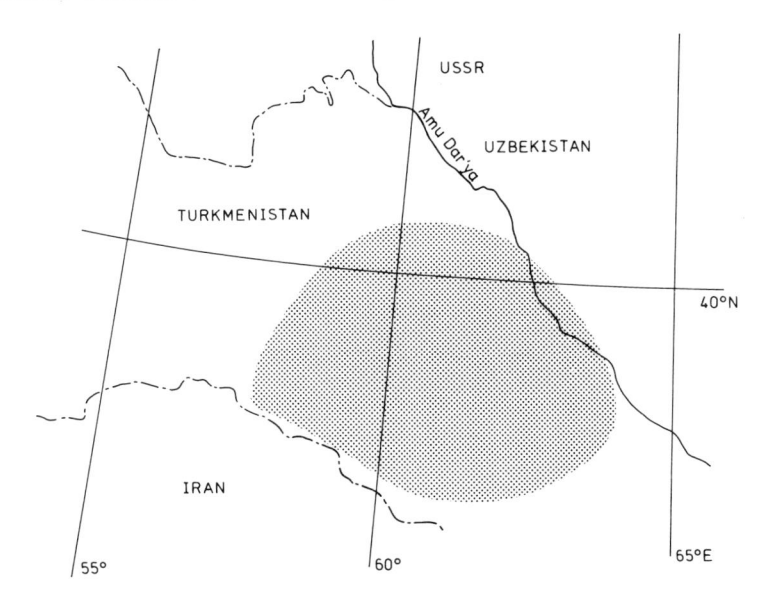

Fig. 90 Range of Saxaul Sparrow P. a. korijewi

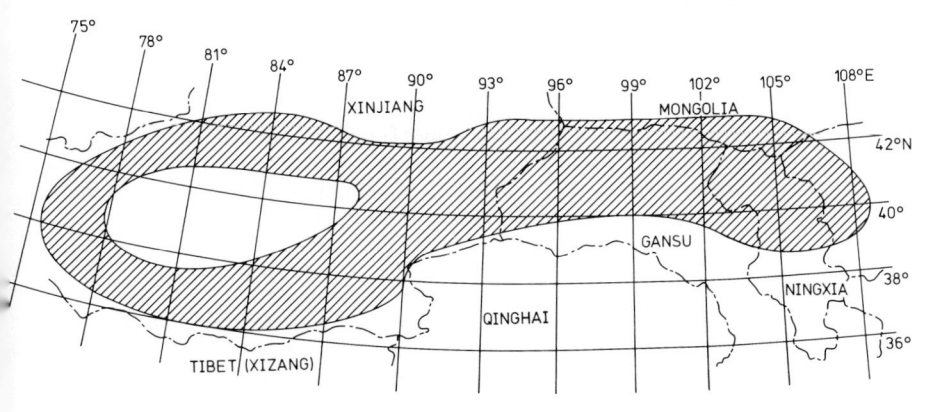

Fig. 91 Range of Saxaul Sparrow P. a. stolickzae

the northern foothills of the Altun Shan, north to Hami, through northern Kansu (Gansu) to the Ordos desert in western Inner Mongolia (Vaurie 1956, Dementiev *et al.* 1970, Zheng 1976) and just penetrates into extreme southern Outer Mongolia (Kozlova 1932). The range is illustrated in Fig. 91. This race is separated from the western population of the nominate race and from *nigricans* by the Tien Shan. According to Ludlow and Kinnear (1933) it is a common resident in Chinese Turkestan.

Passer ammodendri timidus
Once again, the distribution of this race is not clearly defined. It occurs in the western Gobi desert in southern Mongolia (Fig. 92). It appears to have a disjunct range, separated, according to Vaurie (1956), from *stolickzae* by the Gurvan Sayhan mountain range.

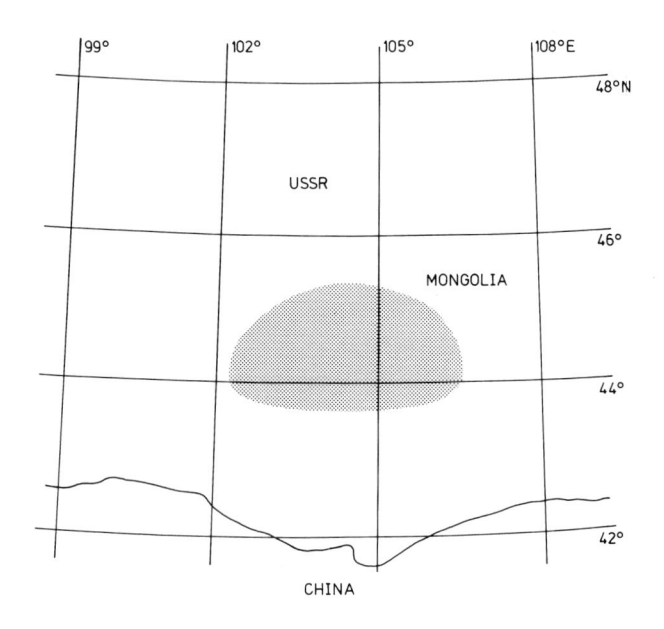

Fig. 92   Range of Saxaul Sparrow P. a. timidus

The range of the separate races are combined in the smaller scale map of Fig. 93; this shows how the six populations are discretely separated from each other. How far this is justified or is merely the result of lack of information from the rather remote parts in which this species occurs must await the result of further studies.

HABITAT
The Saxaul Sparrow is a bird of river valleys in desert or semi-desert country, where there are thickets of saxaul or teterophyllous poplar (*Arthrophytum* sp.), sand acacias (*Ammodendron* sp.), tamarisk (*Tamarix* sp.) or poplars (*Populus* sp.). From there it extends into the foothills of mountains, where there are shrubby areas, and to cultivated parts around oases, penetrating into settlements in winter, though

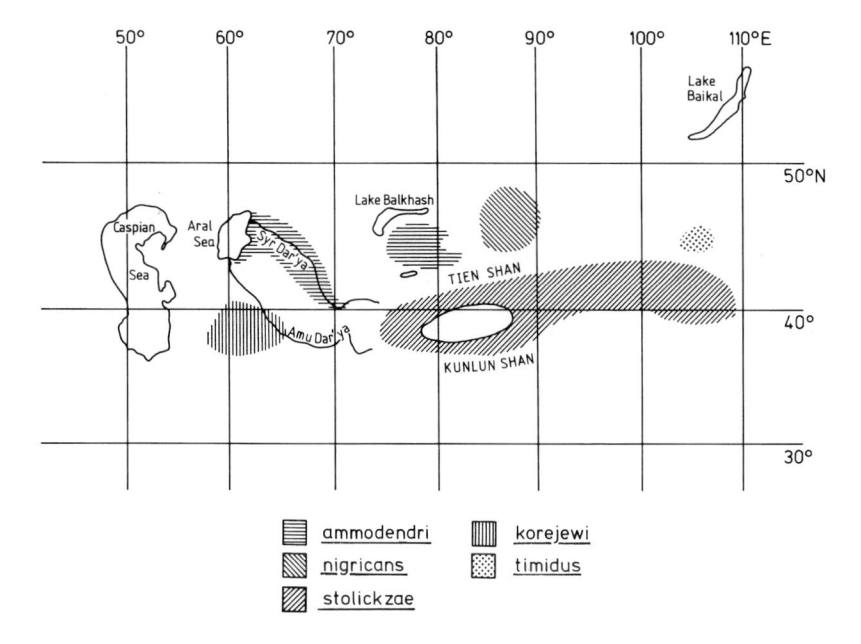

Fig. 93 Distribution of Saxaul Sparrow

according to Carruthers (1949) it is never far from saxaul trees or bushes. It is tied to water and may fly some distance to find it.

BEHAVIOUR

The species is social outside the breeding season. It is largely resident, though flocks may make small local movements during winter months, when numbers at any one place tend to fluctuate widely; at this time of year it may associate with its congeners: Willow Sparrow (Carruthers 1910a), House and Tree Sparrows (Ilyashenko 1979). It is a shy bird, spending much time among the foliage of bushes, making it difficult to see.

BREEDING BIOLOGY

The Saxaul Sparrow is predominantly a hole nester, most frequently using holes or hollows in trees, but also holes in rocky walls, earth banks, isolated uninhabited buildings and in foundations of the large nests of birds of prey. This choice of nest site tends to inhibit social breeding, though occasionally several nests may be found in one tree.

The winter flocks break up about April, individual pairs dispersing to find suitable nest sites. It appears that pairs form in the flock before seeking nest sites.

The height of the nest is determined by the location of the holes; in trees it can be quite close to the ground. The nest is a loose construction of grass, roots and other parts of plants, usually domed over with the entrance either on the side or the top, the structure filling the cavity in which it is placed, though in large cavities it can be left uncovered without a roof. It is lined with feathers and animal fur.

Subspecies	Locality	Month												Reference
		J	F	M	A	M	J	J	A	S	O	N	D	
ammodendri	Kyzylkum (western population)				▬	▬								Dementiev *et al.* 1970
	Semireche (eastern population)				▬	▬	▬							Dementiev *et al.* 1970
stolickzae	Mongolia					▬	▬							Kozlova 1933
	Sinkiang					▬								Ludlow & Kinnear 1933
korejewi	Tedzhen				▬	▬								Dementiev *et al.* 1970

Fig. 94   Breeding season data for Saxaul Sparrow

Published data on the breeding season are summarised in Fig. 94. Clutch size ranges normally from 4–6 eggs, exceptionally 3–7 eggs, with 5–6 most frequent. Two clutches each year are normal. Both sexes take part in incubation and in feeding the young, adults bringing food to the nest every 4–12 minutes. No information is available on either the incubation or the fledging periods. After the young leave the nest the family remains in the breeding area for some time before the juveniles disperse to join post-breeding flocks.

MOULT

The young have a complete post-juvenile moult, beginning in June or July and finishing in August with the last-hatched young. Adults begin their moult in the latter half of July and complete it by late August or early September.

VOICE

Little has been published on the voice. Hartert (1904–22) describes the call as sparrow-like! The song is said to be 'not loud, but pleasantly melodious with fairly diversified intonations' (Shnitnikov 1949). This is confirmed by Ilyashenko (1979).

FOOD

Like other sparrows, seeds form the main part of the diet, particularly seeds of the saxaul, plus insects and insect larvae; the following have been reported: Coleoptera, particularly Curculionidae (weevils), Orthoptera (grasshoppers), Lepidoptera (particularly caterpillars). Ilyashenko (1979), in a study of the food fed to nestlings in the Ili river valley, found that beetles predominated, mainly weevils (60%) and ladybirds (Coccinellidae) (30%), with the remainder made up by caterpillars and grasshoppers.

Because of its small numbers and restriction to desert areas, it is not a pest of agriculture (Dementiev *et al.* 1970).

17: Characteristics and interrelationships

Having summarised the available information on the individual sparrow species, I should now like to look at the genus as a whole, to consider the basic characteristics and the interrelationships between different species, largely as a preamble to discussing their origin and evolution. With the information given in previous chapters we are now in a better position to give a fuller answer to the question: 'What is a sparrow?'.

The sparrows are a very closely related group of birds; very similar in size, have similar, rather stout, conical-shaped bills, basically adapted to feeding on the medium-sized seeds of grasses and other small herbs, with a pronounced tendency towards the larger seeded grasses that are the progenitors of cultivated grains and, through this, to a close association with man the agriculturist, either directly by feeding on cultivated grains themselves or on the seeds of weeds of such cultivation. They are basically birds of grassland savanna, living and nesting in trees, but coming to the ground to search for food. The original sparrows probably built free-standing nests among the branches of trees, forming small colonies for breeding and larger associations outside the breeding season. There has been some radiation from this, most particularly in the use of man's buildings for nesting, that has resulted from the close association with man as a result of specialising on cultivated grains as a source of food, but divergence within the genus as a whole has been rather small.

A common characteristic of the sparrows is the black bill of the male during the breeding season. In most species it reverts to a brownish horn-colour as the male sexually regresses at the conclusion of breeding. In addition, it may also change to black following the post-breeding moult in those species (*eg* the House Sparrow) that show an autumn sexual recrudescence; it then fades to horn again with the advent of winter. It has been shown experimentally in the House Sparrow (Keck 1934) that the change from horn to black is caused by the release of androgens into the blood from the testes: the bills of males in breeding condition quickly revert to a pale ivory following castration, but the black can be restored by the injection of androgens (testosterone, androsterone or dehydroxanderone). The bill of the female House Sparrow also darkens at the height of the breeding season, or after the injection of androgens. This seasonal and sexual pattern is not, however, invariable throughout the genus. For example, in grey-headed sparrows, some adults of both sexes retain the black bill throughout the year, though the proportion appears to vary with the different members of the superspecies group, possibly as a response to some species-isolating mechanism.

The sparrows show their close relationship by the similarity of their pair formation and courtship displays, together with the plumage colour and patterns that are associated with them. Typically the male proclaims ownership of a nest site by calling beside it; this is the well-known *chirrup* call of the House Sparrow and the *chip* call of the Tree Sparrow that have closely similar corresponding calls in the other species. In the presence of a potential mate, the intensity and rate of calling is increased and calls may be strung together to form a rudimentary song. At its highest pitch, calling is accompanied by characteristic posturing. The neck and head are stretched, exposing the black throat patch of the black-bibbed sparrows, the white patch in the case of the Grey-headed Sparrow; the wings are held out slightly from the body, exposing the ruffled-up rump feathers that in most species are conspicuously marked – reddish brown to chestnut in eleven of the species, pale grey-brown to cream or yellowish green in the remainder. In most species the wings are drooped and shivered, but in the Golden, Chestnut and Dead Sea Sparrows they are raised above the back in a shallow V and flicked. The tail is raised and partially spread.

In courtship, the male hops round the female in this stiff posture, alternately bowing deeply and then stretching upwards. If the female is not ready for mating, she lunges at the male with opened bill and then flies off; the male pursues her and when she lands continues his exaggerated bowing display in front of her. In a number of the more social species other sexually active males are attracted by the headlong chase and fly to display in a group in front of the single female, all calling excitedly.

In threat, normally used against rivals of the same sex, the posture is very similar, but the feathers are sleeked rather than ruffled, the head is stretched forward with opened bill, the wings are not shivered, though they may be flicked, and rotated so that the scapular patch is directed towards the opponent; the scapular patch stands out in most species because of its chestnut colour and this is often enhanced by a white wing bar formed by the tips of the lesser coverts and a row of black spots or streaks.

When threatening, the male faces directly towards its opponent; in courtship, in contrast, the male does not face directly towards the female, but presents more of a side view as he parades around bowing in front of her.

Let us now look at some of the species interrelationships in more detail.

THE GREY-HEADED SPARROWS

The five species of grey-headed sparrow described in Chapter 1 are very similar, yet, as shown in Fig. 12, there are a number of areas of sympatry where the species remain separate and do not interbreed. This suggests that there must be subtle differences in habitat selection, food preferences or behaviour. I shall examine the different overlapping species pairs to see if the factors that allow them to co-exist without interbreeding can be determined. Table 45 compares the habitat preferences and sizes of the different species. I have attempted to illustrate the differences in the table by allocating a score of 0–5 for the various habitat types, and 0–2 for the nest-site types. Wing length and overall length are used as indicators of size and details are given of bill dimensions as a possible indicator of adaptations to different types of food. In the absence of sufficient detailed information from the zones of overlap, values in the table are taken generally from the range of the species where, in the absence of the putative superspecies competitor, the choice of habitat and the spectrum of food taken will have been broadened; nevertheless it should be possible to detect differences that could be selected for in the areas of overlap to reduce competition. The actual situation in these areas is described as far as it is known.

P. griseus ugandae/P. swainsonii

The Grey-headed Sparrow and Swainson's Sparrow have a small range of overlap in northern coastal Eritrea (Fig. 95). Over the remainder of Ethiopia they meet, but are segregated by altitude and habitat, the Grey-headed Sparrow occurring in light woodland up to 1,000 m, whereas Swainson's Sparrow is restricted to the plateau above this and occupies open country, cultivated land and villages. There is one apparently extralimital record for *griseus* from Sheikh in Somalia, well inside the range of *swainsonii*, but nearly 1,000 km from the recognised range of *griseus*; this could possibly have been a sport.

It will be seen from Table 45 that in addition to the slight difference in habitat preference, as highlighted by the situation at the zone of contact, Swainson's Sparrow, despite being a larger bird than the Grey-headed Sparrow, has a smaller and finer bill, suggesting that the two species are adapted to different sizes of seeds.

P. griseus ugandae/P. gongonensis

There are two zones of overlap between these species: one in Uganda and extreme western Kenya, Fig. 96, the other along the eastern half of the border between Kenya and Tanzania (see Fig. 12, Chapter 1). Differences between this pair are extreme: the Grey-headed Sparrow is one of the smallest members of the superspecies, the Parrot-billed Sparrow the largest. There are also differences in habitat preference: the Grey-

256 *Characteristics and interrelationships*

Table 45: Comparison of Grey-headed sparrows

		Grey-headed Sparrow (*P.g. ugandae*)	Swainson's Sparrow	Parrot-billed Sparrow	Swahili Sparrow	Southern Grey-headed Sparrow	
Habitat preference[1]:							
villages and built-up areas		2	3	0	$\frac{1}{2}$	0	
cultivated land and separated habitations		2	1	1	1	1	
grassland savanna and bush		$\frac{1}{2}$	1	4	$2\frac{1}{2}$	$1\frac{1}{2}$	
light woodland		$\frac{1}{2}$	0	0	1	$2\frac{1}{2}$	
Nest site[2]:							
holes		1	1	0	2	2	
openly in trees		1	1	2	0	0	
						diffusus	*luangwae*
Approximate size[3]:							
length	mm	150	155	180	150	155	—
wing length							
(males)	mm	83	86	94.5	86	83.5	78
(females)	mm	80	83	91	87	80	78
Bill dimensions:							
culmen length	mm	14.4	14.6	17.4	15.1	14.3	13.4
depth	mm	8.9	8.6	10.0	8.4	8.0	7.4
width	mm	8.5	8.0	9.5	8.3	8.0	7.6
depth:length		0.60	0.59	0.55	0.56	0.56	0.57
width:length		0.57	0.55	0.55	0.53	0.56	0.57
depth:width		1.05	1.08	1.05	1.01	1.00	0.97

[1] scored out of 5   [2] scored out of 2   [3] approximate mean value away from zones of overlap

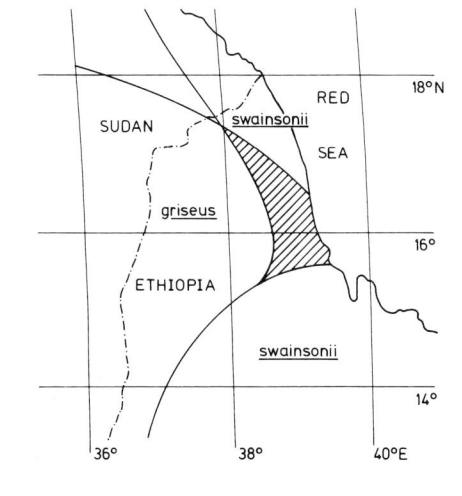

Fig. 95 Zone of overlap between Grey-headed and Swainson's Sparrows

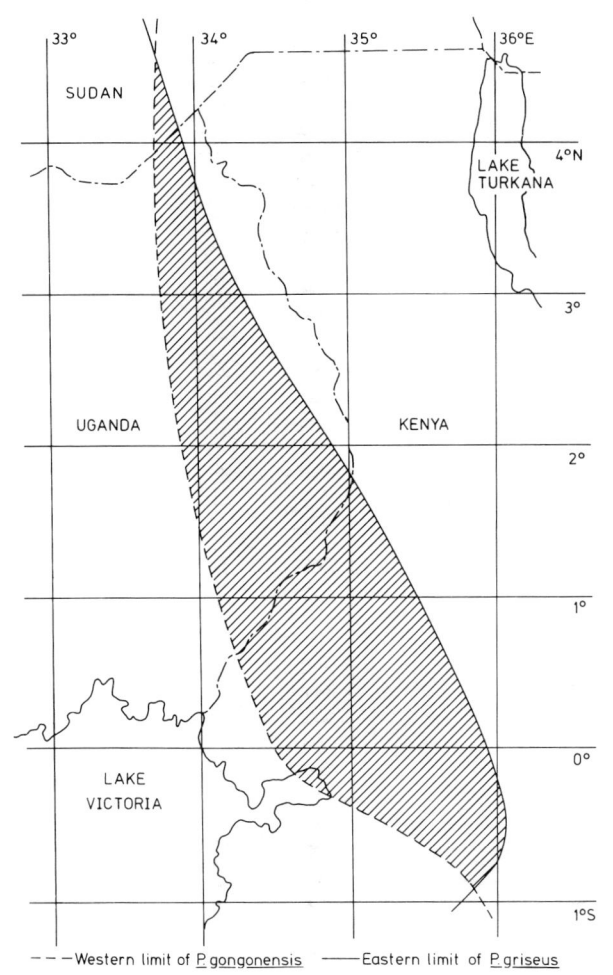

Fig. 96 Zone of overlap between Grey-headed and Parrot-billed Sparrows

headed Sparrow regularly occurs in towns and villages, the Parrot-billed Sparrow more in open country and only rather uncommonly does it come into inhabited areas; in addition, the Parrot-billed Sparrow is generally found in more arid areas than the Grey-headed Sparrow. As the Parrot-billed Sparrow seldom, if ever, uses holes for breeding, there is no reason for competition over nest sites. It is somewhat surprising that the cline in size (both body and bill sizes) in *gongonensis* should be from east to west, with the smallest birds closest to the range of *griseus*, but perhaps the birds are already sufficiently differentiated in other ways for there to be no selective pressure to emphasise the size difference.

P. griseus ugandae/P. suahelicus

The Grey-headed and Swahili Sparrows have a small zone of overlap on the east side of Lake Victoria, possibly extending as far as Nairobi, and a more extensive area in central Tanzania, where the range of the Grey-headed Sparrow extends eastwards

258 *Characteristics and interrelationships*

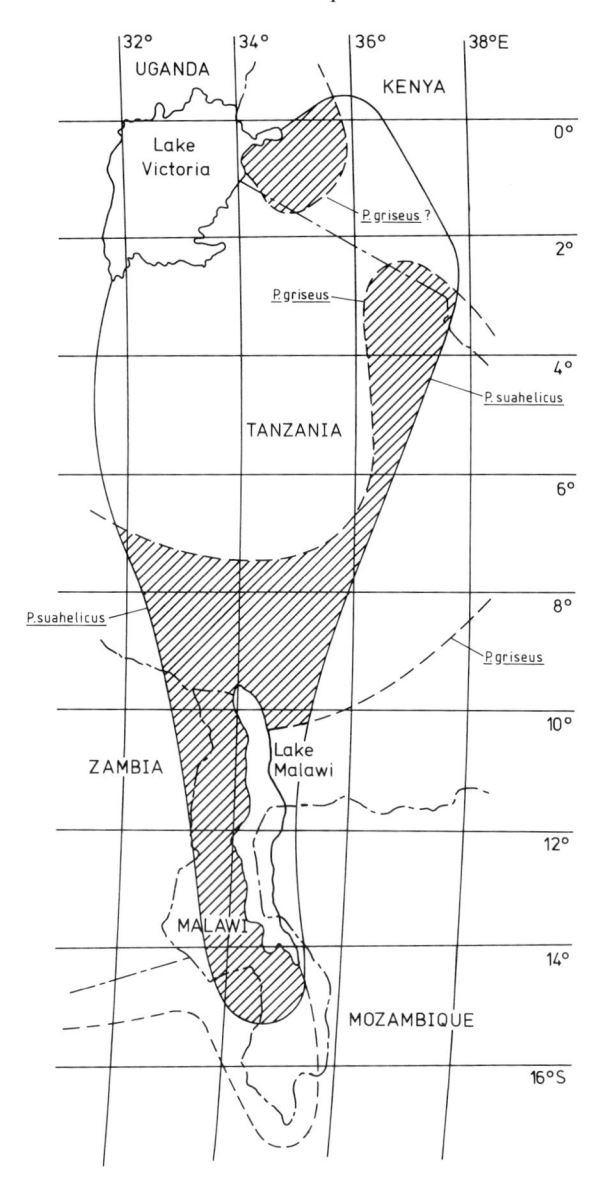

Fig. 97 Zone of overlap between Grey-headed and Swahili Sparrows

to reach the coast at Dar-es-Salaam; the latter has the appearance of a recent extension of range, Fig. 97. Nothing has been reported about the ecology of these two species in the regions of overlap, but, although they are rather similar in size, there are clear differences in habitat preference and nest-site requirements that should allow them to exist sympatrically; differences in bill proportions suggest that they may also be adapted to different seed types.

P. griseus/P. diffusus

There are no less than four zones of overlap between these two species: in central coastal Angola, Fig. 98; in the Luangwa valley in central Zambia, where the isolated *P. diffusus luangwae* population lies completely within the range of the Grey-headed Sparrow, Fig. 99; an extensive area to the south of this on the southern border of Zambia, where the Grey-headed Sparrow meets the northern limit of the main range of the Southern Grey-headed Sparrow, Fig. 100; and, finally, in eastern Tanzania.

Table 45 shows that there are clear differences in habitat preference between the species; this is shown both in Angola, where the Grey-headed Sparrow is confined to the villages and the Southern Grey-headed Sparrow occupies a drier habitat in open country (Traylor 1963), and in the Luangwa valley where the Grey-headed Sparrow is strictly in association with African villages and its congener confined to mopane woodland in the hot valley floor (Benson et al. 1973). The situation in southern Zambia is more complex. Benson (1956), Irwin and Benson (1967), and Hall and Moreau (1970), suggested that, although there is a zone of sympatry in southwest Zambia, the two species hybridise to the east of this in the middle Zambezi valley. On the other hand, Dowsett and Dowsett-Lamaire (1980) in a more recent study involving a larger range of specimens, consider that the two taxa behave as good

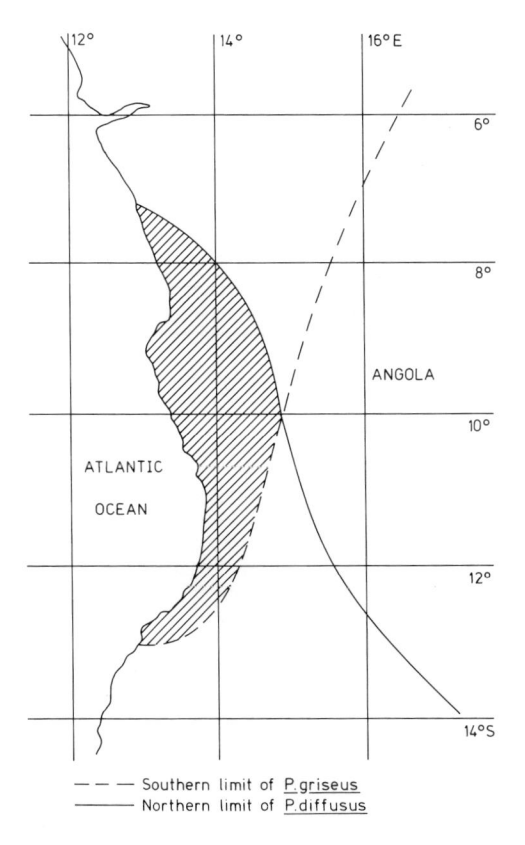

Fig. 98 Zone of overlap between Grey-headed and Southern Grey-headed Sparrows in central coastal Angola

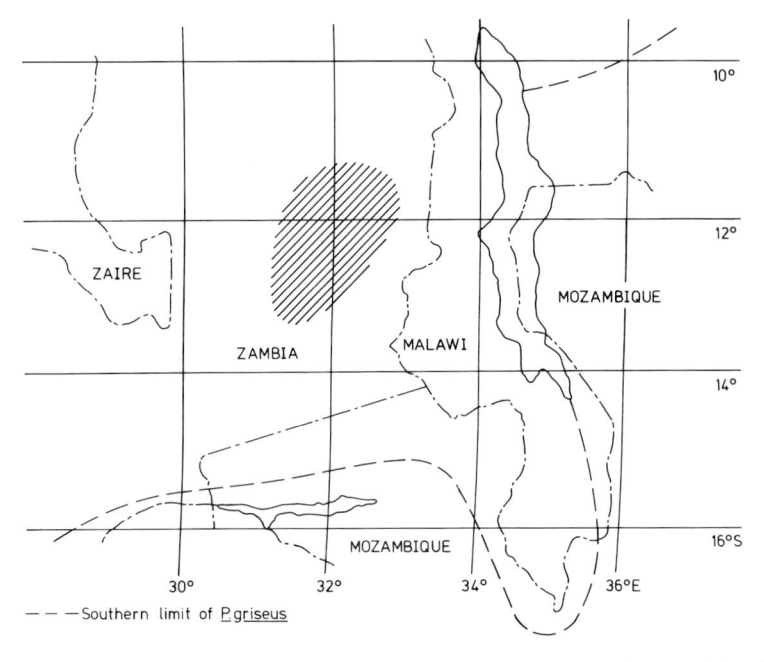

Fig. 99 Zone of overlap of the isolated P. diffusus luangwae *population of Southern Grey-headed with Grey-headed Sparrow*

species over the whole southern province of Zambia, and they point out that in the area of contact one can still recognise more than the 5% of parental phenotypes recommended by Short (1969) as an indication of the lower limit of specific distinctiveness. In the absence of further information I have shown (Fig. 100) a hybrid zone lying to the east of this; it seems likely that if hybridisation does take place there

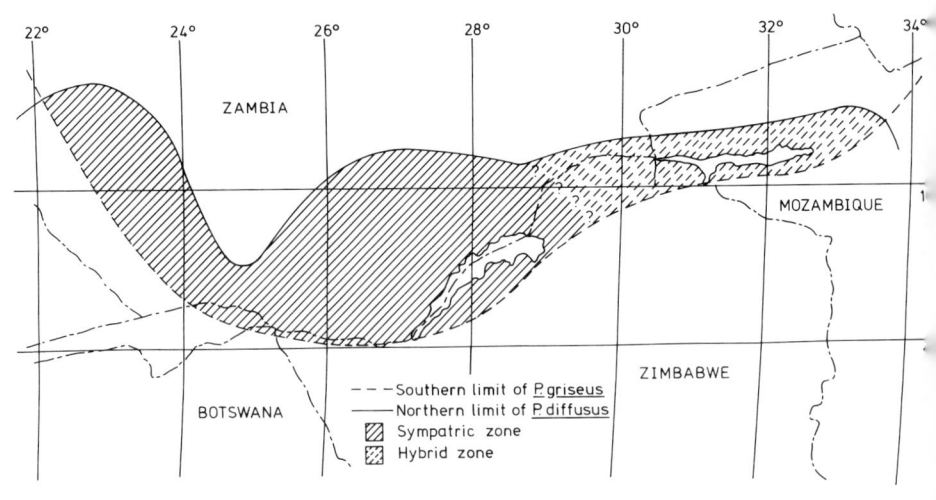

Fig. 100 Zone of overlap between Grey-headed and Southern Grey-headed Sparrows in Zambia, Zimbabwe and Mozambique

it must be the result of lack of ecological diversity in the habitat that is responsible for the breakdown in reproductive isolation. Further study is, however, required to elucidate the situation that has recently become more complicated by the arrival (about 1955) of the House Sparrow, a competitor of the Grey-headed Sparrow for the built-up environment.

Specimens of the two species have been collected in the same area in northern coastal Tanzania (see Fig. 12, Chapter 1), but no study has been made of their respective ecological requirements in this region.

In addition to differences in habitat preference, there are also clear size differences. The isolated race of the Southern Grey-headed Sparrow *luangwae*, is much smaller than the *ugandae* race of the Grey-headed Sparrow, both in overall size and also in bill dimensions (Benson *et al.* 1973) and, as already pointed out in Chapter 1, the Southern Grey-headed Sparrow, though generally larger than the northern species, is abnormally small in the region of overlap in Angola. As White and Moreau (1958) point out, the cline of decreasing size in the Southern Grey-headed Sparrow is in conformity with Bergmann's rule (reduction in size in the warmer parts of the range), whereas the more northerly Grey-headed Sparrow is not modified as would be expected. They suggest from this that the Grey-headed Sparrow is probably a comparatively recent invader. This would have resulted in a selective advantage for the small, Southern Grey-headed Sparrows, which in combination with separation by habitat preferences would help to maintain reproductive isolation. The differences in adult bill coloration (Chapter 1, Tables 1 and 9) may have some significance in this connection.

P. swainsonii/P. gongonensis

There is an extensive zone of overlap between Swainson's Sparrow and the Parrot-billed Sparrow in southern Ethiopia, Fig. 101. In this area Swainson's Sparrow is

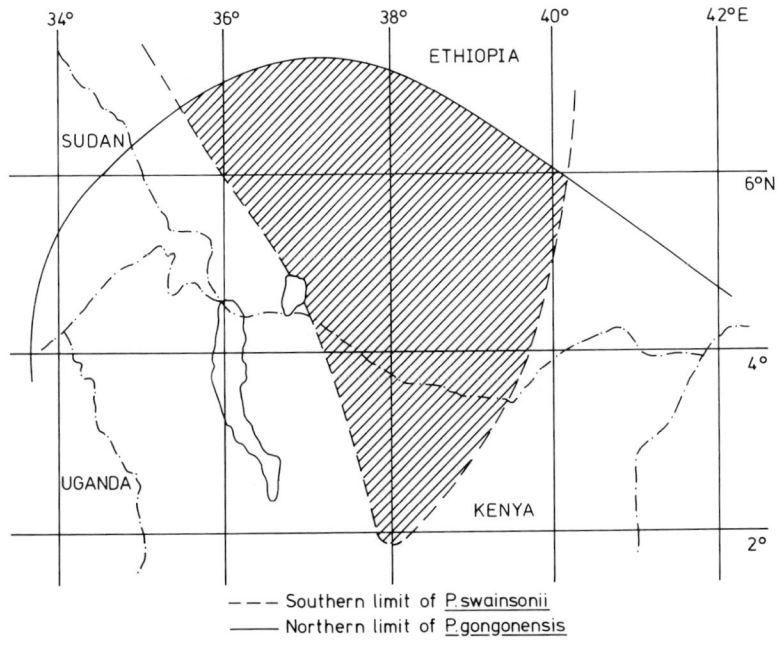

Fig. 101 Zone of overlap between Swainson's and Parrot-billed Sparrows

262 *Characteristics and interrelationships*

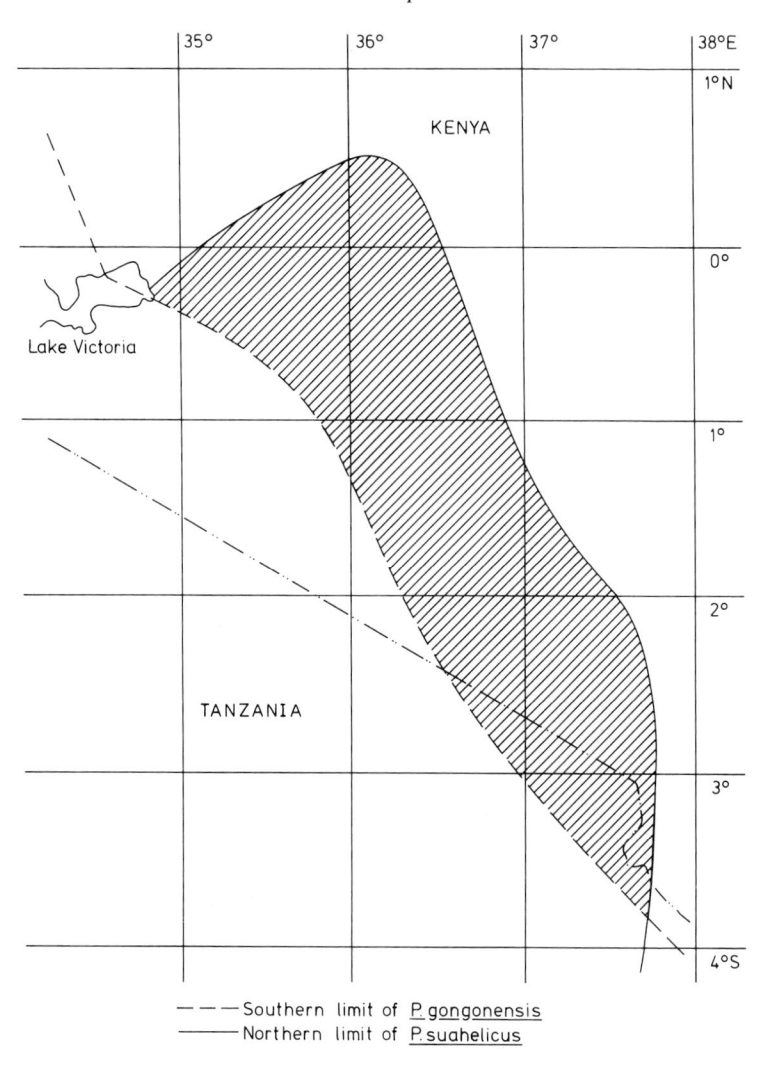

Fig. 102   Zone of overlap between Parrot-billed and Swahili Sparrows

associated with higher altitudes (as it is in western Ethiopia at its contact with the Grey-headed Sparrow) and the Parrot-billed Sparrow is found in the more arid plains. Moreover, the differences between these species are as great as between any pair in the superspecies. Yet in places there appears to be some intergrading for which the subspecific names *turkana* Granvik, *tertale* Benson and *jubaensis* Benson have been proposed, though all these are now recognised as synonyms of *gongonensis*.

P. gongonensis/P. suahelicus

A somewhat similar situation exists between the Parrot-billed Sparrow and the Swahili Sparrow on the southern boundary of the range of the former on the border of Kenya and Tanzania, Fig. 102, though here the two species, with an extreme difference in size, appear to be entirely separated and no intergrades have been reported.

It would appear that the overlap in the ranges of the grey-headed sparrows has been comparatively recent, with the Grey-headed Sparrow perhaps spreading more rapidly. During the period of isolation, sufficient differences in habitat and feeding adaptations, as judged by the bills, not only in overall size, but also in relative proportions, reinforced by rather minor differences in plumage, have allowed them to co-exist with relatively little intergradation in areas of sufficient ecological diversity. This is clearly a dynamic situation that deserves further study for the light it can throw on avian speciation.

GOLDEN AND CHESTNUT SPARROWS

The Golden and Chestnut Sparrows share a number of characteristics that distinguish them from the remaining sparrows. Both are highly social birds that have adopted a nomadic way of life to cope with an irregular food supply that depends on capricious rainfall; they appear to be mutually exclusive, the Golden Sparrow occupying the arid zone to the south of the Sahara and the Chestnut Sparrow replacing it, except for a small area of overlap in west central Sudan, in the arid region of southern Sudan and northern Kenya that would appear to be suitable Golden Sparrow habitat. The two species are similar in size and have a very similar plumage pattern, with yellow and chestnut replacing each other; they also have a similar display at the nest, with wings raised above the back, whereas in all other species, except the Dead Sea Sparrow, the wings are drooped. The raised wing-flicking makes the displaying bird particularly conspicuous. The similarity of the display of the Dead Sea Sparrow is probably a result of convergent adaptation; all three species nest in somewhat similar conditions – a group of trees in a larger area of similar habitat – where a conspicuous display has obvious advantages in forming a breeding aggregation.

The Chestnut Sparrow is tolerant of less arid conditions than the Golden Sparrow, comes more freely into man's settlements and, above all, is a partial nest parasite.

SOUTHERN GREY-HEADED AND CAPE SPARROWS

The Cape Sparrow is largely sympatric with the Southern Grey-headed Sparrow. The latter tends to occupy open acacia woodland, whereas the Cape Sparrow is adapted to open country, regularly occurring in cultivation and fully at home in close association with man, living in gardens and freely nesting on buildings. The Southern Grey-headed Sparrow is not a town or village bird and, though it will nest on isolated habitations, it tends to be displaced from such sites by the more urban-adapted bird.

RUFOUS SPARROWS

The Rufous Sparrow is sympatric over its range with the grey-headed sparrows, but the two are largely separated by habitat, the former occupying more arid areas. The Rufous Sparrow appears to be largely independent of man-altered habitats, seldom occurring in inhabited areas. Unlike the other arid-adapted Afrotropical sparrows, the Golden and Chestnut Sparrows, with which it overlaps in the Sudan, the Rufous Sparrows are largely sedentary and non-social, coping with the arid environment by living at low density, with the pairs widely spaced.

The other member of the rufous sparrow group, the Iago Sparrow, is the only endemic sparrow of the Cape Verde Islands. On islands, where species are not in competition with their congeners, they tend to occupy a greater range of habitat than would otherwise be the case. This is certainly so when one compares the range of habitats occupied by the Iago Sparrow on the Cape Verdes with that occupied by Rufous Sparrows on the African mainland. The situation on the Cape Verdes, however, has been confused by the appearance of the two other sparrows, the Willow Sparrow, that probably arrived in the early years of the 19th century, and the House Sparrow that arrived about 1920. This presents a unique opportunity for studying the relationship between these three sparrows in a region of limited habitat diversity, where they make up no less than 25% of the passerine species that breed in the archipelago.

The Iago Sparrow is clearly well adapted to an arid environment, in contrast to the Willow Sparrow that, as I have shown in Chapter 10, is adapted to much more humid conditions, though it also moves into the built-up environment in the absence of the House and Tree Sparrows. On São Tiago, the largest and most varied of the islands, there is some separation of the two species, the Iago Sparrow occurring in the more arid parts, areas of marginal cultivation with small trees, and the smaller villages, the Willow Sparrow in the richer, cultivated land with larger trees and in the larger built-up areas.

On the smaller islands there does, however, appear to be some competition. This apparently resulted in the extinction of the moisture-loving Willow Sparrow on Brava during the droughts of the 1940s (though it has recently recolonised) and the converse on Fogo, where the Iago Sparrow has never been recorded and the Willow Sparrow reigns supreme.

On Branco, where the Willow Sparrow has only been recorded once (Frade 1976) and on Santo Antão, which it has just colonised and where the situation has not yet had time to stabilise, the Iago Sparrow occupies all available habitats. On Fogo, the Willow Sparrow occurs throughout, except in woodland on the north side. It seems almost certain that the endemic Iago Sparrow must have been present at some time on Fogo, but possibly the Willow Sparrow arrived during a pluvial period, when its arid-adapted congener was at a low level, and was able to displace it completely, just the converse of what occurred in Brava during a period of drought.

The situation on the other islands is less clear cut. On Boa Vista and Maio the Willow Sparrow seems to have decreased in numbers and it is possible that it is now extinct on São Nicolau, with the Iago Sparrow prospering at its expense. The Willow Sparrow is, however, a mobile species that would be expected to recolonise in the event of conditions becoming more suitable for it, as indeed it has done on Brava.

Finally, we have the extraordinary situation of three *Passer* species on São Vicente, a comparatively small island (*ca* 25×20 km), very arid and of limited ecological variety. The House Sparrow appears to do no more than hold out in the town of Mindelo. Unlike other introductions of this species, where, after a period of consolidation, it has spread over unsuitable habitat to found new colonies, in the Cape Verdes it has not even spread to the village of Porto Novo on Santo Antão, within sight and connected by a regular ferry, nor to Praia, the major town on São Tiago, that would be expected to provide a suitable habitat for it. Even on São Vicente its numbers are small, so that the population growth to fill the available habitat and create pressure for dispersion has not occurred; in fact it has not even displaced the endemic Iago Sparrow from the urban habitat of Mindelo.

The occurrence of three sparrow species on São Vicente would appear to be an unstable situation; with the drought conditions that have persisted for the last 20 years it is perhaps not surprising that the Willow Sparrow, the least tolerant of the three species to arid conditions, is the one that first began to hybridise with the House Sparrow (as it does in North Africa in similar conditions) and has even possibly died out. The situation of the sparrows in the Cape Verdes would appear to be a very finely balanced one that has been maintained by alternating cycles of drier and wetter conditions. With continuing drought the endemic Iago Sparrow is likely to hold its own and become the dominant sparrow on all the islands except for Fogo.

SOMALI SPARROW

The Somali Sparrow is yet another arid-adapted species that over much of its range in Somalia is sympatric with none of the other African sparrows. It is a social species, well adapted to living in close association with man. The Kenyan population is sympatric both with the Parrot-billed Sparrow and the Kenyan Rufous Sparrow, but here it tends to occupy the village habitat and there appears to be little competition with its congeners.

THE HOUSE SPARROW IN THE AFROTROPICS

The whole southern African sparrow situation has been disturbed by the comparatively recent introduction and spread of the House Sparrow. In most parts this has led to some displacement of the Southern Grey-headed and Cape Sparrows from inhabited areas, though the situation everywhere is by no means as straightforward as this. In towns an ecological balance seems to have been established between Cape and House Sparrows. In some areas all three species manage to live together and nest on buildings at isolated settlements, both native species successfully defending nest sites against the invader. In contrast, in southern Cape Province there is apparently a different dynamic situation: according to V. Pringle (*in litt.*) the invasion by House Sparrows first resulted in a decrease of Cape Sparrows in Bedford, but in turn the House Sparrow appears to have been displaced as the Southern Grey-headed Sparrow moved in. This may well be another example of the initial dynamism of an invading species that is lacking in the more established parts of its range. A more detailed study is needed to sort out the ecological relationships between these three species.

PALAEARCTIC AND ORIENTAL SPARROWS

The northern hemisphere is dominated by three sparrows: the House and Tree Sparrows, that are perhaps less closely related than some other sparrows of this area, but have evolved separately as urban species in close association with man, and show an interesting relationship in areas in which they are now sympatric; the Willow Sparrow that, although closely related to the House Sparrow, has evolved a different life strategy that enables it to be very successful, despite occurring almost entirely within the range of the latter.

The Willow Sparrow has adopted essentially the same strategy as the Golden Sparrow to exploit a similar, though more structured, food situation that exists in the 'Mediterranean-type' climate zone of the southern Palaearctic. This area is characterised by a short, lush spring, when conditions are suitable for breeding, followed by a long, dry summer when they become increasingly difficult. Like the Golden Sparrow it is an opportunistic breeder that forms large colonies where conditions are suitable, and then moves to another area, usually to the north, where spring comes later, for successive broods. This results in large concentrations that live a nomadic existence to obtain food, though where the climatic succession is more regular this can take on the pattern of a more typical south to north migration. In this way the Willow Sparrow is able to live sympatrically with the sedentary, and closely similar, House Sparrow in an area where the spring flush is greater than can be exploited to the full by its sedentary congener.

Further to the north, however, where the spring flush is not so concentrated and summers are not completely dry, the sedentary House Sparrow population is more fully able to exploit the more evenly distributed seasonal availability of food and the Willow Sparrow does not occur.

In the absence of the House and Tree Sparrows, for example over most of Sardinia, on Madeira, the Canaries and the Cape Verdes, the Willow Sparrow becomes sedentary and expands into urban habitat as well.

As I shall show in the next chapter, I consider the House and Tree Sparrows to have evolved separately in close association with man, the former in the Palaearctic, the latter in the Orient, both subsequently spreading to overlap over much of their ranges. The relationship between these two sparrows, adapted to the same urban habitat, in the area of subsequent overlap is thus of great interest.

The separation of the two urban sparrows, the House and Tree Sparrow, in their range of overlap in the Palaearctic region is well shown by comparison of Tables 28 and 42. The larger House Sparrow (male weight 25–35 g) is clearly dominant over the lighter Tree Sparrow (male weight 20–25 g) in areas of human habitation. The Tree Sparrow then becomes a bird of parkland, farmland and light open woodland, though in many places it is able to retain its urban habitat, not only where it is a resident and well established in its nesting sites before the arrival of a migratory population of House Sparrows (*eg* in Afghanistan, Kazakhstan), but even in many parts where both species are non-migratory. This is particularly the case where, perhaps, both species are approaching the limits of their range – in the high Himalayas, in the high Swiss Alps, in northern Siberia and in northern Europe. Some villages or farm complexes have become the sole preserve of House Sparrows, whereas in others the Tree Sparrow is supreme, nesting in the buildings. The precise ecological relationships between the two species in such areas have still to be established. In some places they even breed harmoniously in close proximity. Schnurre (1930) described both species nesting side by side in the holes of a Sand Martin colony in Germany, and Price (1979) remarked that the small isolated population of Tree Sparrows in the Eastern Ghats of India, bred side by side with House Sparrows in thatched roofs.

The situation in North America, where both species were introduced at about the same time, is of particular interest. Although the Tree Sparrow has survived as a colonist, it has not been particularly successful, its present range being restricted to an area of some 22,000 km^2 north and west of the point of its original introduction. According to Anderson (1978, 1980), in America the two species are syntopic; that is, not only do they overlap in range (sympatric), but also in their utilisation of habitat, with no differentiation in use of food resources, in contrast to the situation in most of Europe where they tend to separate in habitat. Anderson suggests that competition for nest sites may be the major factor in limiting the Tree Sparrow population, with this species only able to maintain a foothold by exploiting smaller items of invertebrate food when feeding its nestlings. It is odd that it has not avoided competition by moving to the more rural type of habitat that it occupies in Great Britain, but perhaps there are endemic native species that have inhibited this. Anderson also draws attention to the possibility that the low genetic variability of the small founding innoculum could have seriously reduced the potential of the Tree Sparrow for adaptation to American conditions. The situation between Tree and House Sparrows in Australia, with the former still largely restricted to the areas of its original introduction, is very similar to that in North America. The failure of the Tree Sparrow to colonise the extreme south of Europe and the Mediterranean islands could well be the result of competition with the House Sparrow, which, as I have pointed out, tends to spread more into a rural habitat away from habitations in these parts.

The Tree Sparrow is unique among the black-bibbed sparrows in that the female has adopted the black-bibbed 'male' plumage. A tempting explanation is that sexual monomorphism in the Tree Sparrow has evolved as a mechanism of species isolation. This will be discussed in the next chapter; here I shall merely consider cases of hybridisation with other species when the species-isolating mechanisms have clearly broken down. There are a number of reports of hybrid Tree × House Sparrow individuals: *viz* Lake District, England, 1892, Suffolk, England, 3.1.1894 (Macpherson 1919–20); Essex, England, 1.5.18 (Nichols 1919); Zwickau, Germany, April 1928 (Meise 1934); Dorset, England, 27.9.55 (Rooke 1957); Norfolk, England, 19.4.56 (Richardson 1957); Białowieża, Poland, 13.12.65 (Ruprecht 1967). Finally, there have been several records from Fair Isle, Scotland: a possible hybrid in 1962 (Davis 1963), a definite one in 1977 (Waterston 1978) and up to five in 1980 (Arnott 1981).

These birds were either shot or trapped and closely examined in the hand. They showed a combination of plumage characters intermediate between males of the two species: crown grey mottled with chocolate, indistinct black cheek patch, double wing bar (Rooke 1957, Ruprecht 1967); ash-grey crown, distinct black cheek patch, single wing bar (Richardson 1957). Although a mixed pair has been bred in captivity (Meise 1951), nothing is known of the paternity of the wild birds referred to above. Ruthke (1930) found a nest with five eggs at Lübeck in Germany that was attended by a male House Sparrow and a female Tree Sparrow; the clutch was abandoned before it hatched and thus it could not be confirmed if this was a genuine case of mixed mating. A few cases of mixed breeding have, however, been observed in the wild. Cheke (1969) induced a pair of Tree Sparrows to foster a clutch of House Sparrow's eggs; the young were ringed before they had fledged. The following year in the same area, he caught a ringed House Sparrow male attending a clutch of Tree Sparrow's eggs that later hatched out three Tree Sparrows and two hybrids; a second clutch in the same nest box hatched out two Tree Sparrows and two hybrids. The female was not seen, but was presumed to be a Tree Sparrow that had been fertilised on both occasions by males of the two species. In 1982, Albrecht (1983) witnessed courtship behaviour between a male Tree Sparrow and a female House Sparrow at Eregli on the Black Sea coast of Turkey; this culminated in an apparently successful copulation and could possibly have resulted in a hybrid brood, though it was not possible for the subsequent fate of the pair to be followed. Lastly, in 1979, Hume (1983) saw a male hybrid and a female House Sparrow feeding young in a nest in North Yorkshire, England.

These observations suggest that assortative mating between Tree and House Sparrows can produce not only viable, but also fertile offspring. Two other hybrids involving the Tree Sparrow have been reported. On Malta, a hybrid Tree × Willow Sparrow was trapped at Hal Far on 7.12.75 (Smith & Borg 1976) and a second was taken at Valletta on 18.12.76 (Sultana & Gauchi 1982).

The examples of hybridisation, or possible hybridisation, from Fair Isle, Turkey and Malta, and the previously discussed case of the hybrid House × Somali Sparrow in Somalia, have one common feature: the occurrence of odd birds of one species in an area where the other is common. In the absence of a partner of its own kind, pairing with the other species becomes more likely, even apparently in the case of the Tree Sparrow overcoming the inhibition imposed by the 'male'-type plumage of the female, in which a female Tree Sparrow would appear as a male to a male House Sparrow; while, conversely, the appearance of a female House Sparrow would be 'wrong' for a male Tree Sparrow. One must assume that, even in the absence of the 'correct' plumage, the behaviour patterns associated with mating are sufficiently close in these related species to be acceptable.

The remaining Oriental and Palaearctic species are closely related to the dominant House and Tree Sparrows and probably owe their rather limited distributions to competition with one or both of the dominant relatives. Two of these species show a tendency to associate with man, no doubt through adaptation to seeds similar to those of cultivated grains, and are able to penetrate the outskirts of inhabited areas. The Pegu Sparrow of Indo-China forms mixed breeding colonies with the Tree Sparrow. For example, at the Bang Phra Nature Reserve in Thailand I saw both Pegu and Tree Sparrows nest building in the same house eaves. At the Horse Serum Farm, about 5 km away, both species were breeding in close proximity, the former in free-standing nests in mango trees, the latter in the nearby office block and stables; the adults were collecting food for their young from different places and did not appear to be in competition. Again, at Hat Karon on Phuket Island, a Tree Sparrow was breeding in a nest box on a beach hut that was close to the sugar palms in which Pegu Sparrows were breeding; once more the two species ignored each other. The Cinnamon Sparrow also comes into built-up areas, but where the more urban House and Tree Sparrows are present it tends to keep to the rather less built-up areas with open gardens, though in villages where its congeners are lacking it becomes the complete 'house sparrow'.

The only northern sparrow that appears to have remained confined to the ancestral savanna-type habitat is the Saxaul Sparrow. This species has not been well studied, but seems to be restricted to arid grassland steppe, where it is not in competition with other sparrows, except possibly for a small range of overlap with the Desert Sparrow in Kazakhstan, though the latter favours an even more arid habitat and tends to be more closely associated with human settlements. The Saxaul Sparrow, however, is likely to be under increasing pressure as modern agricultural techniques allow penetration by cultivation of its traditional ranges in Kazakhstan and Sinkiang.

Finally, we have the radiation of two northern sparrows into riverine or lacustrine habitats: the Dead Sea and Sind Jungle Sparrows. These two sparrows are alone among northern sparrows in that they never make use of holes for nesting, still retaining the ancestral habit of building free-standing nests in trees.

The Dead Sea Sparrow had, until recently, a very restricted distribution, with the nominate race largely confined to a small area in the Rift valley depression of the Levant, where ambient temperatures were so high that only limited incubation of the eggs was necessary and the birds built a nest of coarse twigs in dead trees that allowed cooling of the eggs by free circulation of air during the hottest part of the day. With the growth in cultivation in Israel a plentiful supply of food suitable for the Dead Sea Sparrow was provided by the associated weeds of cultivation. This resulted in the rapid spread of the Dead Sea Sparrow out of the Rift valley into areas of lower ambient temperature. The species has demonstrated the ability to adapt quickly to the need for a more normal pattern of incubation and the use of leafy trees for nesting.

Adaptability is a key word as far as sparrows are concerned. In addition to the striking expansion of the Dead Sea Sparrow, there has been the dramatic change in habitat utilisation by the Cape Sparrow in the vineyards in Cape Province, South Africa, and the remarkable extension of range by the Southern Grey-headed Sparrow, again in Cape Province. The latter is probably also associated with changes in agricultural practices, though the nature of these has not been identified.

Another characteristic in which the sparrows show great adaptability is in the choice of nest site. The situation is summarised in Table 46. No less than 13 (65%) of the 20 species use holes as well as building nests openly in trees; many of them make use of the old nests of other species: in most cases swallows (Hirundinidae), though the Chestnut Sparrow specialises in several species of weaver (Ploceidae). With the exception of the Saxaul Sparrow, only the species least associated with man retain the exclusive use of free-standing nests in trees.

Only three species appear in the table as exclusively using holes; this is unexpected

Table 46: Types of nest sites used by different sparrow species

Region	Species	Free-standing nest in branches of tree	Hole nest (tree, cliff, earth bank, building)
Afrotropical	luteus	only type used	—
	eminibey	only type used, significant nest parasite	—
	griseus	commonly used	commonly used
	swainsonii	commonly used	commonly used
	gongonensis	predominant type	not reported
	suahelicus	—	only type reported
	diffusus	—	only type reported
	melanurus	commonly used	commonly used
	motitensis	predominant type	one race (rufocinctus) also uses holes
	iagoensis	reported	predominant type
	castanopterus	commonly used	commonly used
Palaearctic	simplex	commonly used	commonly used
	domesticus	commonly used	most common
	hispaniolensis	most common	commonly used
	moabiticus	only type used	—
Oriental	pyrrhonotus	predominant type used	has been recorded
	flaveolus	commonly used	commonly used
	rutilans	not common, but used	most common
	montanus	not common, but used	most common
	ammodendri	—	only type used

in the case of the two members of the grey-headed sparrow superspecies and may only reflect the lack of published records. The same could apply to the little studied Saxaul Sparrow.

Most bird species are restricted to one type of nest site: hole nesters have generally greater breeding success (Nice 1957), but the availability of suitable holes may impose a restriction on breeding in otherwise suitable localities. The versatility of sparrows in the use of both types of nest gives maximum flexibility in exploiting breeding possibilities; it is remarkable that so few species, let alone genera, have been able to adopt this breeding strategy.

The breeding strategy of small passerine birds has been the subject of considerable discussion. Some species are single brooded, usually laying large clutches, while others have a prolonged breeding season during which several smaller clutches are laid. The House and Tree Sparrow provide an extreme example of the latter strategy, with up to three or four clutches per year in temperate regions and even six for the House Sparrow in the tropics.

Unlike single brooders, such as the tits (*Parus* sp.), which time their breeding activity to a flush of a particular prey species, the House and Tree Sparrows feed their nestlings on a large variety of invertebrate prey that occurs over a period of many months, but which, particularly in temperate regions, may suffer considerable variations in numbers depending on climatic fluctuations. The most productive clutches place the greatest physiological demands on the adults. McGillivray (1983) found that, with the House Sparrow in Canada, high productivity in one clutch delayed the initiation of a subsequent clutch and also reduced the productivity of that clutch. In general, the second clutch, occurring when conditions are typically most favourable, has the greatest productivity, but in years when conditions are poor at this time, the reduced physiological cost of the failed second clutch increases the chances of successful later broods that together with the early, less successful, first clutches, ensure a high overall annual productivity. The avoidance of the physiological cost of preparation for migration in sedentary populations of sparrows allows the possibility of extending the breeding season to the extreme as well as permitting the earliest possible initiation as soon as conditions become favourable in spring. The prolonged breeding season with multiple broods, with the chances of at least some broods coinciding with favourable conditions, would appear to be the most productive strategy for these sparrows, as indeed their success in colonising new areas bears witness. Golden and Willow Sparrows breeding in more extreme conditions have, as already described, adopted a different strategy with a nomadic way of life adapted to accommodate short, intense periods of food availability.

Adaptability is also shown by sparrows in a number of other ways: feeding methods provide a good example. As well as dehusking seeds with their specially adapted bills, sparrows have evolved a number of techniques for catching the invertebrate prey that is, principally, fed to their young: searching, stalking, pouncing, flycatching. This latter activity occurs not only during the day, but has also been recorded frequently at night, when House Sparrows, at least, collect moths and other insects attracted to lights. In addition, a number of rather more unusual methods of obtaining food have been reported, again mostly for the House Sparrow: searching spider's webs for flies (Rossetti 1983), searching the radiators of parked cars for insects (Hobbs 1955a), turning over stones in gravel paths to locate seeds underneath (Wade & Rylander 1982), kleptoparasitizing digger wasps *Sphex ichneumomeus* as they carried prey back to their burrows (Brockman 1980), kleptoparasitizing Blue Tits that were flying off with bread from a trap set to catch sparrows, but which the sparrows were too cautious to enter (personal observation), carrying a hard crust of bread and dipping it into water to soften it before eating (Clarke 1949), flushing insects from trees and bushes by shaking clusters of leaves (Guillory & Deshotels 1981). Further, at least in the case of the urban sparrows, the considerable ability to learn from other species. Tits (Paridae) were first to exploit milk bottles, delivered to doorsteps in Britain, by pecking through the waxed cardboard closures (and now the aluminium bottle tops), but House Sparrows were quick to copy them, not only in feeding from bottles that the tits had opened, but opening the bottles themselves (Fisher & Hinde 1949). House Sparrows have also learned to copy tits and Greenfinches in clinging to suspended nut feeders; this was first recorded from Germany in the late 1950s (Feldmann 1967, Hill 1977), but was soon followed in England in the late 1960s (Chatfield 1970) and in Canada in 1976 (Boyd 1978). This technique has recently been mastered by the Tree Sparrow.

The one characteristic, however, that sets the sparrows apart from all other bird genera is their widespread association with man. This has developed to a remarkable degree in the two urban sparrows, the House and the Tree Sparrow, that over most of their widespread ranges are almost entirely commensals of man; to such an extent in the House Sparrow that in some places it lives as a wild animal almost entirely under cover in man-made habitats. The birds north of the Arctic Circle in Norway spend the winter days, when there is no natural daylight, inside cattle byres where they find light, warmth and food. Five pairs of House Sparrows even started to breed in November in Dumfriesshire (Dumfries and Galloway), Scotland, in the artificial conditions of a heated turkey house, where they obtained both food and water (Young 1962). Other individuals have taken up residence in large buildings, such as factories, stores, railway stations and airport terminals, and one pair even bred successfully down a coal mine, 640 m below ground (Summers-Smith 1980a).

Many other members of the genus also freely move into settlements and even towns, although not exclusively tied to this habitat. The Willow Sparrow becomes a complete 'house sparrow' in the absence of its two urban congeners. All except three Afrotropical sparrows freely nest in built-up surroundings: the Desert Sparrow uses buildings and wells for nesting, and two of the Oriental sparrows share the outer limits of the built-up environment with the House and Tree Sparrow during the breeding season.

Pair faithfulness is uncommon among small passerine birds. It has been demonstrated in the House and Tree Sparrow, at least in sedentary populations, by colour-ringing. House Sparrows and Tree Sparrows, over much of their range, remain at their nesting places throughout most of the year and it is probable that nest-site faithfulness is the key to mate faithfulness, rather than retention of the mate *per se*. It is not known how far pair faithfulness is characteristic of the genus, though it certainly does not seem to be with migratory Willow Sparrows that breed in Kazakhstan. Gavrilov (1962) found that after the breeding season there birds formed flocks largely composed of one sex, and he concluded that it was probable that they migrated and remained in sexually segregated flocks over winter, basing this on reports by Shulpin (1956), who noted flocks in autumn passage south of Kazakhstan consisting only of males, and by Meinertzhagen (1940) who found the same thing in wintering flocks in Morocco.

A number of studies involving a range of species has shown that pairs that remain together for more than one year have greater breeding success than birds of the same age that are breeding for the first time. This has not been established for the House Sparrow or Tree Sparrow, but it could be another advantage to the species conferred by sedentariness and nest-site retention that assist pair faithfulness.

DISPERSION, COLONISATION AND RANGE EXTENSIONS

Where sufficient data are available on breeding success, as it is for the House Sparrow (Table 27) and to a lesser extent for the Tree Sparrow (Table 41), it is clear that productivity varies considerably, not only from place to place, but also from year to year (emphasising the caution that must be used in placing too much weight on the limited data that are available for the remaining species).

In well-established, settled parts of the range in temperate and subtropical areas, studies show that little change in breeding numbers of House Sparrow occurs over a period of years, suggesting that the population is at the maximum carrying capacity. Lowther (1979b) has shown in a study in Kansas, USA, that the relative population size at isolated farms is determined by 'habitat quality', for which invertebrate food availability during the breeding season was an important parameter. It has to be stressed, however, that we do not know what factors determine the absolute breeding carrying capacity.

In those areas to which the House Sparrow is best adapted, productivity and survival of adults and young is more than sufficient to maintain the numbers at breeding capacity. In addition to the short-term dispersal that occurs after the breeding season (Lowther 1979c; Fleischer *et al.* 1984), the population increase leads to the emigration of a proportion of the surviving young, both as an extension to the autumn dispersal and again at the beginning of the breeding season when birds of the year attempt to take up nest sites. These movements are not obvious in areas where there is a stable House Sparrow population, though they are shown by the appearance of flocks at places where they are not normally seen and for that reason tend to be noticed and recorded; for example, on the coasts and on islands where there is no breeding population. The arrival of small flocks in spring and early summer on offshore islands round the coasts of Britain is a regular occurrence. The surviving young birds provide a non-breeding surplus, which, although not obvious in normal circumstances, is evident by the way in which gaps that occur in the breeding population are rapidly filled.

The situation is different in the more northerly parts of the range where there is a severe winter kill and numbers may be reduced below the summer carrying capacity. For example, Pinowski and Pinowska (1985) found that the Tree Sparrow population in their study area in central Poland ($52°20'N\ 20°50'E$) was correlated with the severity of the preceding winter, in particular the number of days when snow cover was greater than 10 cm and birds were denied access to the weed seeds on which they feed. In such areas the overall population may be lower than the potential breeding capacity and may even depend on immigration from more successful parts of the range. At the northern extremity of the range, breeding colonies may not be permanent, surviving in these marginal areas only for a few years after an invasion. In a similar way, populations in tropical regions may also fall below the potential breeding carrying capacity through limiting conditions at a dry or very hot time of the year when food is difficult to obtain.

With the maximum mortality of adults taking place during the prolonged breeding season (in temperate populations at least), and the lower survival rate of the young from the later broods, it would seem on the face of it disadvantageous to have multiple broods. The fact that this strategy has not been selected against, suggests that it must be more beneficial in terms of recruitment of offspring to the breeding population than the alternative of a single annual brood with presumably the opportunity of prolonging the breeding life.

This combination of high productivity and dispersal gives the possibility of rapid recovery following occasional natural disasters that result in dramatic population decreases: severe storms (see Bumpus 1899, New England, USA; Johnston 1967b, Kansas, USA) and epidemic disease (see Stenhouse 1928 – Fair Isle and Shetland; Menegaux 1919, 1920, 1922 – southern and central France). In addition, it provides an explanation for the somewhat paradoxical rapid expansions of the House Sparrow, that most sedentary of species (sedentary that is in the case of the two races, *domesticus* and *indicus*, that have been involved), following introductions to new areas.

After an introduction, or the expansion into a new area following the creation of suitable conditions, as happened with man's colonisation of Siberia and the setting up of new agricultural communities, the breeding success of new House Sparrow colonies in the absence of competitors and predators, is often dramatic and results in the creation of large population surpluses. It seems probable that these excess young birds form the pioneering bands that lead to rapid expansion. Although in some cases of introduction the expansion has been assisted by the deliberate or involuntary actions of man (in North America, for example, by transplantations and by the birds hitch-hiking on grain cars on the railways), there are frequent examples of colonisation having taken place across long stretches of quite unsuitable country for House Sparrows. This was probably the case in Mexico, where the average annual advance was 200 km (Wagner 1959), and almost certainly so with recent expansions of range in Costa Rica (Reynolds & Stiles 1982), Bolivia (Dott 1986) and the Amazonia region of Brazil (Smith 1973, 1980), as well as in Botswana, Mozambique, Zimbabwe and Zambia (Harwin & Irwin 1966); in Zimbabwe ringed juveniles that had settled down were found more than 100 km from their birthplaces (Irwin 1981).

In some way these wandering flocks must be able to assess the suitability of new areas, not only areas that are inhabited by man and thus probably suitable, though lacking sparrows, but also areas that although suitable are already at their House Sparrow population limit. Craggs (1967), in a detailed study of House Sparrows on Hilbre Island in the Dee estuary, Cheshire, England, over a period of several years, made some very relevant observations as far as the above hypothesis is concerned. Hilbre, a small island (*ca* 1.2 ha), carried a population of 5–7 pairs of House Sparrows when it was inhabited by a farmer who kept a few fowls, the food for which helped to sustain the sparrows through the winter months. Craggs colour-ringed the adult population and many of the young birds raised on the island. Each year there was a sharp fall in numbers in September, when the birds of the year emigrated, and at no time were any new, unringed birds seen on the island, even during the period of autumn and spring movements. The situation changed suddenly when, four years after Craggs had begun his study, the farmer's domestic stock was run down and chicken food was no longer available; not only did the sparrows cease to breed, but following this a regular passage of birds was observed, particularly in autumn (September–November), though also at a lower level in spring (March–June). It is reasonable to suppose that sparrows had flown over the island when the resident birds were breeding, but apparently they assessed it as unsuitable and either flew straight over or only made the briefest of stops, and were not recorded. Once the breeding population had disappeared, however, passing birds made longer stays, but of course did not recolonise the island now that it was no longer suitable for them. An interesting postscript to the Hilbre story is that domestic poultry were re-introduced ten years later and within a few years a breeding colony of House Sparrows re-established itself and built up to seven pairs (Craggs 1976).

In areas that have been newly colonised, there are numerous records of House Sparrows turning up at isolated habitations and then moving on, having apparently assessed them as unsuitable, and of new breeding colonies being established at considerable distances from previously recorded ones; intermediate, less suitable localities being 'in-filled' later as the overall population builds up. Davis (1973) reports how House Sparrows turned up at the Hastings Reservation of the University of California, USA, 24 km from the nearest breeding group: in the period 1939–74 there were four records in October and four in March–April, significantly the times at which dispersal tends to occur.

In addition to deliberate introductions by man and the natural spread that has occurred in the places of introduction, the House Sparrow shows a remarkable ability to make use of man's transport in extending its range. The involuntary hitch-hiking in box cars carrying grain on the American railways has already been mentioned.

There are also several well documented cases of House Sparrows making long journeys on board ships. For example, the Falkland Islands were colonised in 1919 by a number of sparrows that came aboard whaling ships from Montevideo, Uruguay, a journey that lasted five days (Bennett 1926, Hamilton 1944); the Faeroes were probably colonised by birds that were said to have arrived on a ship in about 1933 (Williamson 1945). Lund (1956) reports on sparrows travelling in the holds of ships during the period of spread into Finnmark north of the Arctic Circle in northern Norway, and he considered that, with towns over 200 km apart, this was the most likely way that the colonisation had taken place. Vik (1962) records how six House Sparrows successfully made the 17-day journey from Oslo, Norway, to New York City, USA, in July 1960, being fed and watered by passengers on board ship. The extent to which House Sparrows can survive at sea is shown by the four that joined a ship at Bremerhaven, Germany, on 14th July 1950, and did not disembark until they reached Melbourne, Australia (Gebhardt 1959).

The Tree Sparrow has not been such a successful colonist, but it shows the same characteristics of autumn and spring dispersal. No doubt the expansion in America to the north of St Louis, its point of introduction, has occurred in the way described for the House Sparrow; likewise the more recent colonisations in Sardinia, Sulawesi, Borneo and other parts of southeast Asia. Much of the recent expansion to the islands off the southeast Asian mainland has probably been through ship-assisted passage, though the details have not been documented.

The recent range expansions of the Dead Sea Sparrow in the Near and Middle East, the Pegu Sparrow down the Malay peninsula and the Southern Grey-headed Sparrow in Cape Province, South Africa, are all attributable to the birds rapidly spreading into areas that ecological changes have made suitable for them.

The House Sparrow is remarkable in the way it has successfully colonised a wide variety of environments, though not all colonisations have proceeded at the same rate. We can contrast the rapid spread in North America, South America south of 12–15°S, New Zealand and more recently the Azores, with the induction period in South Africa of 40–50 years between the initial introduction and the rapid expansion since about 1950. Something similar appears to be happening in Australia where there was an interval of about 60 years while the birds consolidated the subtropical southeast corner, before spreading over the Tropic of Capricorn. Again, an early introduction to Belém in the Amazonia region of Brazil, in 1927, failed to prosper and the birds died out, whereas Belém has now been recolonised successfully in the recent spread to northeast Brazil.

A possible hypothesis is that it may be necessary for physiological changes to evolve before the bird is adapted to its new environment. The fact that colonisation is proceeding rapidly in Senegal, by birds probably from South Africa and already adapted to African conditions adds some support to the idea. As previously mentioned, David Wells has suggested that it might be physiological factors that are preventing the colonisation of Malaysia by the Pegu Sparrow consequent on its spread from Thailand. It is even possible that it is the lack of physiological plasticity that has inhibited the successful colonisation of North America and Australia by the Tree Sparrow rather than competition with its introduced congener. Perhaps a key factor in the success of the House Sparrow is its behavioural adaptability that overcomes the physiological constraints in environments well removed from its thermo-neutral point – the insulation from climatic extremes by living indoors in close association with man in winter at high latitudes, the building of roost nests at high altitudes, and in tropical locations the early morning feeding and daytime roosting in shade trees.

Whatever the reasons for the range expansions, there is little doubt that the genus *Passer* is generally a successful and dynamic one. The House Sparrow is still expanding its range, notably in South America and the Afrotropical region; and expanding also within its existing range, despite a possible fall in breeding density in some areas in the early years of this century; and it is most surely increasing its numbers as man continues to modify the landscape in a way that is more favourable for it. The Tree Sparrow is similarly expanding on the islands off southeast Asia. The Willow Sparrow has increased in numbers, if not in range, in the Balkans since the 1950s and has made a significant spread in Semireche, in Kazakhstan, in the last 50 years. The Pegu Sparrow, the Cape Sparrow and the Southern Grey-headed Sparrow are all increasing as has been described. Less spectacularly, though still significantly, the other grey-headed sparrow species are all making modest gains, with forest clearance and agricultural developments creating more suitable habitat. Further north, the Golden Sparrow has expanded from its natural range to drier areas where additional water has been made available through irrigated cultivation.

The only species that appears to be suffering a decline is the Desert Sparrow. It has withdrawn from its range in eastern Sudan as a result of increasing desiccation, and is probably now extinct in its former range in central Iran. This is not the complete story, however; if it were purely a matter of rainfall, the bird would be expected to have moved into the marginal areas on the edges of its range that are suffering desertification. Possibly, expansion there has been inhibited by competition with more vigorous congeners that have managed to retain their position.

The genus *Passer* is of considerable interest as far as evolution is concerned. In Africa we have, on the one hand, the five populations of grey-headed sparrows that had just reached full speciation before climatic changes brought them into contact again; and, on the other, the four populations of African mainland Rufous Sparrows that have not yet come into secondary contact, though with the increasing aridity, partly from natural causes and partly as a result of man's mismanagement of the environment, it seems likely that this will occur in the comparatively near future. Until this happens, it is not possible to be certain about the relationship of these birds to each other, whether they are just well-differentiated subspecies or whether they have evolved far enough to have achieved specific separation. The introduced populations of the House Sparrow and Tree Sparrow provide an excellent opportunity for evolutionary studies as they adapt successfully to their new surroundings, and they are now becoming sufficiently distinct to pose problems in nomenclature. Again, we have the unusual relationship between the House and Willow Sparrows that over much of their sympatric range live side by side as good species, yet in North Africa the species barrier breaks down and they freely interbreed. Finally, there is the possibility that the Chestnut Sparrow is evolving into a brood parasite.

18: Origins and evolution

ORIGIN OF THE GENUS *PASSER*

We are now in a position to discuss the evolution of this closely related group of birds. This must clearly be somewhat speculative, but there are a number of clues that allow us to make reasonably informed guesses.

It seems reasonable to assume that the monomorphic, or at best sesquimorphic, plumage with the absence of a black bib in the male is the more primitive type. All of the birds with this type of plumage occur in the Afrotropical region (7 out of the 11 Afrotropical species), suggesting that this is the centre of origin of the genus. Further, sparrows in the Afrotropical region show a greater diversity than those in the Palaearctic and Oriental regions taken together: in addition to the more typical tree and hole nesting sparrows living in small colonies, such as the grey-headed, Cape, Somali and Iago Sparrows, we find extremes ranging from the solitary Rufous Sparrows to the highly colonial and nomadic Golden Sparrow and the potentially nest parasitic Chestnut Sparrow. This gives further support for the idea that the Afrotropical area has been occupied by sparrows for a longer period of time.

All the sparrows, even those nesting in holes, build domed nests. This is a characteristic of tropical birds, and the fact that the habit still persists with those species using holes suggests that hole-nesting by sparrows is a comparatively recent development and that the open tree site is the more primitive. Table 47 shows the nest sites used in the three geographical regions in which the species occur naturally. I have allocated scores to indicate a preference level for the types of nest sites. Even allowing for the subjectivity that this introduces, it still appears clear that open sites are dominant in the Afrotropical region and holes in the Palaearctic. If the comparison is restricted to the two tropical areas, the African and Oriental ones, the percentage of African species using open tree sites is still the greater, adding further weight to the hypothesis that the Afrotropical region is the origin of the genus and not that open tree sites are simply an adaptation to tropical conditions.

Origins and evolution 277

A characteristic of the sparrows is their post-juvenile moult. This is another feature that is more common in tropical birds (Fogden 1972).

These features all support a tropical origin for the genus with the weight of evidence supporting the Afrotropical region. The most economic hypothesis in my view is that the original sparrow was a rather dull-plumaged bird, sexually monomorphic and

Table 47: *Regional distribution of sparrow nest-site types*

Region	Species	Score*		Comments
		Open Site	Hole Site	
Afrotropical	*griseus*	2	3	
	swainsonii	3	3	
	gongonensis	3	0	
	suahelicus	0	3	
	diffusus	0	3	
	luteus	3	0	
	eminibey	3	0	A significant nest parasite
	melanurus	3	2	
	motitensis	3	0.5	The figures in parenthesis are the scores for the four races for which the nesting behaviour is known; a mean value has been taken
		(3/3/3/3)	(0/2/0/0)	
	iagoensis	1	3	The Cape Verdes are taken as Afrotropical as it seems most likely that *iagoensis* arrived there from the African mainland
	castanopterus	3	3	
Totals	11	24 (73%)	20.5 (62%)	
Palaearctic	*simplex*	3	3	*P. s. simplex* mainly 'holes' *P. s. zarudyni* mainly 'open'
	domesticus	2	3	
	hispaniolensis	3	2	
	moabiticus	3	0	
	rutilans	1	3	
	montanus	1	3	
	ammodendri	0	3	
Totals	7	13 (62%)	17 (81%)	
Oriental	*domesticus*	1	3	'open' sites are used less frequently than by members of the Palaearctic group (Whistler 1941)
	pyrrhonotus	3	1	
	flaveolus	3	2	
	rutilans	0	3	
	montanus	1	3	
Totals	5	8 (53%)	12 (80%)	

* Scoring: 3 = common, 2 = moderately common, 1 = uncommon, 0 = not recorded. Crowns of palms, understory of nests of herons *etc.*, creepers against walls, trees, *etc.* and similar enclosed sites are taken as holes.

278 *Origins and evolution*

living in an open, arid region of the African tropics. The present African, monomorphic, sesquimorphic and dimorphic species with black-bibbed males all evolved from this ancestral sparrow. At a later stage, a black-bibbed sparrow spread up the Nile or Rift valleys into the Palaearctic region and there differentiated into the nine species now found in the Palaearctic and Oriental regions. *Passer luteus* is taken as an Afrotropical species even though it extends to the southwest tip of Arabia; this area is normally placed in the Palaearctic, but it can equally be considered a transition zone lying between the latter and the Afrotropical region, and Lees-Smith (1986) has placed it in the Afrotropical region on the basis of its avifauna.

Another feature of the genus is the strong association with man. Table 48 shows how many species actually nest on man's buildings. Nesting on or in houses occurs in many other species of birds, but in no other genus is this habit so prevalent.

Table 48: Use of houses for nesting by sparrows

Region	Number of species		Total
	Nesting on houses	Not recorded nesting on houses	
Afrotropical	8	3	11
Palaearctic/Oriental	7	2	9
Totals	15	5	20

I shall now consider these two main groups, the Afrotropical and the Palaearctic/Oriental sparrows, separately, beginning with the latter, the more recent in biological terms.

ORIGIN OF THE PALAEARCTIC AND ORIENTAL SPARROWS

A number of fossil remains of *Passer* species dating from the Quaternary have been described from the Levant. The earliest of these comes from the Lower Pleistocene of the Mugharet-Oumm-Qatafa cave in Wadi Khareitoun near Bethlehem in the Judaean hills. Tchernov (1962) identified three species of *Passer* from the Middle Acheulean layer (*ca* 350,000 years B.P.) of the Lower Palaeolithic beds; these he termed as follows:

> *Passer predomesticus* a new species quite distinct from any existing *Passer* species, akin to both *domesticus* and *hispaniolensis*, but closer to the former.
> *Passer cf domesticus* a species closely resembling present-day *domesticus*, but not fully identical with it, *viz* a precursor of *domesticus*.
> *Passer cf moabiticus* similarly a precursor of *moabiticus*.

P. cf domesticus and *P. cf hispaniolensis* were also found in an undetermined Acheulean layer.

Tchernov also mentioned that a *Passer domesticus* precursor has been found in the Upper Levalloiso-Mousterian layer (*ca* 70,000 B.P.) of the Middle Palaeolithic beds in the Kebara cave in the Carmel, and again in the Post-Nutufian layer (*ca* 15,000–20,000 years B.P.) of the Upper Mesolithic in the Abu-Usba Cave, Wadi Falah, also in the Carmel. The approximate datings for the different layers are indicated in parentheses.

In a more recent publication, Tchernov (1984) attributed *Passer predomesticus* to the Yabrudian layers of Oumm-Qatafa, giving the date 140,000 B.P. He also refers to fossil remains from the Hayonim Cave, western Galilee, in Middle Palaeolithic deposits (*ca* 70,000 B.P.) onwards that are indistinguishable from current *Passer domesticus*. These are presumably precursors of present day House Sparrow.

In contrast to Tchernov, Johnston and Klitz (1977) do not consider that *domesticus* and *hispaniolensis* can be distinguished osteologically. Hence, perhaps the most we can infer from the Levantine fossils is that an ancestral type of the *domesticus/hispaniolensis* superspecies was present there about 350,000 years B.P.

As already mentioned, I think it is most probable that the northern hemisphere sparrows originated from a 'black-bibbed' ancestor that spread from the Afrotropical region up the Rift and Nile valleys. There is no evidence on which this can be dated, but, from the fact that some speciation appears to have occurred in the Levant by about 350,000 years B.P., it seems most probable that it took place in the early Pleistocene, possibly triggered off by the change from the Pliocene to the Pleistocene that is dated about 1 million years B.P.

The Pleistocene was a dynamic period of climatic change in the northern hemisphere, the broad succession of vegetational zones of broadleaf forest, dry grassland and boreal forest, oscillating north and south with alternating periods of climatic severity and amelioration. Of these zones only the grassland would have been suitable for sparrows, though no doubt there would have been some penetration of the broadleaf forest zone to the south along the major southwards flowing river systems where the flood plains would inhibit the establishment of permanent forest.

The exact nature of the climatic and vegetational changes in this region during the Pleistocene are complex and by no means fully understood. However, without having to go into the details, it is possible to put forward a plausible hypothesis that would account for the evolution of the nine *Passer* species current in the northern hemisphere, and their distribution and relationships with one another. This is based on the postulate that the ancestral granivorous sparrows (not restricting the term granivorous merely to the seeds of grasses, but including those of small annual herbs) that reached the Nile delta in the early part of the Pleistocene, were able to radiate west in the Palaearctic to the Atlantic seaboard, and east in southern Asia to the Far East in the Oriental Region along the grassland zone (Fig. 103). During severe climatic periods these birds were driven to the southern limits of the regions and separated into isolated populations that were trapped in favourable parts of the Mediterranean basin in Europe, in flood plains and deltas of the major Asian rivers, and in the steppe

280 *Origins and evolution*

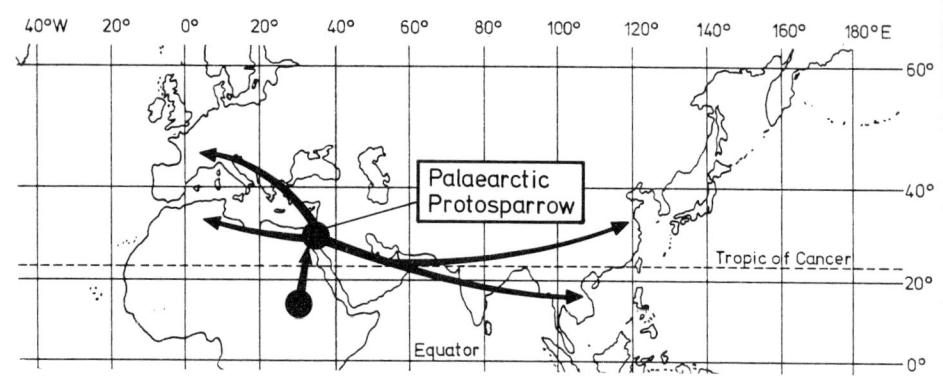

Fig. 103 Proposed spread of sparrows from Afrotropical region to the Levant and subsequent expansion in Palaearctic and Oriental regions during the Pleistocene

region of south-central Asia where conditions, even at the height of the last glaciation (25,000–15,000 years B.P.) remained unglaciated (CLIMAP 1976). These separated populations then evolved to give the present nine species of northern sparrows that subsequently spread to occupy their present ranges during the post-glacial Quaternary and the northern movement of the grassland zone. However, increasingly during this period the broadleaf forest zone has been modified by man, inhibiting its full development to the vegetational forest climax and thus providing more opportunities for sparrows to the south of the climax grassland zone.

Let us consider first the westward spread. Here we are concerned with two present-day species, the House Sparrow and Willow Sparrow. Even so, any historical account has to take into consideration the situation in northwest Africa, Italy and Crete, where there are either stable intermediary forms or the two species occur alongside a complete range of hybrid forms. This complex situation was first studied in detail by Meise (1936), and I have drawn heavily on this extremely valuable study to propose an historical sequence of events (Summers-Smith 1963). Supporting evidence for the datings of the various events is given by electrophoretic enzyme analysis carried out by Parkin (in press) on several populations of House and Willow Sparrows. Parkin's datings are based on my presumption that the House Sparrow arrived in England about 2,000 years ago. It seems reasonable to assume that House Sparrows arrived about the time of the Roman invasion (possibly by deliberate introduction as the Romans appear to have kept sparrows as pets and also as a source of food), though even if this assumption is in error the basic pattern remains, with some modification of the datings.

The basis of the hypothesis is that in the westward expansion during the Pleistocene, the populations in North Africa and southern Europe were exposed to the different climatic conditions existing at that time – humid in North Africa, dry in southern Europe – evolving towards the present-day phenotypes respectively of *hispaniolensis* and *domesticus*. I suggest that the two streams eventually rejoined in central or northern Italy, the southern stream reaching the south of Italy via Pantelleria, Malta and Sicily. Together with the phenetic changes there had also been genetic changes in the populations during the period of spread, but these had not been sufficient to prevent the interbreeding that resulted in offspring with intermediate characteristics. At this point, however, further spread of the parent types – westwards in the case of the northern birds, and eastwards by the southern birds, that had by that time also crossed the Strait of Gibraltar into the Iberian peninsula – was inhibited by the final (Würm) glaciation (*ca* 70,000 years B.P.) that made all except the extreme south of

Europe unsuitable for sparrows. This left a small population in Spain contiguous with the North African birds, another in southern Italy, and a third in the Near and Middle East.

With the retreat of the ice at the end of the Pleistocene, *ca* 12,000–10,000 years B.P., these separated populations began to spread once more, but with the long elapse of time that had taken place further evolution had occurred, giving rise to the Willow Sparrow in northwest Africa and southern Spain, and to the House Sparrow in southeast Europe and the Middle East. The Willow Sparrow was by now primarily adapted to the rather moist conditions that form its preferred present-day habitat; likewise the House Sparrow to the rather drier conditions that were present in the Middle East.

It is suggested that the association of the House Sparrow with man arose in the so-called 'Fertile Crescent', lying largely in the Tigris-Euphrates valley of Mesopotamia (Fig. 104), where man evolved from a nomadic hunter and gatherer to a

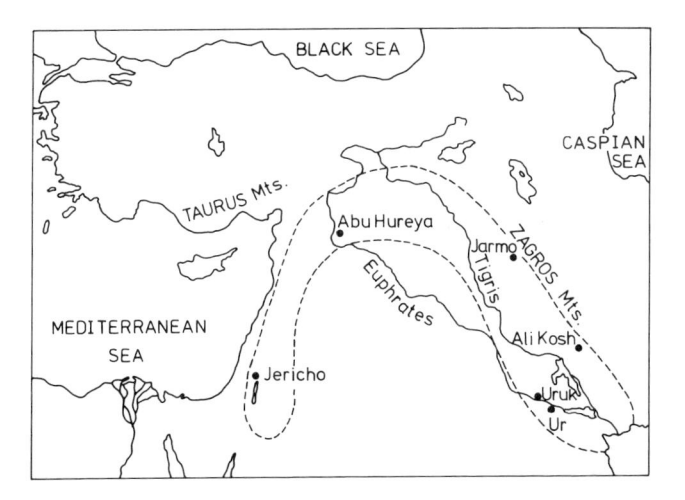

Fig. 104 The 'Fertile Crescent', proposed centre for the origin of the House Sparrow

sedentary agriculturalist. The Middle Eastern sparrows, already specialising on large-seeded grasses, changed their diet from the progenitors of the cultivated grains, such as the wild wheats *Triticum monococcum* (einkorn), *T. aegipiloides* and *T. dicoccoides* and barley *Hordeum vulgare* that grew in this area, to the cultivated grains themselves and then to other foods provided by man, and adopted holes in man's dwellings as their preferred nesting site. The early House Sparrows must, however, have started to spread before this association with man developed, as Parkin's electrophoretic enzyme analyses show that separation of European birds (*P. d. domesticus*) from those in the Middle East (*P. d. biblicus*) had occurred about 40,000 years B.P., presumably in the population trapped in southeast Europe. The same preadaptation to association with man was already present in this population and, when agricultural man spread into this area, the bird developed in the way that already it had done in the Middle East.

Johnston and Klitz (1977) suggest that these early progenitors of the House Sparrow were migrants, retreating to the south in winter as the other *Passer* species in that area, the Willow and Dead Sea Sparrows, still do today; and, from this, the association with man, providing winter food and shelter in his settlements, could

282 *Origins and evolution*

have led to the development of the sedentary behaviour so characteristic of most present House Sparrow populations, with the advantages of living permanently in familiar surroundings and avoiding the hazards of migration.

The combination of genetic and behavioural changes in the two populations meant that, when *hispaniolensis* spread eastwards through the Balkans and *domesticus* westwards to Spain and eventually North Africa, the two taxa were able to live sympatrically as good species. During the period of isolation in the late Pleistocene, the birds in Italy had evolved into a uniform group, intermediate between *domesticus* and *hispaniolensis*. This group spread with the retreat of the ice to northern Italy (and presumably also to Corsica where it is now present) and, although not sufficiently differentiated from either of its progenitors to prevent interbreeding, it was well enough adapted to the local conditions to inhibit significant penetration from the north by birds of the ancestral *domesticus*-phenotype, thereby maintaining the present rather narrow and apparently stable zone of intergradation between *domesticus* and the Italian birds, that I designate as *P. hispaniolensis italiae*. With the end of the Pleistocene, agricultural man spread north and westwards in Europe, following the amelioration in climate that took place from 10,000–5,000 B.P. Johnston and Klitz (1977) reproduce a map from Piggott (1965) showing the distribution of settled agricultural communities in Europe at about 7,000 B.P. (Fig. 105). Assuming that

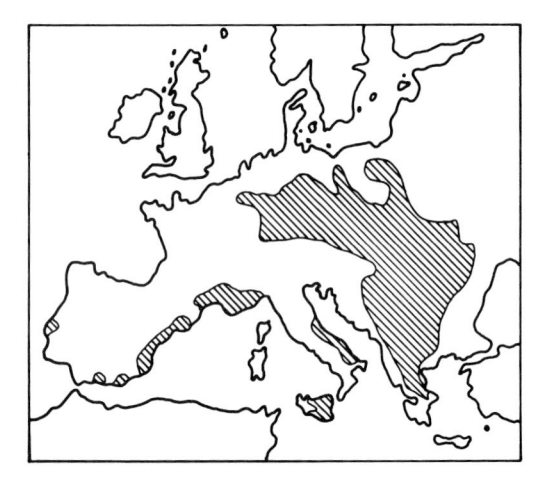

Fig. 105   Distribution of agricultural communities in Europe, approximately 7,000 B.P.

this models the expansion of House Sparrows into Europe, it would have been some time later before they made contact with the Italian Sparrows at their northern limit, allowing still further time for differentiation between the two groups. I do not consider that nominate *hispaniolensis* has ever occurred in this area. (See below for details of its subsequent spread.) *P. hispaniolensis* has, however, been able to make a secondary penetration from the south, where the physical barrier is less extreme, so that there now exists a cline from Malta and Pantelleria, through Sicily to about central Italy.

This evolutionary history tends to support my view that *italiae* is better placed as a race of *hispaniolensis* rather than of *domesticus*, as at the time of the origin of the Italian birds *domesticus* (*sensu stricto* the commensal of man) had not yet evolved, though the preadaptation to associate with man was already present in these early sparrows, as is shown by the way *hispaniolensis* (including *italiae*) readily becomes the urban sparrow in the absence of *domesticus*.

This hypothesis leads to a number of other conclusions. The sparrows breeding on Crete are virtually identical with 'pure' *italiae* occurring in Italy north of about the latitude of Rome. Presumably they have originated in the same way through colonisation by both ancestral types and the production of a stable population of intermediate plumage during the last Pleistocene glaciation. The subsequent spread of the Willow Sparrow to its present distribution across eastern Europe and western Palaearctic Asia, largely between 35° and 40°N (Chapter 10, Fig. 63), would most probably have taken place from Tunisia and Sicily to the Balkans, through Asia Minor to Afghanistan and Kazakhstan, rather than from Spain and round the north of Italy to the Balkans, so it is unlikely that there has ever been the possibility of *hispaniolensis* penetrating the range of *italiae* from the north.

We are now left with the situation in northwest Africa. Following the end of the Pleistocene the climate on the north African coast ceased to be the moist region to which the Willow Sparrow had adapted, and became increasingly drier, particularly from the eastern end. This led to the extinction of the Willow Sparrow from Egypt to Tripolitania with the exception of the isolated population in coastal Cyrenaica. In the meantime the House Sparrow had spread to the Maghreb across the Strait of Gibraltar and eastwards along the North African coast as far as Libya. This route of colonisation is suggested by the fact that, although these birds have been sufficiently isolated from the nominate race that occurs in Spain to form a separate race, *tingitanus*, this race is closer to nominate *domesticus* than to the Egyptian race, *niloticus*, from which, moreover, they are separated by a discontinuity in distribution.

As increasing aridity developed west along the North African coast, so the more moisture-adapted Willow Sparrow was driven further into the habitat occupied by the more arid-adapted House Sparrow and, although the two species can live side by side as good species, they are still very closely related and when forced together into the same biotope are able to breed with each other producing viable and fertile young. This is the situation that has occurred in Tunisia and eastern Algeria, producing a complete range of intermediates between the two phenotypes. In the more isolated oases to the south (*eg* Touggourt), where there is presumably less infusion of new genes, stable intermediate populations have arisen that have been given nomenclatorial status, eg *P. flückigeri* Kleinschmidt 1904, northern Algerian Sahara, and *P. bergeri* Zedlitz 1912, northwest Africa, though these are now recognised as hybrid *domesticus* × *hispaniolensis* forms. These examples of hybridisation occur in a similar situation to that already described for the other *Passer* hybrids and between House and Willow Sparrow on São Vicente in the Cape Verdes; that is, one species in small numbers at the extremity of its range interbreeding with another species that is common in the area.

The very close relationship between House and Willow Sparrows is shown by the fact that hybridisation also occurs in other areas away from those already described. For example, Dr R. F. Johnston informs me (pers. comm.) that hybrids were collected in 1976 at the extreme western end of the range of the Willow Sparrow in eastern Portugal, and I found a small hybrid population at Plavnica on Lake Scutari in southern Yugoslavia in 1965. Meise has similarly interpreted *P. rufipectens* Bonaparte 1950, Egypt, as a *domesticus* × *hispaniolensis* hybrid, and there is also some hybridisation in the isolated populations of *P. d. tingitanus* and *P. hispaniolensis* in Cyrenaica (Hartert 1923, Stanford 1954, Johnston 1969). The similarity in the genotypes of the two species is further demonstrated by the appearance of *hispaniolensis*-type birds in populations well away from the range of *hispaniolensis*; see Bodenstein (1953), who found a *hispaniolensis*-like head pattern in a central European House Sparrow, and

284 *Origins and evolution*

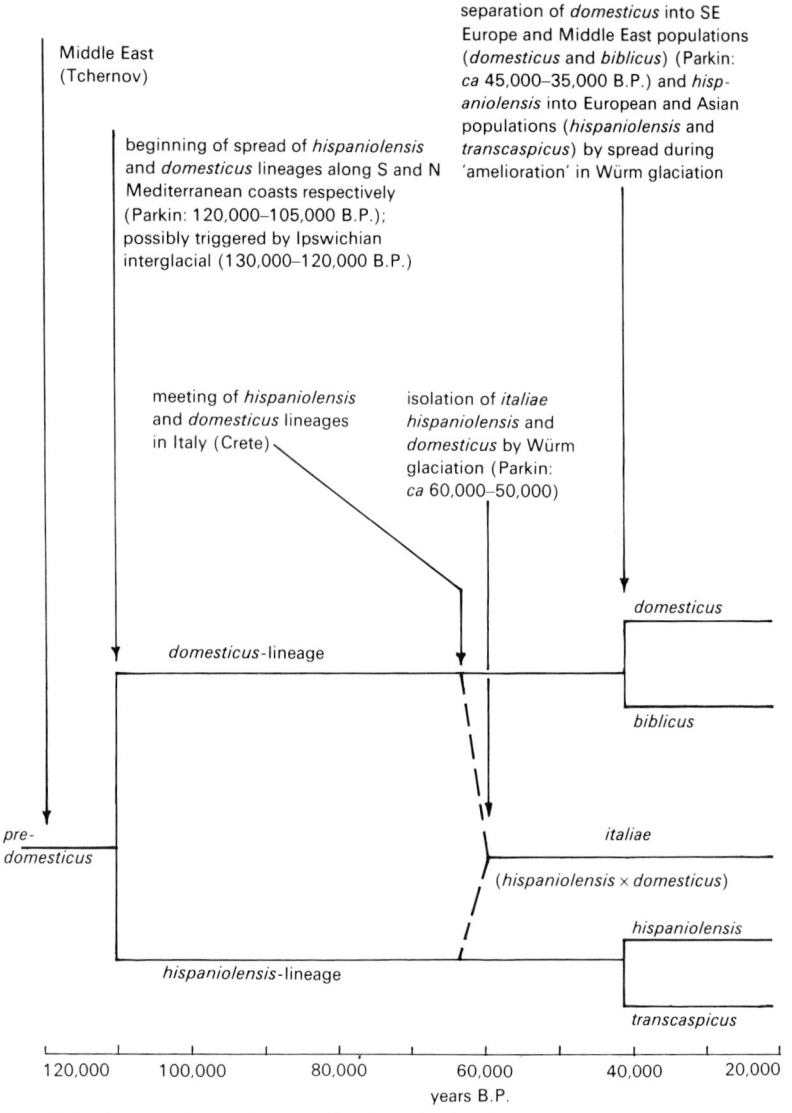

Fig. 106 Scheme for evolution of Willow, 'Italian' and House Sparrows based on proposed spread round Mediterranean; possible causative geological events and datings from Tchernov (1962) and Parkin (in press)

Selander & Johnston (1967), who found a *hispaniolensis*-like breast plumage in 31 out of 1,651 males (*ca* 2%) collected in America.

Fig. 106 shows schematically the proposed course of evolution up to about 30,000 B.P. of *P. domesticus*, *P. hispaniolensis* and the intermediate *P. h. italiae* from an ancestral sparrow, that spread to the Levant from the Afrotropical region at the beginning of the Pleistocene. The datings are no more than approximate.

Baumgart (1984) advances a different hypothesis for the origin of the Italian Sparrow and the other Italian Sparrow-like populations in Crete and the isolated oases in North Africa. He argues that the Willow Sparrow is a summer visitor in most parts of its range, taking advantage of a seasonal surplus of food that allows it to co-exist with the resident House Sparrow, although he considers the two species to have identical ecological requirements. He suggests that the 'Italian Sparrow' populations have originated from Willow Sparrows that have formed sedentary

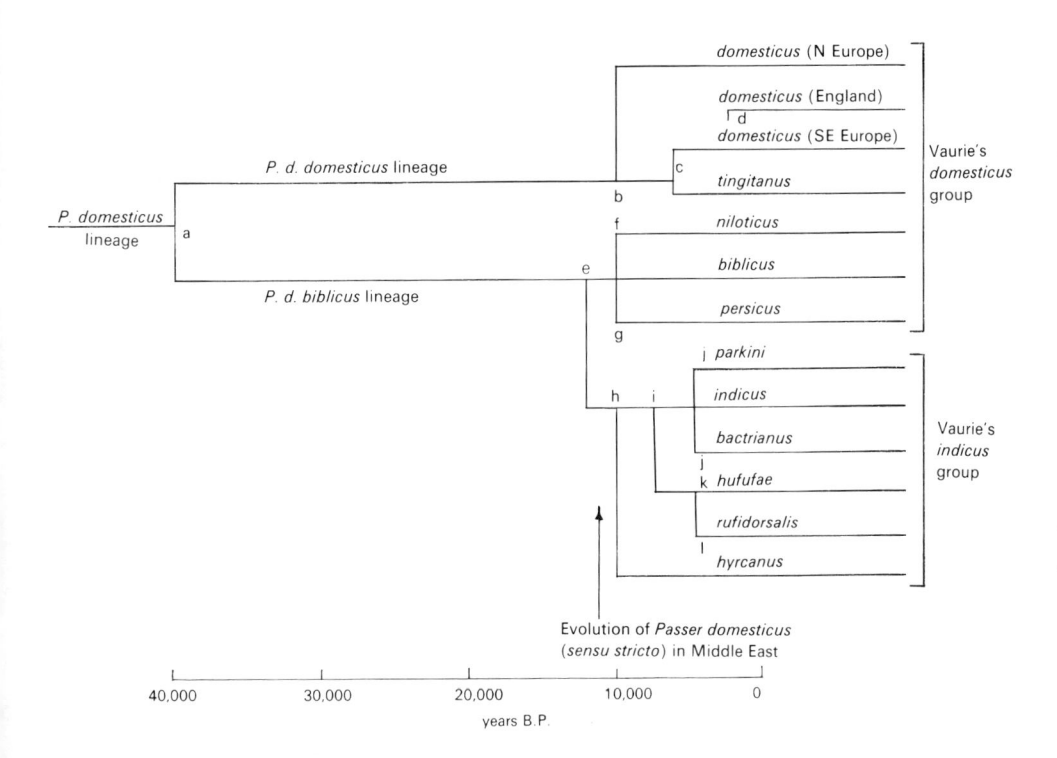

(a) Populations isolated in NE Europe (*domesticus*-lineage) and Middle East (*hispaniolensis*-lineage) at height of Würm glaciation began to spread during a period of climatic amelioration (45,000–25,000 B.P.)
(b) Divergence of northern and southeastern populations of *P. d. domesticus* (Parkin ca 10,000 B.P.)
(c) Spread of *domesticus* from SW Europe to give isolated population, *tingitanus*, in NW Africa
(d) Arrival and separation of *P. d. domesticus* in England (ca 2,000 B.P.)
(e) Spread to India followed by geographical separation giving rise to *indicus*-lineage (Parkin ca 10,000 B.P.)
(f) Separation from *biblicus* by spread to NE Africa to give isolated population in Egypt (*niloticus*)
(g) Secondary spread of *biblicus* to east giving rise to *persicus*
(h) Physical isolation of *indicus* population by Elburz mountains
(i) Secondary spread of *indicus* to SW Asia and NE Africa
(j) Migratory populations derived from *indicus* (*parkini* – Parkin ca 4,000 B.P.)
(k) Climatic separation of *hufufae* in E Arabia
(l) Physical separation of *rufidorsalis* in Sudan

Fig. 107 A tentative scheme for subspeciation of House Sparrow. Datings from Parkin (in press) assume arrival of House Sparrows in England ca 2,000 B.P.

populations, in places where food is adequate to allow overwintering, before the arrival of the House Sparrow. He supports this suggestion by pointing out that these 'Italian Sparrow' populations are all in places that are difficult of access to non-migratory House Sparrows because of physical barriers (the Alps in the case of the Italian population, the Mediterranean in the case of the island populations, and desert terrain in the case of the North African oases populations) and that they have been able to assimilate, or at least prevent displacement by, the occasional infiltrating House Sparrow.

I find this a less satisfactory hypothesis than the one I have advanced above. The co-existence of sympatric resident House and Willow Sparrows in southern Spain, parts of North Africa, Greece and Cyprus speaks against the species having identical ecological requirements. Baumgart's hypothesis does not adequately account for the situation in Italy with the 'pure' Italian Sparrow population in the north and the broad clinal zone of intermediates with the nominate race of the Willow Sparrow, stretching from Sicily to central Italy; nor for the absence of the Willow Sparrow in

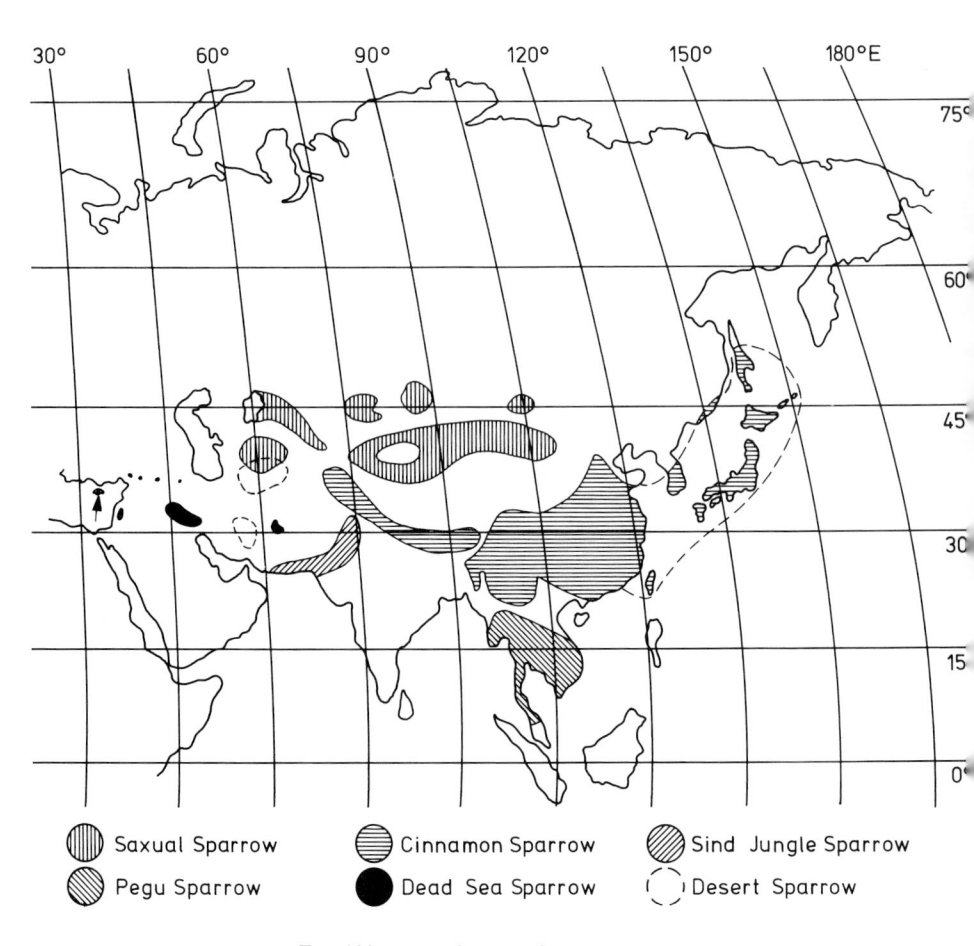

Fig. 108   Distribution of six Asian sparrows

the Balearics and Corfu, and the obviously hybrid region in Algeria and Tunisia.

In addition to the westward spread from the centre of evolution in the Fertile Crescent (Fig. 104), another group of House Sparrows spread eastwards to India and Burma in the Oriental region, giving rise to the *indicus*-lineage that subsequently spread north to Kazakhstan and southwest to Arabia and the Sudan. It would appear that some separation of the *biblicus* (*domesticus*) and *indicus* lineages occurred for a limited period, giving rise to Vaurie's *'domesticus'* and *'indicus'* sub-groups, with a subsequent expansion of range bringing the two groups together, before evolution of genetic isolation, and forming an extensive zone of intergradation in Iran, but only limited intergradation in Arabia and the Sudan. The birds of both groups are closely associated with man, in this case Parkin's dating (Parkin, in press) suggests that the divergence took place about the time of the evolution of the House Sparrow *sensu stricto* (*ie* the commensal of man). Fig. 107 gives a very tentative scheme for the subspeciation of *Passer domesticus* that draws on Parkin's genetic distance datings.

I consider the seven remaining northern sparrows to be Asiatic in origin. One of these is the outstandingly successful Tree Sparrow with its wide Eurasian distribution (Chapter 15, Fig. 82); the remainder are birds of comparatively restricted distributions and are almost completely allopatric (Fig. 108).

Just as in the west, where one species, the House Sparrow, evolved in close association with agricultural man, so did one of the Oriental species, the Tree Sparrow, in the east. I suggest that this occurred in the Huang Ho (Yellow River) valley in China, the cradle of Far Eastern agriculture, where millet and rice were the main cultivated crops. Like the House Sparrow, the Tree Sparrow became a virtual commensal of man, subsequently spreading westwards with sedentary man as far as Burma in the Oriental region, and in a broad sweep northwards into Siberia and west through the Palaearctic to western Europe, to give its present very extensive distribution.

The two urban sparrows are sympatric in Burma, but the House Sparrow has as yet failed to penetrate Indo-China and the Tree Sparrow is absent from the Indian subcontinent, with the exception of an isolated population of the latter in the Eastern Ghats that is probably a recent colonisation. Whether this is a stable situation, determined by interspecific competition, remains to be seen, though there is little historical evidence to suggest recent change. In the north, however, we have seen how the Tree Sparrow has spread into the European homeland of the House Sparrow, largely through a change in its ecology; and more recently how the larger and more aggressive House Sparrow has been able to spread east across Siberia, an area traditionally the preserve of the Tree Sparrow. Again, whether this spread will continue through Manchuria into the heartland of China, the ancestral home of the Tree Sparrow, remains for the future, though the situation here appears to be a more dynamic one than that in Burma.

An extremely puzzling feature is the sexually monomorphic plumage of the Tree Sparrow, the only sparrow in which the female has adopted the black-bibbed 'male' plumage. Could this have evolved as a species-isolating mechanism to inhibit interbreeding between the two urban sparrows? This would have developed in Burma, the only area where the two species have occurred together for a sufficient period of time, and then spread back into the allopatric part of the population in Indo-China and China before the spread of the Tree Sparrow through Siberia to Europe. This hypothesis seems a little improbable, but I can find no other explanation, unlikely as it is. Unlike the bill colour of the male, which is determined by the release of sex hormones (Keck 1934), the 'male' plumage of the House Sparrow appears to be under genetic control as it is not affected by gonadectomy (Amadon 1966b). The 'female' plumage appears to be under separate genetic control and thus adoption of the 'male' plumage by the female Tree Sparrow could have been the result of simple genetic change.

However it evolved, this plumage difference between House and Tree Sparrows must obviously be of some value as a species-isolating mechanism, though as we have seen it is not entirely effective in preventing some interbreeding between these two ecologically, and presumably genetically, very similar sparrow species.

The remaining 'flood plain' populations gave rise to the following species: Chang Jiang (Yangtze) valley, Cinnamon Sparrow; Mekong valley, Pegu Sparrow; Indus valley, Sind Jungle Sparrow; Tigrus-Euphrates delta or Jordan (Rift) valley, Dead Sea Sparrow. The location of the last species is supported by the fossil evidence from the Mugharet-Oumm-Qatafa cave, in Israel, that Tchernov (1962) described as a *moabiticus*-precursor. These four flood-plain species did not evolve to form the same close association with man and, through competition with the two dominant species, House and Tree Sparrows, have only been able to achieve the rather restricted distributions shown in Fig. 108, remaining allopatric with each other, though sympatric with one or both of the two outstandingly successful species.

These four species, though not commensal with man, show their close relationship with the two urban sparrows by having some association with man, living on the outskirts of inhabited places, feeding on grains and two, at least, the Cinnamon and Pegu Sparrows frequently nesting on man's buildings. The Dead Sea Sparrow, originally, may have had a continuous distribution as far east as Sistan, on the borders of Iran and Afghanistan, but is now restricted to a number of rather small, discontinuous populations. Until recently it had the appearance of a relict species, but with its adaptation to changing conditions in the Levant it has made a dramatic resurgence in the last 30–40 years.

The last two northern species, the Desert and Saxaul Sparrows, are desertic forms. The Desert Sparrow is very closely related to the House Sparrow (Hall & Moreau, 1970, place them in a superspecies with the wholly Afrotropical Rufous Sparrow) and tends to replace the latter in inhabited desert regions. The Desert Sparrow probably evolved from a population of House Sparrows that was isolated in the semi-desert area that stretches through the Middle East to southeast Asia. With increasing desiccation the birds died out in the Middle East, spreading into the Sahara about 5,000 years ago as that area entered an arid phase, and giving rise to separate populations in North Africa and southern Asia. The African and Asian subspecies are well differentiated (Vaurie 1956), suggesting that they have been separated for some time. With continuing desiccation, the conditions in the eastern Sahara have become too severe, even for the arid-loving Desert Sparrow, and the Saharan population is still retracting to the west. It died out in the Sudan, from where it was originally described, in the last 100 years and its stronghold is now in the western Sahara. In the same way it has probably become extinct in Iran, leaving the widely separated populations in Russian Turkestan and the western Sahara.

The final Asian sparrow, the Saxaul Sparrow, occurs in Russian and Chinese Turkestans. It is the most distinct in plumage of the northern sparrows and one least associated with man. It is an eremian species that I suggest evolved from a separated population trapped in south-central Siberia during one of the Pleistocene glaciations, in an area of semi-desert steppe that remained free from ice cover (CLIMAP 1976).

Most of the Palaearctic and Oriental sparrows appear to be separated more by geography than by their ecological requirements. This supports the view that they have evolved comparatively recently. Where they are again in contact there is often no clear ecological separation; see, for example, the House Sparrow and the Tree Sparrow in Siberia and northern Europe, the Pegu Sparrow and Tree Sparrow in Indo-China, the House Sparrow and Cinnamon Sparrow in the Himalayas, the hybridisation of the House and Willow Sparrows, not only in the well-recognised hybrid zone in north-west Africa, but also in other parts of their range. Some differences in ecology have also evolved: the preference for a riverine or lacustrine

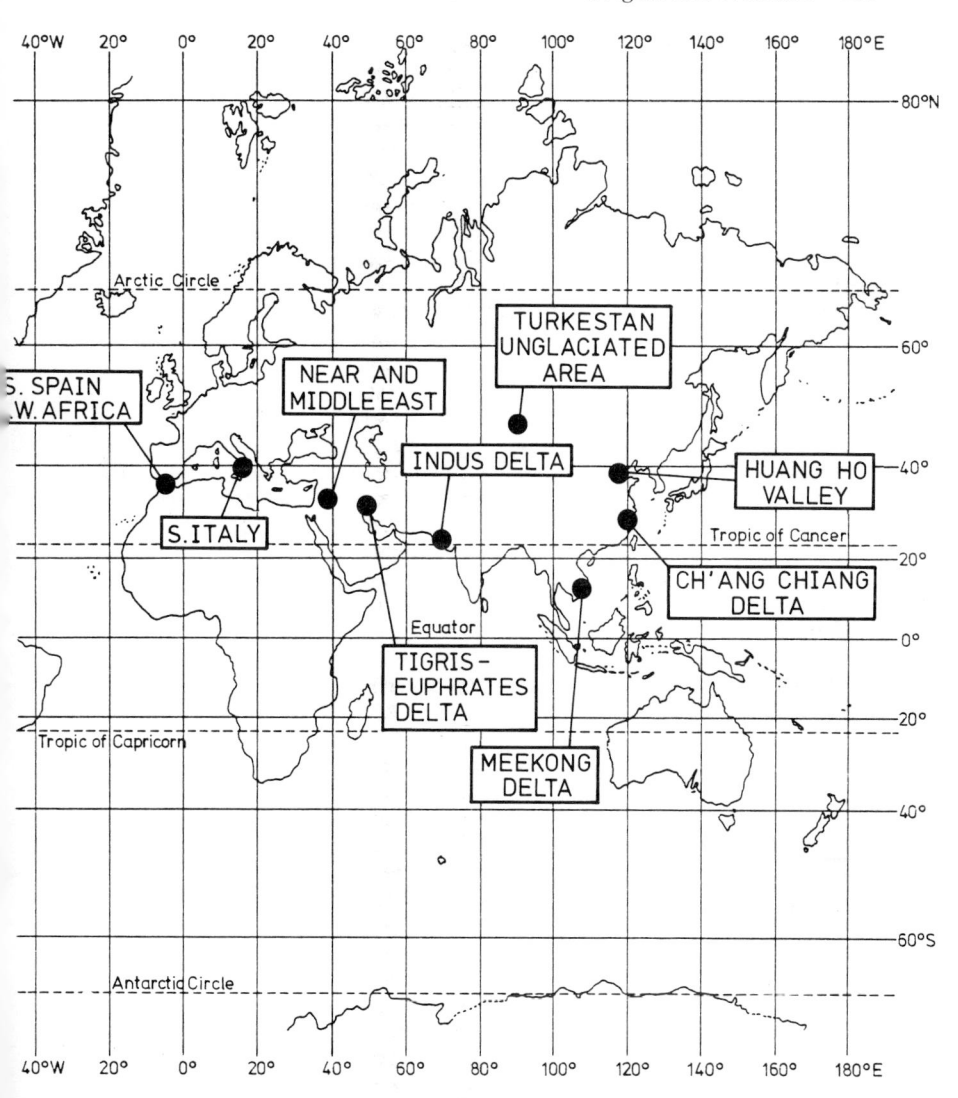

Origins and evolution 289

S. Spain and NW Africa	Willow Sparrow
S. Italy	Italian Sparrow
Near and Middle East	House Sparrow
Tigris-Euphrates Valley	Dead Sea Sparrow
Indus delta	Sind Jungle Sparrow
Meekong delta	Pegu Sparrow
Ch'ang Chiang (Yangtse Kiang) delta	Cinnamon Sparrow
Huang Ho (Yellow river) valley	Tree Sparrow
Turkestan – unglaciated area	Saxaul Sparrow

Fig. 109 Proposed evolution centres for northern hemisphere sparrows

habitat by the Dead Sea Sparrow and Sind Jungle Sparrow, an arid habitat by the Desert and Saxaul Sparrows, the rather moister, lusher habitat by the Willow Sparrow and the montane habitat by the Cinnamon Sparrow.

The proposed evolutionary centres for the sparrows of the northern hemisphere are shown in Fig. 109.

ORIGIN OF THE AFROTROPICAL SPARROWS

The situation in the Afrotropical region is less obvious: sparrows have been there longer than in Eurasia; there is a complete absence of fossil evidence; and the climatic succession is less clear than in the northern hemisphere. However, while a complete picture cannot be offered for all of the species, it is possible to speculate on some of the evolutionary steps.

The Golden Sparrow and Chestnut Sparrow, at first glance very different in appearance, are in fact rather similar; both are small sparrows and the males in breeding plumage, as shown in Table 49, differ only in the relationship between chestnut and yellow. There is actually a direct transition from the subspecies *P. l. euchlorus* which is entirely yellow, through *P. l. luteus* which has chestnut on the back, to *P. eminibey* that is entirely chestnut. The obvious phenetic relationship suggests that the two species may also be phyletically closely related. This is supported by their distributions (Chapter 2, Fig. 14; Chapter 3, Fig. 18), where it almost appears that they replace each other, the Chestnut Sparrow occupying an area of southern Sudan and western Ethiopia that, as already pointed out, would appear to be eminently suitable for the Golden Sparrow. The very small area of overlap in western Sudan suggests a comparatively recent penetration by the Chestnut Sparrow after separation had occurred.

Table 49: Comparison of Golden and Chestnut Sparrows

Feature	*P. eminibey*	*P. luteus*	
		P. l. luteus	*P. l. euchlorus*
Length	115 mm	120 mm	
Head	chestnut	yellow	yellow
Underparts, mantle and back	chestnut	chestnut	yellow

The remaining 'plain' sparrows, the grey-headed sparrows, are a very closely related superspecies group, the members of which have only comparatively recently achieved specific status and, in fact, present an almost impossible situation for tidy classification. This is not surprising in a group of birds that is on the point of evolving into a number of different species, as is well illustrated by the way that zones of sympatry and intergradation occur almost side by side and are difficult to specify precisely.

It seems likely that the original grey-headed sparrow was separated into a number of isolated populations during climatic changes in the Pleistocene. Fig. 110, which is due to Crowe (1978), shows a vegetation map for Africa during a dry climate phase. Grey-headed sparrows, which are adapted to grassland and lightly wooded savanna (mesic savanna), would have had an almost continuous distribution in the Afrotropical region at this time, with the exception of the forest and desert areas. In contrast, during a wet climate phase, Fig. 111 (Crowe 1978), with a massive extension of the areas of tropical and montane forest, they would have become isolated in separated grassland zones. There are five such areas (Fig. 110) corresponding to the number of species in the [*griseus*] superspecies, though this coincidence should not be pushed too far. Nevertheless, it shows how a number of grey-headed sparrow species could have evolved through isolation. Such a pluvial period occurred about 10,000 years ago and it is postulated that the separated populations had just about reached the stage of reproductive isolation when the climate changed at the end of the Pleistocene and the forest retreated to the west, allowing the populations to spread and come into contact once again. An additional factor of importance in the area was man's transition from a hunter to a farmer. This change spread from the north, but came slowly into subequatorial Africa. About 12,000 years ago the Sahara grew colder and less dry, and the land became fertile; agriculture spread there from Egypt, but by 5,000 years B.P. the area began to dry out again and at about the same time agriculture spread to the Ethiopian highlands, possibly through peoples from the desiccating Sahara.

The spread of agriculture into west Africa occurred later, probably not until 2,000–3,000 years B.P., and then gradually progressed further south. It is perhaps significant that the two grey-headed sparrows most associated with man are the two northern species, the Grey-headed Sparrow and Swainson's Sparrow; and the species that is little associated with man is the most southerly, the Southern Grey-headed Sparrow. The Grey-headed Sparrow appears to have increased its range recently; again, this may be a consequence of its close association with man and the effect of man's influence on the habitat.

The population that gave rise to the Parrot-billed Sparrow was presumably isolated in an arid region (perhaps the grassland region shown at the Horn of Africa in Fig. 110) as it is the most arid-loving of the group. This would make it less likely to have come into contact with agriculture and today it is the least associated with man of the five species. It is remarkable to find an absence of sparrows in Kenyan towns such as Isiolo and Nanyuki, though perhaps this is now changing with birds in the towns in southern Somalia and increasingly associating with man in the Kenyan game reserves. As mentioned earlier I think it is probable that the grey-headed sparrows that have recently arrived in Nairobi are more likely to be Grey-headed Sparrows that have spread from the south than Parrot-billed Sparrows.

The rufous sparrows have been subjected to the same climatic conditions as the grey-headed sparrows and, after a more or less continuous sub-Saharan distribution during a dry climate phase (Fig. 110), were separated into a number of isolated populations by a succeeding pluvial phase and the consequent extension of forest cover (Fig. 111). Once again there are five isolated populations, though one of these is on islands lying to the east off the Horn of Africa. A major difference, however, is that rufous sparrows are adapted to much more arid conditions (xeric savanna) than grey-headed and thus expansion of the isolated populations in the recent Quaternary has been much more restricted. As a result, they have not yet come into secondary contact as the present-day dry vegetational zones (Fig. 15, Chapter 2) that are the habitat of this species have still to join up. The relationship of the Rufous Sparrow to the other black-bibbed sparrows is not clear, though it is so different in its social behaviour and its lack of association with man that the relationship does not appear to be particularly close.

In these circumstances, where the populations are still allopatric, it is not possible to determine if evolution has proceeded far enough to prevent interbreeding, as has happened with the grey-headed sparrows, although real differences in behaviour appear to have developed. The lack of intergradation between the parapatric populations, *rufocinctus* and *shelleyi*, is suggestive, compared with the less well-defined boundaries between *motitensis*, *benguellensis* and *subsolanus*. For the present, however, there is no profit in further speculation nor any real merit in treating these populations as other than races of one species.

One obvious anomaly is the absence of a rufous sparrow in the zone of arid savanna

292   *Origins and evolution*

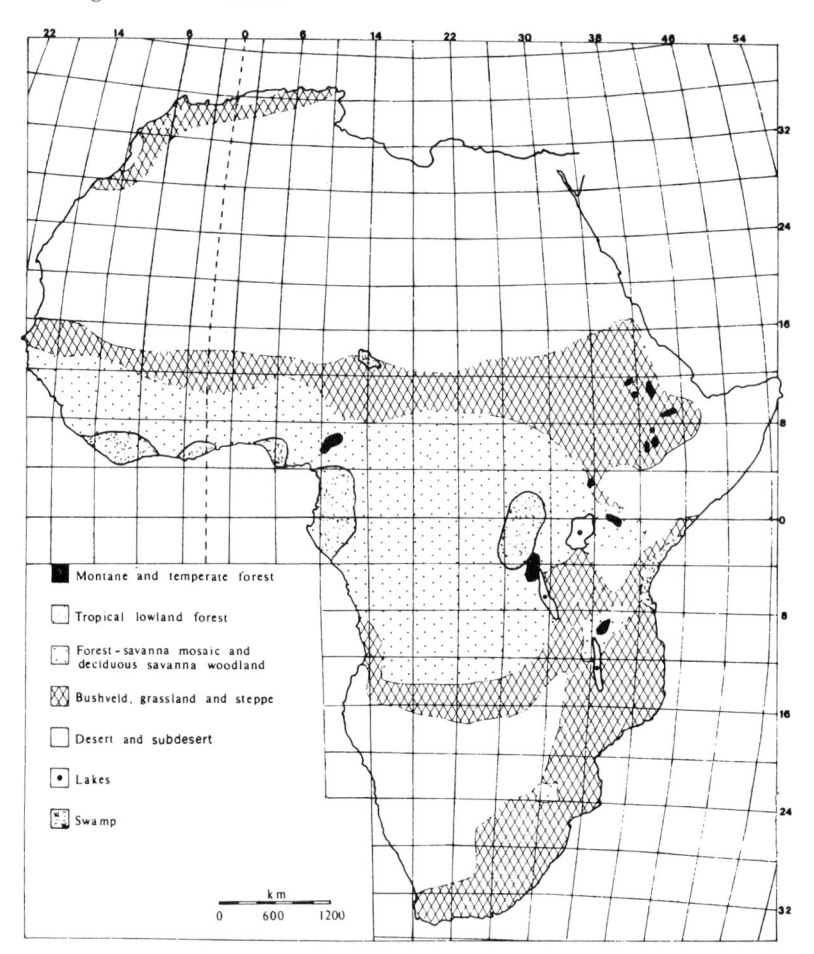

Fig. 110   Proposed vegetation map for Africa during a dry climate phase

in west central Africa, though the similarly-plumaged Iago Sparrow occurs to the west of this on the islands of the Cape Verde Archipelago. This species is obviously closely related to the Rufous Sparrow and would appear to have been derived from a population of rufous sparrows that occurred up to the west African coast, which has since died out, possibly during an earlier pluvial period. Certainly the Iago Sparrow seems to have been separated long enough from the mainland Rufous Sparrow for speciation to have occurred. In the presumed absence of *Passer* congeners at the time of its arrival in the Cape Verdes it has evolved to occupy a much wider niche than has the Rufous Sparrow on mainland Africa.

We are left with the two remaining black-bibbed African species. The Cape Sparrow is very different in appearance from other black-bibbed sparrows, although it shows the characteristic *Passer* tendency to associate with man and breed on his houses. Its origin, for the present, remains obscure. The Somali Sparrow, on the other hand, has very much a conventional 'black-bibbed' plumage pattern; in fact, it so closely resembles one of the Oriental sparrows, the Cinnamon Sparrow, that Meinertzhagen (1951) actually considered it to be a race of the latter (*P. rutilans castanopterus*).

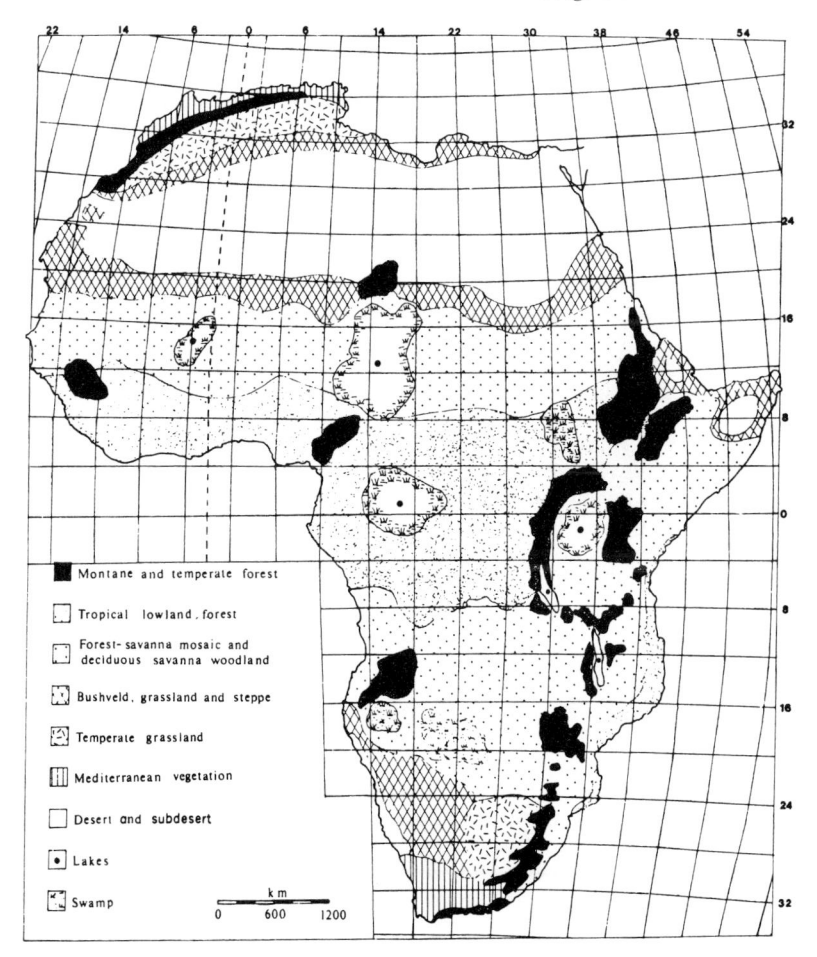

Fig. 111 Proposed vegetation map for Africa during a wet climate phase

When I watched this species in northern Kenya, I was struck by its similarity in all respects to the House Sparrow. One possibility is that both are descendants of the ancestral Palaearctic protosparrow. I feel, however, that the relationship to the House Sparrow is even closer than this and I suggest that the Somali Sparrow is a secondary invader of the Afrotropical region from Eurasia. It is perhaps significant that this is the only Afrotropical sparrow for which hybridisation with the House Sparrow has been reported.

Although we are able to distinguish some relationships between groups of the existing Afrotropical sparrows, the relationship between these groups remains obscure and it is probable that many of the connecting links have become extinct.

THE PRESENT

The problem of unravelling the historical, evolutionary development of a group of birds, such as the sparrows we have been discussing, is the lack of available data to provide a sound basis for the development of a reasonable hypothesis consistent with the known situation. Evolution, however, is not a thing of the past, but a continuous process. The problems in understanding current evolution are quite different from those in the historical situation: the time-scale for which reliable data exists is very short compared with the rate of evolutionary change so that the latter is difficult to assess. However, the introduction of a species to a new, previously unoccupied area makes rapid evolution possible as the species adapts to the new environment. The discontinuity enables one to make comparisons, both with the ancestral stock and also with the founder stock, provided data exists relating to the time of the introduction.

Calhoun (1947) showed that not only had the wings of male House Sparrows in North America increased from a mean of 76.51 mm to 77.60 mm, within 50–60 years of the introduction (*ie* by 1908), but that when birds from four different climatic zones in America were compared, the size (wing length, humerus length, femur length) was inversely correlated with the severity and duration of winter, in conformance with Bergmann's ecogeographical rule (size negatively correlated with winter temperature). Johnston and Selander (1964, 1971, 1973a,b), Packard (1967b) and Johnston (1973), with large samples collected over North America from 1962–65, have shown considerable interlocality differences between North American populations that not only confirm Calhoun's findings with respect to Bergmann's rule, but they also conform with the ecogeographical rules of Allen (extremities increase relative to core size with increasing temperature) and Gloger (individuals in warm and humid areas tend to be more heavily pigmented than those in dry ones). The changes they observed go in parallel with those that occur in native species of birds. Johnston and Selander (1964) found that the greatest colour variation had occurred in Hawaii, where birds introduced from New Zealand in 1870–71 were quite distinct from their European ancestors.

Baker (1980) has also shown that clinal morphometric changes have occurred in the New Zealand population after a similar period of time (*ca* 100 generations), though not only to a lesser extent than has occurred with the American birds – not unexpectedly with the innoculum involving only about 100 birds from England – but in this case in a contrary direction to Bergmann's rule.

The situation in southern Africa is complicated by the fact that there were separate introductions of *P. d. indicus* and *P. d. domesticus*. Crowe, Brooke and Siegfried (1980), however, came to the conclusion that, through differential spread of *indicus*-like birds, southern African House Sparrows now fall within the realm of *P. d. indicus*, though they are morphologically distinct from that race. Dr T. M. Crowe (pers. comm.) informs me that the 'birds from western southern Africa are clearly different from Indian *indicus*: they have greyer and tawnier bellies; are darker overall dorsally; have shorter wings, tarsi and tails; and have narrower bills'. Once more, not only has the introduced population diverged from its progenitors, but there are also differences within the southern African population that correlate significantly with variations in local environment.

While there is thus strong evidence to show that significant changes have occurred in some introduced populations of House Sparrows within 80–100 generations, it does not necessarily follow that these changes have been the result of evolution, *ie* changes in the genotype through natural selection. The changes could merely be phenotypic adaptations to the new environment. As Johnston (1973) himself remarks 'owing to the utility in distinguishing between genetic and environmental influences over morphological variation, the tying of genetic to morphological variation is not trivial for House Sparrows; it is precisely such conjunction that needs to be demonstrated if we wish to speak meaningfully of evolution in American populations of the species.'

There is presumptive evidence, however, to suggest that the changes have, in fact, an evolutionary basis. First, in several of the studies the authors found that the changes were in agreement with the ecogeographical rules of Bergmann, Gloger and Allen, and, moreover, paralleled the changes in native bird populations. Secondly, although the direct evidence of linkage between phenetic characters and specific gene loci is lacking for the House Sparrow, there is circumstantial evidence from biochemical genetic studies that changes in heredity have occurred. Cole and Parkin (1981) and Parkin and Cole (1984), using electrophoretic techniques, have found significant differences in enzymes at polymorphic loci between separate populations of House Sparrows in the English Midlands (incidentally supporting my contention, Summers-Smith, 1963, that the size differences between local populations in Europe had a genetic basis). Similar differences were also found in the introduced populations in Australia and New Zealand (Parkin & Cole 1985), despite the fact that these populations showed a reduction in the number of alleles per locus for the enzymes studied, as might be expected from the reduced gene pools of small innoculi – the 'Founder Effect' (Mayr 1963). More interestingly, however, the introduced populations, despite the reduction in alleles from the founder effect, showed a higher rate of differentiation than the British and European Continental populations, suggesting the effect of natural selection in new populations subjected to new environments.

These House Sparrow studies have been restricted by lack of complete data about the founding populations. The recent formation of new colonies in the Azores (1960) and in Senegal (1970–80) provides a fresh opportunity for more objective treatment. It is pleasing to be able to report that both these populations are already being thoroughly investigated by the Departmento de Biologia, Universidade dos Açores and by Dr M.-Y. Morel, respectively.

It should be pointed out that, in contrast to the changes that have occurred in introduced House Sparrow populations, Barlow (1973, 1980) could find no significant variations in the introduced Tree Sparrow in America that correlated with the ecogeographical rules, and Bannerman (1963) was unable to detect any significant difference between Willow Sparrows of the Canary Islands and western European birds, despite the former having, earlier, been given subspecific status (*P. hispaniolensis canariensis* Tschusi 1914). Barlow attributed this lack of differentiation in the introduced Tree Sparrow to the restricted gene pool of the initial innoculum of only 20 birds. Details of the colonisation of the Canaries by the Willow Sparrow are not known, but this, too, was probably a small founding population; moreover, I have suggested that it could have occurred by natural spread and hence further admixtures of north African genotypes could occur from time to time in the same way and so restrict differentiation.

Some of the populations of introduced House Sparrows are now sufficiently distinct from all other populations to meet the requirements of racial status. This is certainly the case with the birds in southern Africa and Hawaii, though Johnston and Selander (1973a) point out that as yet the American continental birds show only modest differentiation from the putative English stock, insufficient to justify nomenclatorial recognition, particularly as the situation in America is a dynamic one with the population not yet having attained its immediate evolutionary potential.

It seems inevitable in the absence of further introductions that all of the introduced populations will diverge sufficiently from the ancestral stocks to warrant racial status, merely on grounds of nomenclatorial convenience, quite independently of whether these divergences can be ascribed to evolution, *sensu stricto*. It is suggested that, when introduced populations are accorded subspecific names, a consistent nomenclatorial system should be used that will preserve the known nature of their origin. This could be the population locality with the prefix *novo* (=new, recent), *eg P. d. novosudafricanus, P. d. novohawaiiensis*.

I am now in a position to answer the question I originally posed about the reasons for the success of the House Sparrow. Belonging to a genus of birds that had evolved to feed on the large-seeded savanna and steppe grasses, it was eminently preadapted to become a commensal of man once the latter became a cultivator of these plants. By chance *Passer domesticus*, or its progenitor *sensu stricto*, was present where this development took place, but a similar association could easily have developed with several other *Passer* species had they been so situated, as in fact did occur with *P. montanus* in the Far East. The greater success of the House Sparrow compared with the Tree Sparrow has been a consequence of its association with western, colonising man.

Having given the reason for the success of the House Sparrow, we are led to ask what limits its success. Why is it not commoner than it is? Bird populations can be limited by a number of factors: suitability of the biotope, in other words availability of food and suitable nesting places, predation and disease. As we have seen, the House Sparrow is capable of maintaining a production rate far above that required to maintain its numbers, as is shown by the rapid increases in population in places to which it has been introduced. Nor do nest sites appear to be a limiting factor– when there is a shortage of suitable holes it regularly takes to the branches of trees and even the tops of telegraph poles. Although I have suggested earlier that Sparrowhawks could be having an effect on House Sparrows numbers in certain parts of Great Britain, this is somewhat abnormal and there is little real evidence of predators generally exerting a limiting influence. We are left with food. Clearly the House Sparrow is in competition with other species, and in the different environments in which House Sparrows occur there will be a number of such competitors that interact to maintain a reasonably stable balance. The strength of the House Sparrow lies in its ability to exploit a wide range of environments through its omnivorous diet, and its adaptability to a large range of circumstances and climatic conditions.

19: The systematic position of the sparrows

Classification is essential to bring some order into a situation where there are a large number of items. It is implicit that animal classification should show phylogenetic relationships, either in the traditional Linnaean system with a series of subgroupings, or by an evolutionary tree (cladogram). Traditional systematics has been based on morphological characters. But are similarities evidence of genuine relationship or merely of convergent adaptation? The question has been particularly difficult to answer for sparrows and opinions still differ about their relationship to other groups of birds.

In this chapter I shall be concerned with two problems: first, the relationship of sparrows to other genera and other seed-eating passerines; secondly, the relationship between species in the genus *Passer*.

Four families of seed-eating oscines are generally recognised: the Buntings (Emberizidae), the Finches (Fringillidae), the Waxbills (Estrildidae) and the Weavers (Ploceidae). Sharing some characteristics with finches and weavers, the sparrows have alternately been placed in the Fringillidae in a grouping known as the 'passerine-finches' that contains the genera *Petronia* (Rock and Bush Sparrows) and *Montifringilla* (Snow Finches), or with the Ploceidae in an assemblage known as the 'sparrow-weavers'. The latter includes some or all of the following genera in addition to those mentioned above: *Plocepasser* (4 species), *Histurgops* (1 species), *Pseudonigrita* (2 species), *Philetairus* (1 species) and *Sporopipes* (2 species).

Rather than attempt an historical review of the various opinions that have been advanced with respect to the systematic relationships of *Passer*, which have altered almost every time a new character has been investigated as a taxonomic criterion, I propose to examine the arguments associated with each of the characters that have been used, with the hope that this will give a clearer picture.

Review of taxonomic characters and classification of sparrows

SKELETAL STRUCTURE AND MYOLOGY (MUSCULATURE)

Chapin (1917), although noting that *Passer* were essentially alike in skeletal characters with the ploceines and estrildines, retained them in Fringillidae. He placed *Sporopipes, Histurgops, Plocepasser* and *Philetairus* in the Ploceinae, a subfamily of Ploceidae. Sushkin (1927) considered the similarities in certain skeletal characters (largely unspecified) between *Passer* and the ploceines to be evidence for their relationship, though this view was criticised by Sibley (1970) and Bock and Moroney (1978b) on the grounds that little factual evidence has been produced and that what evidence was produced was not dealt with in a systematic manner. Beecher (1953), on the basis of musculature, placed *Passer* (*domesticus* and *griseus* examined) in the subfamily Passerinae of the Ploceidae. *Dinemellia* was included by Beecher in this subfamily, though without much conviction; *Dinemellia* is more usually placed with *Bubalornis* in the subfamily Bubalornithinae of the Ploceidae.

Pocock (1966) pointed out that *Passer* and *Petronia* have certain apertures (foramina G) in the posterior wall of the orbit (eye socket) that are lacking in Ploceinae, *Bubalornis, Plocepasser, Amblyospiza* and *Sporopipes* (all normally recognised as belonging to the family Ploceidae), but are present in *Carduelis* (Fringillidae) and many other families. Although he recognised that the aperture of the foramina in the orbit was rather a poor taxonomic criterion, nevertheless he favoured the recognition of a separate, and new, family, Passeridae, for the sparrows as they were not in his opinion closely related to *Plocepasser* and the rest of the Ploceidae, considering the resemblance between *Passer* and the Ploceidae to be the result of convergence.

Bock and Moroney (1978a, 1978b) reported that the preglossale muscle (*M. hypoglossus anterior*) is present in *Passer, Montifringilla* and *Petronia*, but absent in Ploceinae, Fringillidae and other nine-primaried oscines. They considered that the 'passerine-finches' (Passeridae) could only have evolved from an ancestor possessing a *M. hypoglossus anterior* and that they had acquired their seed-eating adaptations quite independently of the ploceines. They argued that this character had considerable taxonomic significance 'because it is a feature of complex structure and possessing a complex series of morphological connections with surrounding features.' On this evidence they considered that the genera *Passer, Montifringilla* and *Petronia* comprised a monophyletic and extremely close-knit group, worthy of separate family rank and of obscure phylogeny. They found no relationship with the 'sparrow-weavers' and favoured either the re-establishment of the subfamily Plocepasserinae in Ploceidae, containing the genera *Plocepasser, Histurgops, Pseudonigrita, Philetairus* and *Sporopipes*, or the uniting of these genera with Bubalornithinae.

In contrast, Bentz (1979), on the basis of appendicular myology, retained Passerinae as a subfamily of Ploceidae, containing the genera *Plocepasser, Pseudonigrita, Philetairus, Passer, Petronia, Montifringilla* and *Sporopipes* (following Moreau (1962)), although according to his own results *Passer, Petronia* and *Montifringilla* showed features not present in the other ploceids examined; he treated these, however, as derived characters. To some extent this classification was by default, as in his conclusions he admits that the exact position of *Passer* has still not been conclusively determined.

FEEDING MECHANISMS

The form of the ridges on the palatal surface of the horny bill in *Passer* and *Montifringilla* is similar to that of the Ploceinae (Sushkin 1927). This was confirmed by Ziswiler for *Passer* (1965) and for *Montifringilla* (1967c).

Also, according to Ziswiler (1967a), the sparrows are similar to the Ploceidae in histology of the alimentary tract, but different from Fringillidae and Emberizidae. On the other hand, they differ from the weavers in the more complicated structure of the mucous glands of the oesophagus (Ziswiler 1967b).

PTEROLOGY

Sushkin (1927) considered that possession of an obvious, though sometimes minute, outermost primary by the sparrows showed their relationship to weavers rather than to finches (Fringillidae), which have only nine obvious primaries, the 10th, outer,

Fig. 112 Wing of House Sparrow (dorsal view); primaries numbered descendantly

being vestigial, concealed and displaced by the next (Fig. 112). This view was strongly opposed by Bates (1934) on the grounds that in *Passer* this feather is rotated dorsally, whereas in the weavers, although it may also be much reduced in length, it is in its normal position. For this reason he favoured the retention of the sparrows in the Fringillidae; in this view he was supported by Bannerman (1963). Sushkin (1927), however, considered that the sparrows (*Passer, Montifringilla* and including with these *Pyrgolauda, Onychostruthus*, Petronia, Gymnoris* and *Sorella*) show an affinity with Ploceidae in having a complete post-juvenile moult, whereas moult is only partial in the Fringillidae. Mayr (1927) suggested that the complete juvenile moult in *Montifringilla* showed this genus to be 'passerine' rather than 'fringilline' in character. Clench (1970), as the result of an extensive study of passerine pterylosis, considered that this was an adequate criterion for distinction at the family level. On this basis she separated *Sporopipes* and *Passer*, though unfortunately her material for *Pseudonigrita* and *Plocepasser* was inadequate to reach a decision on their relationships. A further distinction is that young *Passer* are born naked, a feature they share with the Ploceidae (Collias & Collias 1971), whereas young Fringillidae have down.

BEHAVIOUR

As early as 1850, Lafresnaye (1850) suggested that, in contrast to the widely held view at the time, sparrows might be more closely related to weavers (Ploceidae), in particular to the genus *Plocepasser*, than to finches (Fringillidae), based on characters of nest construction and building.

* This genus does not appear in modern bird lists and I have not been able to find any information about it.

Sushkin (1927) also considered that the sparrows' domed nest and side entrance showed that they were closer to weavers than to finches with their open-cup nests. Collias and Collias (1964), however, state that the weaving ability of sparrows is much less well developed and the nests of Passerinae do not bear any close resemblance to those of the other subfamilies of Ploceidae (*viz* Bubalornithinae, Plocepasserinae, Ploceinae, Sporopipinae and Estrildinae) and concluded that their relationship to weavers is uncertain.

Harrison (1965) suggests that the wing-flicking, nest-advertising display of *Passer moabiticus* shows an affinity between *Passer* and Ploceidae. I think this is unlikely as this display is restricted to *Passer luteus, eminibey* and *moabiticus*, whereas the normal display in the genus is a shivering of the drooped wings. The similarity of this display to that of some ploceids is more likely to be the result of convergent adaptation to a similar type of nesting habitat than of genuine relationship.

Both *Passer* and *Petronia* tend to be social species, with most species of *Passer* nesting in loose colonies. In this they show a resemblance to weavers. Against this, both sexes in the sparrows, with the possible exception of *P. luteus* and *P. eminibey*, feed the young, whereas normally this is done in weavers only by the female. Also in contrast, Moreau (1967) pointed out that sparrows differ from Ploceinae in that they dust bathe, whereas the latter do not.

The taxonomic criteria discussed so far have the disadvantage that they are largely concerned with adaptive characters – feeding mechanisms, flight mechanisms, behaviour – and this makes it difficult to differentiate between primitive (phylogenetic) relationships and convergence. This has led to the use of more fundamental methods – cytology, serology, and biochemical techniques involving the genetic material itself – to determine fundamental relationships.

CYTOLOGY

Bulatova (1973) concluded that the karyotypes of *Pyrgolauda* (= *Montifringilla*), *Montifringilla* and *Petronia* are much closer to those of the finches (Fringillidae) than to *Passer*.

SEROLOGY

Sibley (1970) has shown that the electrophoretic patterns of the egg-white proteins of *Passer* differ strikingly from those of ploceines and are closer to emberizines. In contrast, *Montifringilla* show a much better match for *Ploceus* than for *Passer*. *Philetairus*, which most authors, who treat Passerinae as a subfamily of Ploceidae, include in Passerinae, was placed by Sibley as closer to *Passer* than to the ploceines, though he considered the degree of its relationship to *Passer* remained to be determined.

The composition of the secretion of the uropygial (preen) gland of the sparrow is more like that of Fringillidae and Emberizidae than that of Ploceidae (Poltz & Jacob 1974).

DNA ANALYSIS

Using a technique known as DNA × DNA hybridisation, Sibley and Ahlquist (1983) have produced, over a number of years, a series of avian phylogenies of increasing refinement. In recent years their ideas on the immediate relationships of sparrows to weavers and finches have changed. In 1981 (Sibley & Ahlquist 1981) all three of the taxa in question have been given family status (Passeridae, Ploceidae,

Fringillidae) in the superfamily Fringilloidea. In a later treatment (Sibley & Ahlquist 1983) the family Passeridae disappeared and Passerinae was joined with Ploceinae in the family Ploceidae, the *Passer* and Ploceid lines separating about 36 M.Y.A. (million years ago). The most recent phylogeny (Sibley & Ahlquist 1985, 1986; Sibley, Ahlquist & Monroe in press) adheres to this basic scheme, though Passeroidea has been substituted for Fringilloidea as the superfamily name and the family name has been changed from Ploceidae to Passeridae, with the branching dates respectively at 42 and 36 M.Y.A. As Sibley and Ahlquist (1985) say themselves 'these divergences are so close together* that they cannot be certain of the exact sequence of the branchings.' It is clear, however, that in their view sparrows are more closely related to weavers than to finches.

PROPOSED CLASSIFICATION OF SPARROWS AND THEIR ALLIES

As we have seen, most systematists of the last century (*eg* Sharpe 1888) placed sparrows and the following genera, that are usually grouped with them, *Montifringilla, Petronia, Gymnor(h)is, Carpospiza, Auripasser, Sorella*, with finches, largely on account of wing structure, but no recent authors have included them in Fringillidae. Sushkin (1927) was the first to argue strongly that they should be placed as a subfamily, Passerinae, of Ploceidae, and this has had the support of many authorities since then (*eg* Sclater 1930, Mayr & Amadon 1951, White & Moreau 1958, Moreau 1962), though differences occur as to whether the plocepasserine genera should be included in Passerinae.

Sushkin's arguments for placing sparrows and their allies in Ploceidae have come increasingly under criticism (*eg* Bates 1934, Sibley 1970, Bock & Moroney 1978b). A significant number of authorities has expressed doubts about the affinity of sparrows and has considered them sufficiently different from Ploceidae to warrant full family rank, Passeridae (Grönvold, quoted by Bannerman 1948, Pocock 1966, Sibley 1970, Bock & Moroney 1978b). Voous (1977), after a full consideration of all the facts, came to the conclusion, in his *List of Recent Holoarctic Bird Species*, that Old World Sparrows (*Passer*), Rock Sparrows (*Petronia*) and Snow Finches (*Montifringilla*) were best treated as a taxonomic unit (Passeridae) of unknown, though probably distant, relationship to Ploceidae. Voous was concerned only with Holarctic species and thus expressed no opinion on the African genera *Plocepasser, Histurgops, Pseudonigrita* and *Philetairus* that have usually been placed with sparrows in the subfamily Passeridae. Bock and Moroney (1978b), with new evidence, came to the same conclusion regarding *Passer, Petronia* and *Montifringilla*.

For a non-specialist the taxonomic evidence is very conflicting and it is difficult to know how much weight should be put on the different characters that have been used in discussing the relationships of Old World Sparrows: how far these are ancestral and how far derived and attributable to convergence. Zoological classification is more than merely a means of communication – 'simply to provide a convenient practical means by which zoologists may know what they are talking about' (Simpson 1961) – it is primarily concerned with relationships (phylogeny) – 'avian classification should express the content of natural groups, that is taxa that are our best estimates of genealogically related groups of species' (Cracraft 1981). For this reason I do not consider it helpful to treat Old World Sparrows as a subfamily of Ploceidae as long as there are reasonable doubts that suggest otherwise.

Pending further evidence that confirms the position of Old World Sparrows as a subfamily, Passerinae, of the family Ploceidae (or Passeridae, see Sibley 1986) that also includes the weavers, Ploceinae, and the waxbills, Estrildinae, I prefer to keep them as a separate family, Passeridae, of the superfamily Fringilloidea (or Pas-

* In terms of accuracy of the dating techniques (author).

seroidea) that includes all the seed-eating oscines (*ie* the species currently placed in Fringillidae, Ploceidae and Estrildidae). This follows the conclusion reached by Friedmann (1960), when he considered the relationship of the sparrows to Fringillidae and Ploceidae, by Voous (1977) in his *List of Recent Holoarctic Bird Species*, by Cracraft (1981) in his recent attempt at a natural phylogenetic classification, and by the Committee on Classification and Nomenclature of the American Ornithologists' Union (American Ornithologists' Union 1983).

GENERA IN THE FAMILY PASSERIDAE

The following genera have at one time been included in the subfamily Passerinae: *Passer* (including *Auripasser* and *Sorella*), *Petronia*, *Pyrgilauda*, *Onychostruthus* (?), *Plocepasser*, *Histurgops*, *Pseudonigrita*, *Philetairus* and *Sporopipes*.

Sushkin (1927) placed *Sporopipes* closer to Ploceinae than to Passerinae. Pocock (1966) also considered that *Sporopipes* and *Passer* were not closely related. Ziswiler (1968) concluded that *Sporopipes* is a distinct, specialised member of the Ploceidae that should be placed as a monotypic subfamily, Sporopipinae. Clench (1970) reached the same conclusion on the basis of the body pterylosis.

The evidence for the four plocepasserine genera (*Plocepasser*, *Histurgops*, *Pseudonigrita* and *Philetairus*) is less clear cut, but on the basis of their tongue musculature, obviously a key factor in the phylogeny of the seed-eating oscines, Bock and Moroney (1978a) considered that they were more closely related to ploceids than to passerine-finches (genera *Passer*, *Montifringilla* and *Petronia*). Until fresh studies are made that throw further light on this question, I consider that the plocepasserine genera are best retained as a subfamily, Plocepasserinae, of Ploceidae, for which the English term 'sparrow-weavers' can continue to be applied.

Montifringilla is more difficult to place. Bulatova (1938) and Sibley (1970) both separate it from the Passeridae, the former author even dissociating *Petronia* from *Passer*, placing this genus and *Montifringilla* with the finches. Against this, Bock and Moroney (1978b) strongly argue for the inclusion of *Montifringilla* in Passeridae.

In addition to similarities in wing structure and the complete post-juvenile moult, *Montifringilla* also resembles *Passer* and *Petronia* in that the bill of the male changes from horn-coloured to black as the bird comes into breeding condition. For these reasons I favour the retention of *Montifringilla* in Passeridae, for which I retain the old term 'passerine-finches'.

The classifications for the passerine-finches and the sparrow-weavers are given in Figs. 113 and 114. On the basis of observations of the breeding behaviour of *Petronia*

PASSERIDAE

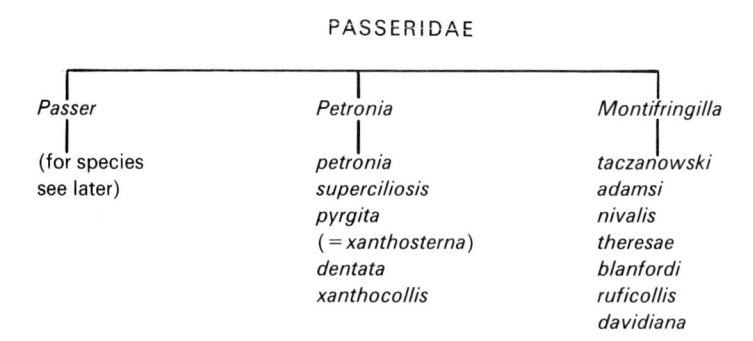

Passer	*Petronia*	*Montifringilla*
(for species see later)	petronia superciliosis pyrgita (=xanthosterna) dentata xanthocollis	taczanowski adamsi nivalis theresae blanfordi ruficollis davidiana

Fig. 113   Proposed classification for passerine-finches, family Passeridae

PLOCEIDAE

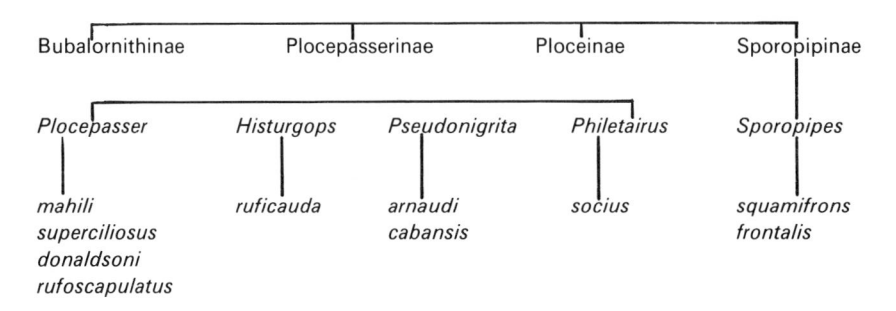

```
Bubalornithinae      Plocepasserinae           Ploceinae              Sporopipinae

Plocepasser    Histurgops      Pseudonigrita     Philetairus         Sporopipes

mahili         ruficauda       arnaudi           socius              squamifrons
superciliosus                  cabansis                              frontalis
donaldsoni
rufoscapulatus
```

Fig. 114   Proposed classification for sparrow-weavers, subfamily Plocepasserinae

brachydactyla (= *Carpospiza brachydactyla*), Prof. Mendelssohn and his co-workers at the University of Tel Aviv consider that this species is a cardueline finch and should thus be transferred to the Fringillidae in the monotypic genus *Carpospiza* (see Bock & Moroney 1978b). For this reason *brachydactyla* has been excluded from *Petronia*.

The species in Passeridae mostly build covered nests. This is always the case if the nest is in the branches of a tree or bush, where it is domed with an entrance hole on the side, sometimes with a short entrance tunnel. In a cavity, such as a hole in a tree or rock face, the nests of *Passer* and *Petronia* may be domed over, completely filling the space, though where the cavity is small the nest may be reduced to a lining on the bottom. Some species use both types; *eg Passer domesticus* and *Petronia xanthocollis* nest both openly in trees and also in holes; in the latter site the nest may be domed or a mere pad in the bottom of the hole. The species of *Montifringilla* nest normally in holes in the ground, usually in animal burrows. Little information is available on the nest construction, but again it appears that it can either be a bulky mass, domed over to fill the hole completely, or merely a lining at the bottom of the hole. In all cases with the three genera the nests are loose bundles of material, not woven as is characteristic of true weavers, Ploceinae, or even with 'thatched roofs' as is the case with sparrow-weavers, Plocepasserinae (Collias & Collias 1964).

THE GENUS *PASSER*

Having given my views on the position of the True Sparrows, I should now like to look at the relationships of the taxa within the genus *Passer*. Here I have used a wider range of categories with the view that the nomenclature used should be appropriate to reveal the subtle differences in relationship that exist. The use of nomenclature in this way is not new. Dr R. F. Johnston, one of the leading workers on sparrows, when considering the complex sparrow situation in the Mediterranean basin (Johnston 1969a), advocated the elevation of the Italian Sparrow, the intermediate between the House Sparrow and the Willow Sparrow that I place as a subspecies of the latter, to full specific rank, *P. italiae*, on the grounds that they 'have resulted from events long since gone and since today most of them ... are in poor genetic contact with the parental species'. In contrast, the same author, in his studies on the evolution of the House Sparrow since its introduction to North America, found greater variation in morphology among American House Sparrows than in the whole *domesticus-italiae-hispaniolensis* complex in Europe, and in this context was more inclined to treat them

Table 50: *Classification of the genus* Passer

Genus	Subgenus	Superspecies*	Species	Semi-species†	Subsp.
Passer	Passer	[*griseus*][1]	*griseus*		3
			swainsonii		—
			gongonensis		—
			suahelicus		—
			diffusus		3
		[*iagoensis*][2]	*iagoensis*[2]		—
			motitensis[3]	*motitensis/benguellensis/subsolanus*	3
				cordofanicus	—
				shelleyi	—
				rufocinctus	—
				insularis	—
			melanurus[4]		3
		[*domesticus*][5]	*predomesticus*[6]		
			domesticus[7]	'*domesticus*'-group	5
				'*indicus*'-group	6
			hispaniolensis[8]	*hispaniolensis/transcaspicus*	2
			italiae[9]		—
			simplex[10]	*simplex*	—
				zarudyni	—
			castanopterus[11]		—
			moabiticus[12]	*moabiticus*	—
				yatii	—
			pyrrhonotus[13]		—
			flaveolus		—
			rutilans[14]		3
			ammodendri		5
			montanus		6
	Auripasser	[*auripasser*][15]	*luteus*[16]	*luteus*	—
				euchlorus	—
			eminibey		—

* 'A group of entirely or essentially allopatric taxa that were once races of a single species but which now have achieved species status.' (Amadon 1966a.)

† '... the term semispecies ... designates populations which have part way completed the process of speciation. Gene exchange is still possible among semi-species, but not as freely as among conspecific populations.' (Amadon 1966a.)

NOTES:
  1. The five grey-headed sparrows are separated into five species on the basis of small zones of sympatry, though in other areas they intergrade. It seems best to treat them as members of a superspecies.
  2. *P. iagoensis* closely resembles *P. motitensis*, but is separated by behaviour and size (Summers-Smith 1984a). I place them as members of a superspecies.
  3. *P. motitensis* consists of five allopatric populations. No detailed comparative study has been made of these birds, though some observations suggest that considerable differences exist,

indicating some divergence. Ecological differences are almost certainly developing and it seems best at this stage to treat the five populations as emerging species (*viz* semi-species).

4. The relationship of *P. melanurus* is obscure.

5. The [*domesticus*] superspecies includes the extinct *predomesticus* (see 6, below), *P. domesticus*, *P. hispaniolensis*, *P. simplex* and *P. castanopterus*.

6. *P. predomesticus* is an extinct species described from fossil remains from Palestine (Tchernov 1962). It appears to have been a precursor of *P. domesticus* and *P. hispaniolensis*.

7. *P. domesticus* can be divided into two distinct groups: a Palaearctic and an Oriental one (using the latter term to reflect evolutionary origin rather than present distribution, which includes Palaearctic forms), which Vaurie (1956) suggests may have had separate evolutionary centres. The Palaearctic or '*domesticus*'-group includes the subspecies *domesticus*, *tingitanus*, *biblicus*, *persicus* and *niloticus*; the Oriental or *indicus*-group, members of which are brighter coloured and mostly rather smaller, includes *indicus*, *parkini*, *bactrianus*, *hufufae*, *hyrcanus* and *rufidorsalis*. A number of authors have argued that the Oriental birds are sufficiently distinct from the Palaearctic ones to merit specific separation. For example, Gavrilov and Korelov (1968) found that not only do *domesticus* and *bactrianus* differ morphologically, but that they overlap extensively in Kazakhstan without interbreeding; Yakobi (1979), also, considered that the Indian Sparrow (*bactrianus*) should be separated from the House Sparrow (*domesticus*), basing his conclusion on behavioural grounds.

I do not consider the differences justify specific separation, but recognise that there are differences by placing the '*domesticus*'- and '*indicus*'-groups as semi-species.

8. All authorities are agreed that the House and Willow Sparrows are very closely related. Baumgart (1984) describes them as 'time-differentiated species' with identical ecological requirements. The extensive hybrid zone in North Africa and the regular occurrence of hybridisation in other parts where they are sympatric justify including *P. hispaniolensis* in the [*domesticus*] superspecies.

9. *P. h. italiae* is a stabilised hybrid between *P. hispaniolensis* and *P. domesticus* that a number of authors have suggested merits specific rank (*eg* Johnston 1969a). Because of the zones of intergradation with the parent species, I do not consider this view can be sustained. In my view it is better placed as a subspecies of *hispaniolensis*, rather than the conventional one that it is a subspecies of *domesticus*.

10. Hall and Moreau (1970) place *P. simplex* in the [*domesticus*] superspecies. It appears to be basically a desert form of *domesticus*. The two subspecies *simplex* and *zarudnyi* are so widely separated, both geographically and almost certainly in time, that I consider them as semi-species.

11. I find *P. castanopterus* closely related to *P. domesticus* on behavioural grounds; also wild hybrids between the two species have been reported. This, I consider, warrants including it in the [*domesticus*] superspecies. I think it is quite possibly a secondary invader of the Afrotropical region, following its evolution from a common ancestor with *P. domesticus* in the Levant.

12. When first described, *P. yatii* and *P. mesopotamicus* were treated as separate species. The latter is now considered to be synonymous with *P. moabiticus*. A recent comparative study of the morphology and plumage of *moabiticus* and *yatii* by Boros and Horvath (1954) led them to the view that *P. m. yatii* should be treated as a full species, though apparently they did not consider any specimens from southern Iraq and southwest Iran, from where *mesopotamicus* was described; such birds tend to be intermediate between *moabiticus* and *yatii*. I consider *moabiticus* and *yatii* are best treated as semi-species.

13. *P. pyrrhonotus* is a small replica of *P. domesticus*, but otherwise does not show any particularly close relationship.

14. Meinertzhagen (1951) considered that *P. rutilans* was so alike *P. castanopterus* that it was a race of the latter. Hall and Moreau (1970) placed *P. rutilans* in the [*domesticus*] superspecies; Johnston and Klitz (1977) also found them very close phenetically. I do not find any grounds to support this view and prefer to keep *P. rutilans* separate.

15. *P. luteus* and *P. eminibey* are very distinct from the remainder of the genus; both are very small and show, in males, a stepwise transition in plumage from all yellow (*P. l. euchlorus*) through yellow with a chestnut back (*P. l. luteus*) to all chestnut (*P. eminibey*). I consider this justifies placing them together in a superspecies.

16. *P. l. luteus* and *P. l. euchlorus* appear sufficiently distinct to warrant treating them as semi-species.

as all belonging to one species (Johnston 1973). Both courses are equally valid and appropriate in the circumstances in which they were considered. Gavrilov and Korelov (1968) have discussed another problem in the classification of the species, the position of the Indian sparrows. As already discussed in the nomenclature section of Chapter 9, this group appears to lie somewhat between subspecies of the House Sparrow and a fully independent species.

My proposed classification of the genus is given in Table 50. This table, together with the appended notes, is designed to highlight the relationships of the twenty taxa already treated as full species in the text, and also to indicate those taxa (semi-species) that appear to me to have evolved further than merely to racial status and are likely to become full species in future.

Appendix A

A KEY TO THE NAMES OF THE SPARROWS

abyssinicus	*Passer griseus abyssinicus* Neumann 1908. Ghadi-Saati, Mareb River, Eritrea. Synonym of *Passer swainsonii*.
aegyptiaca	*Pyrgita aegyptiaca* Brehm 1831. Egypt. See *aegyptiacus*.
aegyptiacus	*Passer aegyptiacus* (Brehm) 1831. Egypt. Synonym of *P. hispaniolensis*.
ahasvar	*Passer ahasvar* Kleinschmidt 1904. Marrakech. Synonym of *P. domesticus tingitanus*; intergrades with *P. hispaniolensis*.
albiventris	*Passer albiventris* Madarász 1911. Sudan. Synonym of *P. griseus ugandae*.
alexandrinus	*Passer alexandrinus* Madarász 1911. Alexandria. Synonym of *P. domesticus niloticus*.
ammodendri	*Passer ammodendri* Gould 1872. Turkestan; Djulek above Kyzl Orda on the Syr Darya according to Hartert 1904. Species of *Passer*; subspecies of *P. ammodendri*.
annectans	*Passer rutilans annectans* Koelz 1952. Mawryngkneng, Khasi Hills. Synonym of *P. rutilans intensior*.
Arboreus	*Passer Arboreus* Foster 1817. Substitute name for *Fringilla montana* Linnaeus. See *montana*.
arboreus	*Passer arboreus* Bonaparte 1850. Sennar, eastern Sudan. Synonym of *Passer domesticus rufidorsalis*.
arcuata	*Fringilla arcuata* Gmelin 1788. Cape of Good Hope. See *arcuatus*.
arcuatus	*Passer arcuatus* (Gmelin) 1788. Cape of Good Hope. Synonym of *Passer melanurus melanurus*.
arrigoni	*Passer hispaniolensis arrigoni* Tschusi 1903. La Maddalena, Sardinia. Synonym of *P. h. hispaniolensis*.
assimilis	*Passer assimilis* Walden 1870. Tonghoo [Toungoo], Burma. Synonym of *P. flaveolus*.
Auripasser	*Auripasser* Bonaparte 1851. Subgenus of *Passer*.
bactrianus	*Passer domesticus bactrianus* Zarudny & Kudashev 1916. Merv [Mary], Tashkent; the type is from Tashkent according to Meinertzhagen 1938. Subspecies of *P. domesticus*.

Appendix A

baicalicus	*Passer domesticus baicalicus* Keve 1943. Kultuk, southern Lake Baikal. Synonym of *P. d. domesticus*.
balaeroibericus	*Passer domesticus balaeroibericus* von Jordans 1923. Valldemosa, Mallorca. Synonym of *P. d. domesticus*.
batanguensis	*Passer rutilans batanguensis*. Batang, Sichuan. Synonym of *P. r. rutilans*.
benguellensis	*Passer iagoensis benguellensis* Lynes 1926. Huxe, Benguella [southern Angola]. Subspecies of *P. motitensis*.
bergeri	*Passer italiae bergeri* Zedlitz 1912. North west Africa. Intergrades between *P. domesticus tingitanus* and *P. h. hispaniolensis*.
biblicus	*Passer domestica biblicus* Hartert 1904.
	= *Passer domesticus biblicus* Hartert 1904. Sueme, Palestine. Subspecies of *P. domesticus*.
boetticheri	*Passer montanus boetticheri* Stachanow 1933. Namsky Ulus [Namskiy], 120 km below Yakutsk on the Lena. Synonym of *P. m. montanus*.
bokotoensis	*Passer montanus bokotoensis* Yamashina 1933. Mako, Bokoto [Hoko Island, Pescadores]. Synonym of *P. montanus dilutus*.
brancoensis	*Passer brancoensis* Oustalet 1883. Branco Island, Cape Verde Islands. Synonym of *P. iagoensis*.
brutius	*Passer italiae* (Vieill.) var. *brutius* de Fiori 1890. Calabria. Synonym of *P. hispaniolensis italiae*; intergrades between *P. hispaniolensis italiae* and *P. h. hispaniolensis* in southern Italy (Apulia), Sicily and Malta.
buryi	*Passer domesticus buryi* Lorenz & Hellmayr 1901. Yeshbum, Aden Protectorate [Yashbum, South Yemen]. Synonym of *P. domesticus indicus*.
canariensis	*Passer hispaniolensis canariensis* Tschusi 1915. La Oliva, Fuerteventura, Canaries. Synonym of *P. h. hispaniolensis*.
carnicus	*Passer domesticus carnicus* Vallon 1912. North Friaul [Friuli], Italy. Intergrades between *P. d. domesticus* and *P. hispaniolensis italiae*.
castanopterus	*Passer castanopterus* Blyth 1855. Somaliland. Species of *Passer*; subspecies of *P. castanopterus*.
catellatus	*Passer catellatus* Kleinschmidt 1935. England; the type is from Sussex according to Clancey 1948. Synonym of *P. m. montanus*.
caucasicus	*Passer domesticus* var. *caucasicus* Bogdanov 1879. Caucasus. Synonym of *P. d. domesticus*.
cheprini	*Passer domesticus cheprini* Phillips 1913. Giza [El Giza], near Cairo. Synonym of *P. domesticus niloticus*.
cinnamomea	*Pyrgita cinnamomea* Gould 1836. *apud montes Himalayenses;* restricted to NW Himalayas by Ticehurst 1927. See *cinnamomeus*.
cinnamomeus	*Passer rutilans cinnamomeus* (Gould) 1936. [NW Himalayas]. Subspecies of *P. rutilans*.
cisalpina	*Fringilla cisalpina* Temminck 1820. Italian Alps. See *cisalpinus*.
cisalpinus	*Passer domesticus cisalpinus* Schlegel 1844.
	= *Passer domesticus cisalpinus* (Temminck) 1820. Italian Alps. Synonym of *P. hispaniolensis italiae*.
ciscaucasicus	*Passer montanus ciscaucasicus* Buturlin 1929. Near Vladikavkauz [Ordzhonikidze], northern Caucasus. Synonym of *P. m. montanus*.
colchicus	*Passer domesticus colchicus* Portenko 1960. Artvin, Turkey. Synonym of *P. domesticus biblicus*.
confucius	*Passer confucius* Bonaparte 1853. China [error: Rangoon]. Synonym of *P. domesticus indicus*.
cordofanicus	*Passer cordofanicus* Heuglin 1871. Melspez, Korodofan, Sudan. Subspecies of *P. motitensis*.
Corospiza	*Corospiza* Bonaparte 1850. Synonym of *Passer*.
damarensis	*Passer arcuatus damarensis* Reichenow 1902. Brakwater, 12 miles south of Windhuk, Damaraland [Windhoek, Namibia]. Subspecies of *P. melanurus*.
debilis	*Passer rutilans debilis* Hartert 1904. Sind-Tal in Kashmir [Sind Valley in Kashmir]. Synonym of *P. rutilans cinnamomeus*.
diffusa	*Pyrgita diffusa* A. Smith 1836. Between the Orange river and the tropic; restricted to 'near Kuruman' by Macdonald & Hall 1957. See *diffusus*.
diffusus	*Passer diffusus* (A. Smith) 1836. [near Kuruman]. Species of *Passer*; member of [*griseus*] superspecies.
diniz	*Passer domesticus diniz* Floericke 1926. Portugal. Synonym of *P. d. domesticus*.
domestica	*Fringilla domestica* Linnaeus 1758. Sweden. See *domesticus*.

domesticus	*Passer domesticus* Pallas 1811. = *Passer domesticus* (Linnaeus) 1758. Sweden. Superspecies and species of *Passer;* subspecies of *P. domesticus.*
dybowskii	*Passer montanus dybowskii* Domaniewski 1915. Ussiri Valley and Korea; restricted to latitude 48°N on the Ussiri by Vaurie 1956. Synonym of *P. m. montanus.*
emini	*Sorella emini* Hartert 1881. = *eminibey qv.*
eminibey	*Sorella eminibey* Hartlaub 1880. = *Passer eminibey* (Hartlaub) 1880. Lado, northern Uganda. Species of *Passer;* member of *Auripasser* subgenus.
enigmaticus	*Passer enigmaticus* Zarudny 1903. Hurmuk [Hormak] and Kamschar, Persian Baluchestan. Synonym of *P. domesticus indicus;* specimens are intersexes according to Mayr 1949.
eritrea	*Passer griseus eritrea* Zedlitz 1911. Tacasse [Tekeze], Eritrea. Synonym of *P. swainsonii.*
erythrophrys	*Passer erythrophrys* Temm. Senegal [error: Cape Verde Islands]. Synonym of *P. iagoensis.*
euchlora	*Auripasser euchlora* Bonaparte 1850. Kunfuda [Al Qunfidhah], southern Arabia. See *euchlorus.*
euchlorus	*Auripasser euchlorus* Bonaparte 1850. = *Passer luteus euchlorus* (Bonaparte) 1850. [Al Qunfidhah, Saudi Arabia]. Subspecies of *P. luteus.*
flaveolus	*Passer flaveolus* Blyth 1844. Arakan, Burma. Species of *Passer.*
flückigeri	*Passer flückigeri* Kleinschmidt 1904. Northern Algerian Sahara. Intergrades between *P. domesticus tingitanus* and *P. h. hispaniolensis.*
fulgens	*Passer castanopterus fulgens* Friedmann 1934. Indunumara Mts., Kenya. Subspecies of *P. castanopterus.*
galliae	*Passer italiae* var. *galliae* Chigi 1904. Upper Italy. Synonym of *P. hispaniolensis italiae.*
georgicus	*Passer diffusus georgicus* Reichenow 1904. Damaraland, South West Africa [Otjimbingwe, Namibia]. Synonym of *P. d. diffusus.*
gongonensis	*Pseudostruthus gongonensis* Oustalet 1890. = *Passer gongonensis* (Oustalet) 1890. Gongoni, near Mombasa. Species of *Passer;* member of [*griseus*] superspecies.
gobiensis	*Passer montanus gobiensis* Stachanow 1933. Chuoy-ho, Gobi du Sud [Ch'ingshui Ho, Ningsia]. Synonym of *P. montanus dilutus.*
grisea	*Fringilla grisea* Vieillot 1817. United States [error: Senegal, Lafresnaye 1839]. See *griseus.*
griseigularis	*Passer griseigularis* Sharpe 1888. Kandahar. Probably synonym of *P. domesticus bactrianus,* but not identifiable as to subspecies.
griseogularis	mis-spelling of *griseigularis qv.*
griseus	*Passer griseus* (Vieillot) 1817. [Senegal]. Superspecies and species of *Passer;* subspecies of *P. griseus.*
guasso	*Sorella eminibey guasso* van Someren 1922. Archer's Post, northern Kenya. Synonym of *P. eminibey.*
gularis	*Pyrgita gularis* Lesson 1839. Senegal. Synonym of *P. g. griseus.*
halfae	*Passer domesticus halfae* Meinertzhagen 1921. Wadi Halfa, northern Sudan. Intergrades between *P. domesticus niloticus* and *P. domesticus rufidorsalis.*
hamburgia	*Loxia hamburgia* Gmelin 1789. Hamburg. Synonym of *P. m. montanus*
Hansmanni	*Passer Hansmanni* Bolle 1856. Cape Verde Islands. Synonym of *P. iagoensis.*
hemileucus	*Passer hemileucus* Ogilvie-Grant & Forbes 1900. Abd el Kuri Island. Synonym of *P. motitensis insularis.*
hepaticus	*Passer montanus hepaticus.* Tezu, Mishmi Hills, NW Assam. Synonym of *P. montanus malaccensis.*
Hispaniae	*Passer montanus Hispaniae* von Jordans 1933. Oreposa, Castellon, eastern Spain. Synonym of *P. m. montanus.*
hispaniolensis	*Fringilla hispaniolensis* Temminck 1820. = *Passer hispaniolensis* (Temminck) 1820. Algeciras, southern Spain. Species of *Passer;* member of [*domesticus*] superspecies; subspecies of *P. hispaniolensis.*

hostilis	*Passer hostilis* Kleinschmidt 1915. Tring, Herts, Great Britain. Synonym of *P. d. domesticus*.
hufufae	*Passer domesticus hufufae* Ticehurst & Cheesman 1924. Hufuf town, Hasa Province, eastern Arabia. Subspecies of *P. domesticus*.
hyrcanus	*Passer domesticus hyrcanus* Zarudny & Kudashev 1916. Astrabad [Gorgān], Gilan and Mazandaran, northern Iran. Subspecies of *P. domesticus*.
iagoensis	*Pyrgita iagoensis* Gould 1838. = *Passer iagoensis* (Gould) 1838. São Tiago, Cape Verde Islands. Species of *Passer*.
ignoratus	*Passer rutilans ignoratus* Deignan 1948. Mount Omei, Szechwan [Emei Shan, Sichuan]. Synonym of *P. r. rutilans*.
indicus	*Passer indicus* Jardine & Selby 1835. India; restricted to Bangalore by Kinnear 1925. Subspecies of *P. domesticus*; recognised as full species by some authorities.
insularis	*Passer insularis* Sclater & Hartlaub 1881. Socotra. Subspecies of *P. motitensis*.
intensior	*Passer rutilans intensior* Rothschild 1922. Mekong valley [Yunnan]. Subspecies of *P. rutilans*.
italiae	*Fringilla italiae* Vieillot 1817. Italy. Subspecies of *P. hispaniolensis*; recognised as subspecies of *P. domesticus* by some authorities, by others as full species, *Passer italiae*.
iubalaeus	*Passer montanus iubalaeus* Reichenow 1907. Caucasus to Tsingtao [Qindao, Shandong]; type locality is Tsingtao according to Hartert & Steinbacher 1932. Synonym of *P. montanus dilutus*.
jagoensis	see *iagoensis*.
jubaensis	*Passer griseus jubaensis* Benson 1942. Mandera, Juba River [Giuba River]. Synonym of *P. gongonensis*.
jubalaeus	see *iubalaeus*.
kaibatoi	*Passer montanus kaibatoi* Munsterhjelm 1916. Kaiba Island, southern Sakhalin. Synonym of *P. montanus saturatus*.
kansuensis	*Passer montanus kansuensis* Stresemann 1932. Heitsuitse, northern Kansu; above Sining on the Sining river, northeastern Tsinghai according to Vaurie 1959. Synonym of *P. montanus dilutus*.
kikuchi	*Passer rutilans kikuchi* Kuroda 1924. Horisha, Nanto [Nan-t'ou] district, central Formosa [Taiwan]. Synonym of *P. r. rutilans*.
kleinschmidti	*Passer griseus kleinschmidti* Grote 1922. Ngaundere [French] Cameroons [N'gaoundéré, Cameroun]. Synonym of *P. g. griseus*.
kordofanicus	see *cordofanicus*.
korejewi	*Passer ammodendri korejewi* Zarudny & Härms 1902. Eastern part of Transcaspia between the foothills of the Paropamisus and the Amu Dar'ya. Subspecies of *ammodendri*.
laeneni	*Passer griseus laeneni* Niethammer 1955. Bol, east bank of Lake Chad. Subspecies of *P. griseus*.
lisarum	*Passer rutilans lisarum* Stresemann 1940. Mount Victoria [Chin Hills, southern Burma]. Synonym of *P. rutilans intensior*.
loangwae	see *luangwae*.
luangwae	*Passer diffusus luangwae* Benson 1956. Mupamadzi River, Mpika area, Luangwa Valley, Northern Rhodesia [Zambia]. Subspecies of *P. diffusus*.
lutea	*Fringilla lutea* Lichtenstein 1823. Dongola, northern Sudan. See *luteus*.
lutea	*Auripasser lutea* Bonaparte 1850. Kunfuda [Al Qunfidhah, Saudi Arabia]. See *euchlorus*.
luteus	*Passer luteus* (Lichtenstein) 1823. Dongola, northern Sudan. Species of *Passer*; member of *Auripasser* subgenus; member of [*auripasser*] superspecies; subspecies of *P. luteus*.
malaccensis	*Passer montanus malaccensis* Dubois 1885. Malacca [Melaka]. Subspecies of *P. montanus*.
maltae	*Passer hispaniolensis maltae* Hartert 1902. Malta, Sicily. Intergrades between *P. hispaniolensis italiae* and *P. h. hispaniolensis*.
manillensis	*Passer montanus manillensis* Hachisuka 1941. Manila. Synonym of *P. montanus saturatus*.
margaretae	*Passer montanus margaretae* Johansen 1944. Western Siberia. Synonym of *P. m. montanus*.

maximus	*Passer montanus maximus* Schäfer 1938. Jyekundo (Yü-shu], southern Tsinghai. Synonym of *P. montanus tibetanus*.
melanura	*Loxia melanura* Müller 1776. Cape of Good Hope; restricted to Cape Town by Macdonald 1957. See *melanurus*.
melanurus	*Passer melanurus* (Müller) 1776. [Cape Town]. Species of *Passer*, subspecies of *P. melanurus*.
melitensis	*Passer domesticus melitensis*. Malta. Intergrades between *P. hispaniolensis italiae* and *P. h. hispaniolensis*.
mesopotamicus	*Passer mesopotamicus* Zarudny 1904. Mochammera [Khorramshahr, Khuzestān], southwestern Iran. Synonym of *P. m. moabiticus*.
minor	*Pyrgita minor* Brehm 1842. Egypt and Buchara. Synonym of *P. hispaniolensis*.
moabiticus	*Passer moabiticus* Tristram 1864. Palestine [southern end of Dead Sea]. Species of *Passer*; subspecies of *P. moabiticus*.
montana	*Fringilla montana* Linnaeus 1758. 'In Europe'; restricted to Bagnacavallo, Ravenna, Italy, by Clancey 1948. See *montanus*.
montanus	*Passer montanus* Pallas 1811. = *Passer montanus* (Linnaeus) 1758. [Bagnacavallo, Ravenna, Italy]. Species of *Passer*; subspecies of *P. montanus*.
mosambicus	*Passer griseus mosambicus* van Someren 1921. Lumbo, Portuguese East Africa [Mozambique]. Subspecies of *P. diffusus*.
motitensis	*Pyrgita motitensis* A. Smith 1836. = *Passer motitensis* (A. Smith) 1836. Old Lakatoo, 60 miles south of Orange River [Error: Motita, near Old Lakatoo, 135 miles north of Orange river, Winterbottom 1966]. Species of *Passer*.
mozambicus	see *mosambicus*.
neumanni	*Passer griseus neumanni* Zedlitz 1908. Salamona, about 16 miles west of Massawa, eastern Eritrea. Synonym of *P. swainsonii*; intergrades with *P. griseus ugandae*.
nigricans	*Passer ammodendri nigricans* Stepanyan 1961. Northern Sinkiang. Subspecies of *P. ammodendri*.
nigricollis	*Pyrgita nigricollis* Burton 1838. South India. Synonym of *P. domesticus indicus*.
nikersoni	*Passer nikersoni* Madarász 1911. Chor-em-Dul, Sennar district, Sudan. Synonym of *P. griseus ugandae*.
niloticus	*Passer domesticus niloticus* Nicoll & Bonhote 1909. El Faiyûm, Egypt. Subspecies of *P. domesticus*.
obscuratus	*Passer montanus obscuratus* Jacobi 1923. Central China, Hupeh and Szechwan [between Wanhsien and I-ch'ang]. Intergrades between *P. montanus dilutus* and *P. m. malaccensis*.
occidentalis	*Passer occidentalis* Shelley 1883. West Africa; Lokoja, south Nigeria according to Lynes 1926. Synonym of *P. g. griseus*.
orientalis	*Pyrgita orientalis* Brehm 1855. 'In the East'. Synonym of *P. hispaniolensis*.
orientalis	*Passer montanus orientalis* Clark 1910. Hakodate and Fusan [Hokkaido and Korea]; restricted to Korea by Deignan 1952. Synonym of *P. montanus saturatus*.
pagorum	*Pyrgita pagorum* Brehm 1831. Central Germany. Synonym of *Passer domesticus*.
pallidissimus	*Passer montanus pallidissimus* Stachanow 1933. 'Harma Bouroung', eastern Zaidam, northern Tsinghai [Qaidam, Qinghai]. Synonym of *P. montanus dilutus*.
pallidus	*Passer montanus* Brisson var. *pallidus* Zarudny 1904. Eastern Iran [Eastern Khorāsān]. Synonym of *P. montanus dilutus*.
parkini	*Passer domesticus parkini* Whistler 1920. Srinagar, Kashmere [Kashmir]. Subspecies of *P. domesticus*.
parvirostris	*Pyrgitopsis rutilans parvirostris* Momiyama 1927. Quelpart [Cheju] Island. Synonym of *P. r. rutilans*.
Passer	*Passer* Brisson 1760.
payni	*Passer italiae payni*. Corsica. Synonym of *P. hispaniolensis italiae*.
persicus	*Passer domesticus persicus* Zarudny & Kudashev 1916. Arabistan [Khuzestān], southwestern Iran. Subspecies of *P. domesticus*.
predomesticus	*Passer predomesticus* Tchernev 1962. Oumm-Qatafa Cave, Wadi Khareitoun, near Bethlehem, Israel. Extinct species of *Passer*.

312 Appendix A

Pseudostruthus	*Pseudostruthus gongonensis* Oustalet 1890. = *Passer gongonensis* (Oustalet) 1890. Gongoni, near Mombasa. Species of *Passer*.
Pyrgita	*Pyrgita* Cuvier 1817. Synonym of *Passer*.
pyrrhonotus	*P*[*asser*] *pyrrhonotus* Blyth 1844. Bahawalpur, Sind. Species of *Passer*.
rikuzenica	*Passer montanus rikuzenica* Kumagai 1928. Wakayanagi, Hondo. Synonym of *P. montanus saturatus*.
romae	*Passer italiae* var. *romae* Chigi 1904. Rome, Italy. Synonym of *P. hispaniolensis italiae*.
ruficinctus	see *rufocinctus*.
rufidorsalis	*P*[*asser*] *rufidorsalis* Brehm 1855. Northeast Africa; restricted by Khartoum, Sudan, by Vaurie 1959. Subspecies of *P. domesticus*.
rufipectus	*Passer rufipectus* Bonaparte 1850. Egypt. Hybrid *P. domesticus* × *P. hispaniolensis*.
rufocinctus	*Passer rufocinctus* Finsch & Reichenow 1884. Lake Naivasha, Kenya. Subspecies of *P. motitensis*.
russatus	*Passer russatus* Temminck & Schleg, 1850. Japan. Synonym of *P. rutilans*.
rustica	*Pyrgita rustica* Brehm 1831. Griefswald, Germany [GDR]. Synonym of *P. domesticus*.
rutilans	*Fringilla rutilans* Temminck 1835. = *Passer rutilans* (Temminck) 1835. Japan. Species of *Passer*; subspecies of *P. rutilans*.
saharae	*Passer simplex saharae* Erlanger 1899. Tunisian Sahara [Jebel Dekanis according to Hilgert 1908]. Synonym of *P. s. simplex*.
Salicaria	*Pyrgita Salicaria* Bonaparte 1838. Clerical mistake for *salicicola* according to Hartert 1904. Synonym of *P. hispaniolensis*.
salicarius	*Passer salicarius* Keys & Blas 1840. Synonym of *P. hispaniolensis*.
salicicola	*Fringilla salicicola* Vieillot 1828. Provence, France, and Spain. Synonym of *P. hispaniolensis*.
saturatus	*Passer saturatus* Stejneger 1885. Riu Kius [Ryūkyū Retto]; Okinawa according to Phillips 1947. Subspecies of *P. montanus*.
scandens	*Loxia scandens* Hermann 1783. France. Synonym of *P. m. montanus*.
schaeferi	*Passer rutilans schaeferi* Stresemann 1939. Süd-Tibet; Shigatse. Synonym of *P. rutilans cinnamomeus*.
schiebeli	*Passer italiae schiebeli* Rokitansky 1934. Canea, Crete. Synonym of *P. hispaniolensis italiae*.
semiretschiensis	*Passer domesticus semiretschiensis* Zarudny & Kudashev 1916. Verny [Alma-Ata], Djarkent [Panfilov] and Przhevalsk, Russian Turkestan. Synonym of *P. d. domesticus*.
senckenbergianus	*Passer italiae senckenbergianus* Hartert 1904. Egypt. Intergrades between *P. domesticus niloticus* and *P. hispaniolensis*.
septentrionalis	*Pyrgita septentrionalis* Brehm 1831. Denmark. Synonym of *P. m. montanus*.
shansiensis	*Passer montanus shansiensis* Yamashina & Kiyosu 1943. Shansi. Synonym of *P. montanus dilutus*.
shelleyi	*Passer shelleyi* Sharpe 1891. Lado, southern Sudan. Subspecies of *P. motitensis*.
sibiricus	*Passer domesticus sibiricus*. Tyumen, Tomsk. Synonym of *P. d. domesticus*.
simplex	*F*[*ringilla*] *simplex* Lichtenstein 1823. = *Passer simplex* (Lichtenstein) 1823. Ambukol [Ambikol] on the Nile, Sudan. Species of *Passer*; subspecies of *P. simplex*.
simplex	*Corospiza simplex* Bonaparte 1858. Algeria. As above.
sititoi	*Passer montanus sititoi* Momiyama 1940. Seven Islands of Izu [Izu-Shotō]. Synonym of *P. montanus saturatus*.
Sorella	*Sorella* Hartlaub 1880. Synonym of generic name *Passer* and subgeneric name *Auripasser*.
soror	*Passer domesticus soror*. Nikawella State Farm, Rattota, Matale district, Ceylon [Sri Lanka]. Synonym of *P. domesticus indicus*.
stegmanni	*Passer montanus stegmanni* Dementiev 1933. Yakutsk. Synonym of *P. m. montanus*.
stolickzae	*Passer stolickzae* Hume 1874. About 4 miles east of Kashgar. Subspecies of *P. ammodendri*.
stygiceps	*Passer diffusus stygiceps* Clancey 1954. Umzinyatti Falls, Inanda, near Durban, Natal. Synonym of *P. d. diffusus*.

Appendix A   313

suahelicus — *Passer griseus suahelicus* Reichenow 1904. Bussissi [Mwanza district, northern Tanzania]. Species of *Passer*; member of [*griseus*] superspecies.
subsolanus — *Passer motitensis subsolanus* Clancey 1964. Ingwezi Ranch, Syringa, Matabeleland, Southern Rhodesia [Zimbabwe]. Subspecies of *P. motitensis*.
swainsonii — *Pyrgita swainsonii* Rüppell 1840.
= *Passer swainsonii* (Rüppell) 1840. Northern Abyssinia [Ethiopia]. Species of *Passer*; member of [*griseus*] superspecies.
taivanensis — *Passer montanus taivanensis* Hartert 1904. Taihoku, Formosa [Taiwan]. Synonym of *P. m saturatus*.
tauricus — *Passer domesticus tauricus* Portenko 1960. Simferopol, Crimea. Synonym of *P. d. domesticus*.
terekius — *Passer hispaniolensis terekius* Buturlin 1929. Malaia, Areshevka [Bol'shaya Areshevka], Kizlar district, Terek delta, northern Caucasus. Synonym of *P. hispaniolensis transcaspicus*.
tertale — *Passer griseus tertale* Benson 1942. 30 miles west of Yavello, south Abyssinia [Yabelo, Ethiopia]. Synonym of *P. gongonensis*; intergrades with *P. swainsonii*.
thierryi — *Passer diffusus thierryi* Reichenow 1899. Mangu [Sansanné, Mango], Togo. Synonym of *P. g. griseus*.
tibetanus — *Passer montanus tibetanus* Baker 1925. Khumbajong, Tibet. Subspecies of *P. montanus*.
tilmenensis — *Auripasser luteus tilmenensis* Bates 1932. Taberréshat [northeast of Bourem], French Sudan [Mali]. Synonym of *P. l. luteus*.
timidus — *Passer timidus* Sharpe 1888. Gobi Desert. Subspecies of *P. ammodendri*.
tingitanus — [*Passer domesticus*] *A. Tingitanus* Loche 1867.
= *Passer tingitanus* v. Homeyer 1869. Algeria. Subspecies of *P. domesticus*.
tokungai — *Passer montanus tokungai* Kuroda & Yamashina 1935. Chihfeng, Jehol, southern Manchuria. Synonym of *P. montanus dilutus*.
transcaspicus — *Passer hispaniolensis transcaspicus*. Transcaspia eastwards; [the type is from Iolotan' according to Tschusi 1903]. Subspecies of *P. hispaniolensis*.
transcaucasicus — *Passer montanus transcaucasicus* Buturlin 1906. Akhalzykh [Akhaltsikke], Transcaucasia. Subspecies of *P. montanus*.
turkanae — *Passer griseus turkanae* Granvik 1934. Lotonok, Turkana, northwestern Kenya. Synonym of *P. gongonensis*; intergrades with *P. swainsonii*.
ugandae — *Passer diffusus ugandae* Reichenow 1904. Uganda (Manjonga). Subspecies of *P. griseus*.
valloni — *Passer italiae* (Vieill.) var. *valloni* Chigi 1906. East Frial [Friuli], Italy. Intergrades between *P. d. domesticus* and *P. hispaniolensis italiae*.
vicinis — *Passer melanurus vicinis* Clancey 1958. Bethlehem, eastern Orange Free State. Subspecies of *P. melanurus*.
volgensis — *Passer montanus volgensis* Ognev 1913. Volga delta. Synonym of *P. montanus*.
yatii — *Passer yatii* Sharpe 1888. Dedadi, western Afghanistan [Sistan]. Subspecies of *P. moabiticus*.
yunnanensis — *Passer rutilans yunnanensis* La Touche 1923. Lotukow, SE Yunnan. Synonym of *P. rutilans intensior*.
zaissanensis — *Passer montanus zaissanensis* Poliakov 1911. Kara Irtysh, in Zaissan Nor region [Lake Zaysan, Kazakhstan]. Synonym of *P. montanus dilutus*; intergrades with *P. m. montanus*.
zarudnyi — *Passer simplex zarudnyi* Pleske 1896. Transcaspia. Subspecies of *P. simplex*.
zedlitzi — *Passer griseus zedlitzi* Gyldenstolpe 1922. Near Benguella town, Portuguese West Africa [Angola]. Synonym of *P. g. ugandae*.

Appendix B

GAZETEER

List of place names and geographical locations not shown in either the *Times Atlas of the World: Comprehensive Edition*, Times Books, 5th Edition, 1975, or *Zhonghua Renmin Gongheguo Fen Sheng Dituji*, Zhongguo Beijing: Ditu Chubanshe, 1980.

Adung Valley, Burma   28°15′N 97°37′E
Akrotiri Salt Lake, Cyprus
                      34°37′N 32°58′E
AMAZONIA: the basin of the river Amazon, Brazil.
ARABIA: peninsula of SW Asia, including Saudi Arabia, Yemen, South Yemen, Oman, Bahrain, Qatar and the United Arab Emirates.
ASIA MINOR: Turkey in Asia.
Baba Garage, Senegal   14°57′N 16°31′W
BALKANS: countries of the Balkan peninsula, including Yugoslavia, Romania, Bulgaria, Albania, Greece, Turkey in Europe.
Bang Phra, Thailand   13°12′N 100°57′E
Bar Cenisio, Italy   45°11′N   6°59′E
Brenkhausen, FGR   48°23′N   7°47′E
Bulow, GDR   53°30′N 11°30′E
Buraskanda, Garhwal, India
                   30°25′N 78°14′E
Bzenec, Czechoslovakia 49°01′N 17°10′E

Cadipietra (Steinhaus), Italy
                   49°00′N 11°59′E
Chersónisos, Crete   35°20′N 25°23′E
Chokpakskii Pass, Kazakhstan, USSR
                   42°30′N 72°27′E
Cloverlake, Texas, USA 34°20′N 101°43′W
Col de Bretolet, Canton Valais, Switzerland
                   46°09′N   6°48′E
Coldspring, Wisconsin, USA
                   42°53′N 88°19′W
Dhanaulti, Garhwal, India 30°25′N 78°17′E
Dida Galgalla Desert, Kenya 3°N 38°15′E
Dziekenów Leśny, Poland 52°20′N 20°50′E
'El Seranillo', Guadalajara, Spain
                   40°45′N   3°06′W
EURASIA: the land mass of Europe and Asia; equivalent to the combined Palaearctic and Oriental regions, but excluding North Africa.
FAR EAST: Japan, China, North and South Korea, East Siberia and Indo-China.

Fitzroy, Falkland Islands 51°45′S 58°14′W
Gaglioni, Italy 45°08′N 7°01′E
Ghadi-Saati, Mareb river, Ethiopia
15°35′N 39°18′E
Glenn Dale, Maryland, USA
38°59′N 76°49′W
Göksu delta, Turkey
(near Silifke 36°22′N 33°57′E)
Green Patch, Falkland Islands
51°34′S 58°08′W
Hal Far, Malta 35°49′N 14°31′E
Hal Hambo, Somalia 1°54′N 45°05′E
Halle, GDR 51°28′N 11°53′E
Hat Karon, Phuket Island, Thailand
7°52′N 98°18′E
Hilbre Island, Cheshire, England
53°23′N 3°13′W
HORN OF AFRICA: Somalia and adjacent territories in NE Africa.
Inanda, Natal, South Africa
29°45′S 30°52′E
INDO-CHINA: the region between India and China, comprising Burma, Thailand, Laos, Cambodia (Kampuchea), Vietnam and peninsular Malaysia.
Indunamara Mts., Kenya 2°12′N 36°56′E
Kamschar, Persian Baluchestan
(near Sarbäz 26°40′N 61°20′E)
Kangasala, Finland 61°27′N 24°03′E
Karen Hills, Kawthoolei State, Burma
ca 18°N 97°30′E
KAROO: high, arid plateaux in Cape Province, South Africa.
Kuer Momer Sarr, Senegal 15°52′N 15°59′W
Kungrad-Kul, Uzbekistan, USSR
42°30′N 59°38′E
Lado, Uganda 3°48′N 31°53′E
LAKE DISTRICT: region of lakes and mountains in Cumbria, England
Lammasinghi, Visakhapatnam, Eastern Ghats, India 17°42′N 83°24′E
'La Bomba', Guadalajara, Spain
40°30′N 3°11′W
Ladhowai, Punjab, India 31°00′N 75°51′E
Lanslevillard, France 45°17′N 6°55′E
La Paz, Department, El Salvador: (capital Zacatecoluca) 13°29′N 88°51′W
La Rosière du Montvalescan, France
45°39′N 6°51′E
LEVANT: eastern part of Mediterranean with its islands and neighbouring countries.
Lone Oak, Kentucky, USA
37°00′N 88°47′W
MAGHREB: northwest Africa including Morocco, Algeria and Tunisia.
Malamfatori, Nigeria 13°37′N 13°23′E
Mannchar Lake, Pakistan 26°25′N 67°40′E
Manovo-Goundi-St. Floris National Park, Central African Republic 3°37′N 17°01′E
Mara Serena Lodge, Kenya 1°25′S 35°02′ E
Massif du Mont Cenis, France
45°14′N 6°52′E

MESOPOTAMIA: the tract of land between the rivers Tigris and Euphrates.
MIDDLE EAST: countries from Egypt to Iran inclusive.
Montgenèvre, France 44°56′N 6°43′E
Motzener See, GDR 52°13′N 13°27′E
NEAR EAST: Turkey and the Balkan states bounded by the Adriatic, Aegean and Black Seas.
NEW ENGLAND: NE USA, consisting of Maine, New Hampshire, Vermont, Massachusetts, Rhode Island and Connecticut.
Oktibbeha Co., Mississippi, USA
33°28′N 88°48′W
Old Lakatoo, Cape Province, South Africa
27°04′ S 23°50′E
Pékesse, Senegal 15°07′N 16°25′E
Phi Phi Le Island, Krabi, Thailand
7°41′N 98°46′E
Pressenger, Austria 46°38′N 13°27′E
Ross Island, Andaman Islands
11°40′N 92°44′E
Rottenau, GDR 52°00′N 12°50′E
SAHEL: stretch of arid land in W Africa south of the Sahara Desert, comprising parts of Mauritania, Senegal, Mali, Niger and Chad.
San Domingos, São Tiago, Cape Verde Islands
15°00′N 23°40′W
SEMIRECHE: region of eastern Kazakhstan, USSR, extending from Targabatay south to the Tien Shan and east of Lake Balkhash.
Shaweishan Island, China 31°25′N 122°14′E
Sleské Rudoltice, Czechoslovakia
50°13′N 17°43′E
Steckby, GDR 51°58′N 12°06′E
Teal Inlet, Falkland Islands
51°35′ S 58°26′W
Thathur, Garhwal, India 30°30′N 78°06′E
TRANSCASPIA: former Russian Republic, east of Caspian Sea.
TRANSCAUCASIA: region of Soviet Union lying south of the Caucasus between the Black and Caspian Seas; includes the Georgian, Azerbaijan and Armenian SSRs.
Trins, Austria 47°04′N 11°22′E
TURKESTAN: region of central Asia, divided into Russian (West) Turkestan that includes Turkmenistan, Uzbekistan, Tadzhikistan, Kirghizia and parts of Kazakhstan, and Chinese (East) Turkestan, the modern Autonomous Region of Sinkiang in western China.
Umzinyata Falls, Inanda (qv)
Valldemosa, Majorca 39°42′N 2°37′E
Wadi Akwandra, Egypt 23°15′N 34°18′E
Wat Phai Lom, Thailand 13°33′N 100°15′E
West Point Island, Falklands
51°21′ S 60°40′W
Wieniec, Poland 54°20′N 18°56′E

References

Abdulali, H. 1964. The birds of the Andaman and Nicobar Islands. J. Bombay Nat. Hist. Soc. 61: 483–571.
Abé, M. T. 1969. [Ecological studies on *Passer montanus kaibatoi* Munsterhjelm.] Bull. Govt. Forest. Expt. Station. No. 220: 11–57. (In Japanese with English summary.)
Abé, M. T. 1970. Some factors influencing growth of nestlings of *Passer montanus kabatoi* Munsterhjelm. Int. Studies on Sparrows 4: 7–10.
Agostinho, J. 1963. Variations dans l'avifauna des Açores. Alauda 31: 305–306.
Albrecht, J. S. M. 1983. Courtship behaviour between Tree Sparrow and House Sparrow in the wild – a possible case of hybridisation. Sandgrouse 5: 97–99.
Alexander, B. 1898a. An ornithological expedition to the Cape Verde Islands. Ibis (7) 4: 74–118.
Alexander, B. 1898b. Further notes on the ornithology of the Cape Verde Islands. Ibis (7) 4: 277–285.
Alexander, B. 1899. An ornithological expedition to the Zambezi river. Ibis (7) 5: 549–583.
Ali, S. 1949. *Indian Hill Birds*. London: Oxford Univ. Press.
Ali, S. 1962. *The Birds of Sikkim*. London: Oxford Univ. Press.
Ali, S. 1963. A note on the Eastern Spanish Sparrow, *Passer hispaniolensis* Tchusi, in India. J. Bombay Nat. Hist. Soc. 60: 318–321.
Ali, S. 1977. *Field Guide to the Birds of the Eastern Himalayas*. London: Oxford Univ. Press.
Ali, S. & Ripley, S. D. 1974. *The Handbook of the Birds of India and Pakistan*, Vol. 10. Bombay: Oxford Univ. Press.
Ali, S. & Whistler, H. 1933. The Hyderabad Ornithological Survey, Part III. J. Bombay Nat. Hist. Soc. 36: 898–919.
Allan, R. G. & Jackson, J. J. 1973. Notes on Golden Sparrow nesting near Khartoum, Sudan. U.N. Development Programme. Regional Project 'Research into the control of grain-eating birds.' Internal Report No. 412, Project Quelea, Dakar, Senegal.
Allouse, B. E. 1953. The Avifauna of Iraq. Iraq Nat. Hist. Mus. Publ. 3: 1–166.
Alonso, J. C. 1984a. Estudio comparado de los principales parametros reproductivos de *Passer hispaniolensis* y *Passer domesticus* in España centro-occidental. Ardeola 30: 3–21.
Alonso, J. C. 1984b. Zur Mauser spanischer Weiden- und Haussperlinge (*Passer hispaniolensis* und *domesticus*). J. Ornith. 125: 209–223.
Alonso, J. C. 1986. On the status and distribution of the Spanish Sparrow (*Passer hispaniolensis* (Temm.)) in Iberia. Int. Studies on Sparrows 13: 35–45.
Amadon, D. 1966a. The superspecies concept. Syst. Zool. 15: 245–249.
Amadon, D. 1966b. Avian plumages and molts. Condor 68: 263–278.
American Ornithologists' Union. 1983. *Check List of North American Birds*. Lawrence, Kansas, USA: Allen Press Inc.
Anderson, T. R. 1973. A comparative ecological study of the house sparrow and the tree sparrow near Portage des Sioux, Missouri. PhD. Thesis, St. Louis University, Missouri, USA. (Quoted by Dyers *et al.* 1977.)
Anderson, T. R. 1975. Fecundity of the house sparrow and the tree sparrow near Portage des Sioux, Missouri, USA. Int. Studies on Sparrows 8: 6–23.
Anderson, T. R. 1978. Population studies of European Sparrows in North America. Occ. Papers Mus. Nat. Hist. Univ. Kansas 70: 1–58.
Anderson, T. R. 1980. Comparison of nestling diet of sparrows, *Passer* spp., within and between habitats. Symposium on genus *Passer*. Proc. 17th I.O.C. 1978: 1162–1170. Deutsch. Orn. Gesell. R. Nohrung (Ed.). Berlin.
Archer, G. & Godman, E. M. 1961. *Birds of British Somaliland and the Gulf of Aden*, Vol. 4. Edinburgh & London: Oliver & Boyd.
Argyle, F. & Gel, B. 1978. Birds in Eilat, Spring 1978. Eilat Field Study Centre, Society for the Protection of Nature in Israel.
Arnott, J. (Ed.) 1981. Fair Isle Bird Observatory Report, 1980. (33): 59.
Ash, J. S. & Colston, P. 1981. A House × Somali Sparrow *Passer domesticus* × *P. castanopterus* hybrid. Bull. Br. Orn. Cl. 100: 291–294.
Ash J. S. & Miskell, J. E. 1983. The Birds of Somalia: their habitat, status and distribution. Scopus Special Suppl. No. 1.

Ash, J. S. & Shafeeg, A. (in preparation). A checklist of the birds of the Maldives archipelago.
Ashford, R. W. 1978. First record of House Sparrow from Papua New Guinea. Emu 78: 36.

Bachkiroff, Y. 1953. Le moineau steppique au Maroc. Service de la Défense des Végétaux No. 3: 1–135 (Maroc).
Baker, A. J. 1980 Morphometric differentiation in New Zealand populations of the House Sparrow (*Passer domesticus*). Evolution 34: 638–653.
Baker, E. C. S. 1921. Handlist of the 'Birds of India', Part III. J. Bombay Nat. Hist. Soc. 27: 692–744.
Baker, E. C. S. 1926. *The Fauna of British India*. Birds Vol. III (2nd Edn.). London: Taylor & Francis.
Balat, F. 1970. Einige Erkenntnisse aus der Brutbionomie des Feldsperlings, *Passer montanus*, in Mahren, Tschechoslowakei. Int. Studies on Sparrows 4: 10–13.
Balat, F. 1971. Clutch size and breeding success of the Tree Sparrow, *Passer montanus* L., in central and southern Moravia. Zoologicke Listy 20: 265–280.
Balat, F. 1972. Zur Frage des Legebeginns bei dem Feldsperling, *Passer montanus* L. Zoologicke Listy 21: 235–244.
Balat, F. 1974a. Zur Frage des Nistkonkurrenz des Feldsperlings *Passer montanus* L. Zoologicke Listy 23: 123–135.
Balat, F. 1974b. Gelegungrösse und Brutverluste des Haussperlings *Passer domesticus* (L.) in Mittelmahren. Zoologicke Listy 23: 229–240.
Balat, F. 1976. Dispersionsprozesse und Brutortstreue beim Feldsperling *Passer montanus*. Zoologicke Listy 25: 39–49.
Balat, F. & Touskova, I. 1972. Zur Erkenntnis der Biomasse-Production der Nachkommenschaft des Feldsperlings, *Passer montanus* L. Zoologicke Listy 21: 325–335.
Bannerman, D. A. 1912. The birds of Gran Canaria. Ibis (12) 6; 567–627.
Bannerman, D. A. 1948. *The Birds of Tropical West Africa*, Vol. 6. London: Crown Agents for the Colonies.
Bannerman, D. A. 1951. *The Birds of Tropical West Africa*, Vol. 8. London: Crown Agents for the Colonies.
Bannerman, D. A. 1953. *The Birds of West and Equatorial Africa*, Vol. 2. Edinburgh & London: Oliver & Boyd.
Bannerman, D. A. 1963. *History of the Birds of the Canary Islands and of the Salvages*. Edinburgh & London: Oliver & Boyd.
Bannerman, D. A. & Bannerman, W. M. 1965. *The Birds of Madeira*. Edinburgh & London: Oliver & Boyd.
Bannerman, D. A. & Bannerman, W. M. 1968. *History of the Birds of the Cape Verde Islands*. Edinburgh & London: Oliver & Boyd.
Barlow, J. C. 1973. Status of the North American population of the European Tree Sparrow. In: Hardy & Morton (Eds.) 1973.
Barlow, J. C. 1980. Adaptive responses in skeletal characters of the New World population of *Passer montanus*. Symposium on the genus *Passer*. Proc. 17th Int. Orn. Congr. 1978: 1143–1149. Deutsch. Orn. Gesell. R. Nohrung. (Ed.) Berlin.
Barnard, C. J. 1979. Birds of a feather. New Scientist 13 Sep.: 818–820.
Barnard, C. J. 1980a. Flock feeding and time budgets in the House Sparrow (*Passer domesticus* L.). Animal Behaviour 28: 295–309.
Barnard, C. J. 1980b. Equilibrium flock size and factors affecting arrival and departure in feeding House Sparrows. Animal Behaviour 28: 503–511.
Barnard, C. J. 1980c. Factors affecting flock size mean and variance in a winter population of House Sparrows. Behaviour 74: 114–127.
Barnard, C. J. 1980d. Flock organisation and feeding budgets in a field population of House Sparrows (*Passer domesticus*). Symposium on the genus *Passer*. Proc. 17th Int. Orn. Congr. 1978: 1117–1121. Deutsch. Orn. Gesell. R. Nohrung (Ed.). Berlin.
Barnard, C. J. 1983. So you think you know everything about sparrows? BBC Wildlife, Dec.: 90–93.
Barrows, W. B. 1889. The English Sparrow (*Passer domesticus*) in North America. U.S. Dept. Agric. Div. Economic Orn. and Mamm. Bull. No. 1.
Bäsecke, K. 1949. Phragmites Samen als Nahrung für Feldsperling. Vogelwelt 70: 21.
Bates, G. L. 1934. Birds of the southern Sahara and adjoining countries in French West Africa. Ibis (13) 4: 685–717.
Bates, G. L. (with notes by H. St. J. Philby) 1936. Birds in Jidda and central Arabia collected in 1934 and early in 1935, chiefly by Mr. Philby. Ibis (13) 6: 531–556.

Bauer, Z. 1975. The biomass production of the Tree Sparrow *Passer m. montanus* (L.) population in the conditions of the flood forest. Int. Studies on Sparrows 8; 124–139.
Baumgart, W. 1980. Einige brutbiologische Besonderheiten des Weidensperlings. Der Falke 27: 78–86.
Baumgart, W. 1984. Zur Charakterisierung von Haus- und Weidensperlings *Passer domesticus* und *Passer hispaniolensis* als 'zeitdifferente Arten'. Beitr. Vogelk. 30: 217–242.
Beaman, M. (Ed.). 1978. Ornithological Society of Turkey Bird Report No. 4, 1974–75.
Beaman, M., Porter, R. F. & Vittery, A. (Eds.) 1975. Ornithological Society of Turkey: Bird Report No. 3, 1970–73.
Beaven, R. 1866. Letter to Editor. Ibis (2) 2: 419–420.
Beecher, W. J. 1953. A phylogeny of the Oscines. Auk 70: 270–333.
Beesley, J. S. S. & Irving, N. S. 1976. The status of birds of Gaberone and its surroundings. Botswana Notes and Records 8: 231–261.
Beimborn, D. A. 1967. Population ecology of the English Sparrow in North America. MSc. Thesis, University of Wisconsin, Milwaukee. (Quoted by Pinowski & Kendeigh (Eds.). 1977.)
Beimborn, D. A. 1976. Sex ratios in the House Sparrow. Bird Banding 47: 13–18.
Bennett, A. G. 1926. A list of the birds of the Falkland Islands and Dependencies. Ibis (12) 2: 306–333.
Benson, C. W. 1947. Notes on the birds of southern Abyssinia. Ibis 89: 29–50.
Benson, C. W. 1953. A Check List of the Birds of Nyasaland. Blantyre & Lusaka: The Nyasaland Society & The Publications Bureau.
Benson, C. W. 1956. The relationship of *Passer griseus* (Vieillot) and *Passer diffusus* (Smith), with a description of a new race of the latter. Bull. Br. Orn. Cl. 70: 38–42.
Benson, C. W. & Benson, F. M. 1977. *The Birds of Malawi*. Limbe, Malawi: D. W. K. Macpherson (sponsor).
Benson, C. W., Brooke, R. K., Dowsett, R. J. & Irwin, M. P. S. 1973. *The Birds of Zambia*, 2nd. Edn. London: Collins.
Bent, A. C. 1965. Article on English Sparrow, *Passer domesticus domesticus* (Linnaeus) in: *Life Histories of North American Weaver Finches, Blackbirds, Orioles and Tanagers*. New York: Dover Publications.
Bentz, G. D. 1979. The appendicular myology and phylogenetic relationships of the Ploceidae and Estrildidae (Aves: Passeriformes). Bull. Carnegie Mus. Nat. Hist. No. 15: 1–25.
Berck, K.-H. 1961, 1962. Beiträge zur Ethologie des Feldsperlings (*Passer montanus*) und dessen Beziehung zum Haussperling (*Passer domesticus*). Vogelwelt 82: 129–173; 83: 8–26.
Berg, A. B. van den & Roever, J. W. de 1984. Plumages and bare parts of Desert Sparrow. Dutch Birding 4: 139–140.
Bergtold, W. H. 1921. The English Sparrow (*Passer domesticus*) and the motor vehicle. Auk 38: 244–250.
Bethune, G. de 1961. Notes sur le moineau friquet, *Passer montanus* (L.). Gerfaut 51: 387–398.
Betts, F. N. 1966. Notes on some resident breeding birds of S. W. Kenya. Ibis 108: 513–530.
Bezzel, E. 1957. Beiträge zur Kenntnis der Vogelwelt Sardiniens. Anz. Orn. Ges. Bayern 4: 589–707.
Bibby, C. 1975. Observations on the moult of the Tree Sparrow. Ringing & Migration 1: 148–157.
Blair, H. M. S. 1936. On the birds of East Finmark. Ibis (13) 6: 280–308.
Blem, C. R. 1973. Geographic variation in the bioenergetics of the House Sparrow. In: Hardy & Morton (Eds.). 1973.
Blem, C. R. 1975. Geographic variation in wing-loading of the House Sparrow. Wilson Bull. 87: 543–549.
Blyth, E. 1867. The Ornithology of India. A commentary on Dr. Jerdon's 'The Birds of India'. Ibis (2) 3: 1–48.
Bock, W. J. & Moroney, J. T. 1978a. The preglossale of *Passer*: a skeletal neomorph. J. Morphology 155: 99–110.
Bock, W. J. & Moroney, J. T. 1978b. Relationships of the passerine finches (Passeriformes: Passeridae). Bonn. zool. Beitr. 29: 122–145.
Bodenstein, G. 1953. Auftreten einer *hispaniolensis*-ähnlichen Kopfzeichnung bei einem mitteleuropaischen Haussperling. Orn, Mitt. 5: 72.
Bond, J. 1960. *Birds of the West Indies*. London: Collins.
Boros, I. & Horvath L. 1954. Clarification of the position of the *Passer moabiticus moabiticus* Tristr. and *Passer moabiticus yatii* Sharpe. Acta Zool. Acad. Sci. Hungary 1: 43–51.
Bortoli, L. 1973. Sparrows in Tunisia. In: Kendeigh & Pinowski (Eds.). 1973.

Bortoli, L. & Bruggers, R. 1976. Nidification de *Quelea quelea* dans la delta Nigerien en 1976. P.N.U.D./Recherche pour la lutte contre oiseaux granivores *'Quelea quelea'*. SR 258.
Boswall, J. & Kanwanach, S. 1978. The birds of Phi Phi Le Island, Krabi, Thailand. Nat. Hist. Bull. Siam Soc. 27: 83–97.
Bourdelle, M. E. & Gabin, M. J. 1950/51. Bull. Stations Françaises de Baguage (7): 1–38 & 43.
Bourne, W. R. P. 1955. The birds of the Cape Verde Islands. Ibis 97: 508–556.
Bourne, W. R. P. 1957a. The breeding birds of Bermuda. Ibis 99: 94–105.
Bourne, W. R. P. 1957b. Additional notes on the birds of the Cape Verde Islands with particular reference to *Bulwaria mollis* and *Fregata magnificens*. Ibis 100: 275–276.
Bourne, W. R. P. 1966. Further notes on the birds of the Cape Verde Islands. Ibis 108: 425–429.
Boyd, A. W. 1932. Notes on the Tree Sparrow. Brit. Birds 25: 278–285.
Boyd, A. W. 1933. Notes on the Tree Sparrow 1932. Brit. Birds 26: 273–274.
Boyd, A. W. 1934. Notes on the Tree Sparrow 1933. Brit. Birds 27: 250–260.
Boyd, A. W. 1935. Notes on the Tree Sparrow 1934. Brit. Birds 28: 347–349.
Boyd, A. W. 1949. Display of the Tree Sparrow. Brit. Birds 42: 213–214.
Boyd, G. 1978. House Sparrows feeding at suspended feeders. Brit. Birds 71: 318.
Brackbill, H. 1969. Two male House Sparrows copulating on the ground with the same female. Auk 86: 146.
Britton, D. 1981. Spanish Sparrows in the Isles of Scilly. Brit. Birds 74: 150–151.
Britton, P. L. (Ed.). 1980. *Birds of East Africa, their Habitat, Status and Distribution*. Nairobi: E.A.N.H.S.
Brockman, H. J. 1980. House Sparrows kleptoparasitize digger wasps. Wilson Bull. 92: 394–398.
Brooke, R. 1973. House Sparrows feeding at night in New York. Auk 90: 206.
Brosset, A. (in press). Ecology and behaviour of the Birds of the Ivindo Basin, N. E. Gabon.
Broun, M. 1972. Apparent migratory behaviour in the House Sparrow. Auk 89: 187–189.
Brown, L. H. & Britton, P. L. 1980. The Breeding Season of East African Birds. Nairobi: EANHS.
Brown, L. H., Urban, E. K. & Newman, K. 1982. *The Birds of Africa*, Vol. 1. London: Academic Press.
Browne, P. W. P. 1981. Breeding of six Palaearctic birds in south west Mauritania. Bull. Br. Orn. Cl. 101: 306–310.
Bruggers, R. L. 1977. Summary of the present research on the Golden Sparrow. P.N.U.D./Recherche pour la lutte contre les oiseaux granivores. *'Quelea quelea'*. SR 255.
Bruggers, R. L. & Bortoli, L. 1976. Dry season nesting of the Golden Sparrow near Richard Toll, Senegal. Terre et Vie 30: 521–527.
Brunel, J. & Thiollay, J. M. 1969. Liste préliminaire des oiseaux de Côte d'Ivoire. Alauda 37: 315–337.
Bulatova, N. S. 1973. A cytological study of three related families of birds: Fringillidae, Emberizidae, Ploceidae. J. Syst. Evolutionsforschung 11: 233–239.
Bumpus, H. C. 1899. The elimination of the unfit as illustrated by the introduced sparrow, *Passer domesticus*. Biol. Lectures, Marine Biol. Lab. Woods Hole: 209–226.
Bundy, G. 1976. *The Birds of Libya*. London: Brit. Orn. Union.
Bundy, G. & Morgan, J. H. 1969. Tripolitanian birds, Part II. Bull. Br. Orn. Cl. 89: 151–159.
Butler, A. L. 1899. The birds of the Andaman and Nicobar Islands. J. Bombay Nat. Hist. Soc. 12: 555–571.

Calhoun, J. B. 1947. The role of temperature and natural selection in relationship to the variations in the size of the English Sparrow in the United States. Am. Nat. 81: 203–228.
Carruthers, D. 1910a. On the birds of the Zarafschan basin in Russian Turkestan. Ibis (9) 4: 436–475.
Carruthers, D. 1910b. On a collection of birds from the Dead Sea and N. W. Arabia, with contributions to the ornithology of Syria and Palestine. Ibis (9) 4: 475–491.
Carruthers, D. 1949. *Beyond the Caspian*. Edinburgh: Oliver & Boyd.
Cawkell, E. W. & Hamilton, J. E. 1961. The birds of the Falkland Islands. Ibis 103a: 1–27.
Chapin, J. P. 1917. The classification of the weaver-birds. Bull. Am. Mus. Nat. Hist. 37: 243–280.
Chapin, J. P. 1954. The birds of the Belgian Congo, Vol. 4. Bull. Am. Mus. Nat. Hist. 75B.
Charlwood, R. H. 1981. Spanish Sparrows in the Isles of Scilly. Brit. Birds 74: 150.
Chatfield, D. G. P. 1970. House Sparrow feeding from nut bag. Brit. Birds 63: 345.
Cheesman, R. E. 1919. Exhibition of nest and eggs of *Passer moabiticus mesopotamicus*. Bull. Br. Orn. Cl. 40: 59.

Cheesman, R. E. & Sclater, W. L. 1936. On a collection of birds from north western Abyssinia. Ibis (13) 6: 163–197.
Cheke, A. S. 1969. Mechanism and consequences of hybridisation in sparrows *Passer*. Nature 222: 179–180.
Cheke, A. S. 1973. Movements and dispersal among House Sparrows, *Passer domesticus* (L.), at Oxford, England. In: Kendeigh & Pinowski (Eds.) 1973.
Chia, H.-K., Bei, T. H., Chen, T. Y. & Cheng, T. 1963. [Preliminary studies on the breeding behaviour of the tree sparrow (*Passer montanus saturatus*).] Acta Zool. Sinica. 15: 527–536. (In Chinese with English summary)
Christison, A. F. P. 1941. Notes on the birds of Chagai. Ibis (14) 5: 531–556.
Churcher, P. B. & Lawton, J. H. 1987. Predation by domestic cats in an English village. J. Zool., Lond. 212: 439–455.
Cink, C. L. 1976. The influence of early learning on nest site selection in the House Sparrow. Condor 78: 103–104.
Clancey, P. A. 1948. Seasonal bill variation in the Tree Sparrow. Brit. Birds 41: 115–116.
Clancey, P. A. 1959. Notes on Grey-Headed Sparrows from Kenya Colony. Bull. Br. Orn. Cl. 79: 131–132.
Clancey, P. A. 1964a. New subspecies of the Greater Sparrow *Passer iagoensis* Gould and the Black-cheeked Waxbill *Estrilda erythronotes* (Vieillot) from South Africa. Durban Mus. Novit. 7: 137–139.
Clancey, P. A. 1964b. On the original description of *Passer iagoensis motitensis* Smith. Bull. Br. Orn. Cl. 84: 110.
Clancey, P. A. 1964c. *The Birds of Natal and Zululand*. Edinburgh & London: Oliver & Boyd.
Clancey, P. A. 1965. Further on *Passer motitensis* (Smith). Bull. Br. Orn. Cl. 85: 41.
Clancey, P. A. 1971. A handlist of the birds of southern Moçambique. Lourenço Marques: Instituto de Investigação Cientifica de Moçambique.
Clark, J. H. 1903. A much-mated House Sparrow. Auk 20: 306–307.
Clarke, G. C. W. 1949. Ingenious sparrows. Country Life 105: 1131.
Clench, M. H. 1970. Variability in body pterylosis with special reference to the genus *Passer*. Auk 87: 650–691.
CLIMAP, 1976. The surface of the ice-age earth. Science 191: 1131–1137.
Cole, D. J. 1962. House Sparrows in northern Cape and Bechuanaland Protectorate. Ostrich 33: 54–55.
Cole, S. R. & Parkin, D. T. 1981. Enzyme polymorphism in the House Sparrow *Passer domesticus*. Biol. J. Linnaen Soc. 15: 13–22.
Collias, N. E. & Collias, E. C. 1964. Evolution of nest building in the weaver birds (Ploceidae). Univ. Calif. Publ. Zool. 73: 1–239.
Collias, N. E. & Collias, E. C. 1971. Ecology and behaviour of the Spotted-backed Weaverbird in the Kruger National Park. Koedoe 14: 1–27.
Common, A. M. 1956. Anting by House Sparrow. Brit. Birds 49: 155.
Cottam, C. 1929. The fecundity of the English Sparrow in Utah. Wilson Bull. 41: 193–194.
Courtney-Latimer, M. 1955. The English Sparrow at East London. Bokmakierie 7: 32.
Cracraft, J. 1981. Toward a phylogenetic classification of the recent birds of the world (Class Aves). Auk 98: 681–714.
Craggs, J. D. 1967. Population studies of an isolated colony of House Sparrows (*Passer domesticus*). Bird Study 14: 53–60.
Craggs, J. D. 1976. An isolated colony of House Sparrows. Bird Study 23: 281–284.
Craig, A. J. F. K., Every, B. & Summers-Smith, D. 1987. The spread of the Southern Greyheaded Sparrow in Cape Province, South Africa. An. Cape Prov. Mus. (Nat. Hist.) 16: 191–200.
Cramp, S. 1969. Dead Sea Sparrow. In: Gooders, J. (Ed.) 1969.
Cramp, S. 1971. The Dead Sea Sparrow: further breeding places in Iran and Turkey. Ibis 113: 244–245.
Creutz, G. 1949. Untersuchungen zur Brutbiologie des Feldsperlings (*Passer m. montanus* L.). Zool. Jahrb. 78: 133–172.
Crowe, T. M. 1978. The evolution of guineafowl (Galliformes, Phasianidae, Numidae): taxonomy, speciation and biogeography. Ann. S. Afr. Mus. 76: 43–136.
Crowe, T. M., Brooke R. K. & Siegfried, W. R. 1980. Evolution and adaptive radiation in southern African House Sparrows *Passer domesticus*. Abstract: 5th Pan-African Orn. Congr., Malawi.
Currie, A. J. 1909–10. The Rufous-backed Sparrow (*Passer pyrrhonotus*) nesting in the Punjab. J. Bombay Nat. Hist. Soc. 19: 259–260.
Currie, A. J. 1915–16. The birds of Lahore and its vicinity. J. Bombay Nat. Hist. Soc. 24: 561–577.

Darwin, C. 1841. *The Zoology of the Voyage of H.M.S. Beagle*, Vol. 2, Birds. London: Smith, Elder.
Davis, J. 1954. Seasonal change in bill length of certain passerine birds. Condor 56: 142–149.
Davis, J. 1973. Field notes concerning dispersal of house sparrow at Hastings Natural History Reserve. Ms. (Star Rt., Box 80, Carmel Valley, California 93924, USA.) (Quoted by Johnston & Klitz 1977.)
Davis, P. E. (Ed.) 1963. Fair Isle Bird Observatory Bull. 1962. (15): 24.
Davison, W. 1883. Notes on some birds collected in the Nilghiris and in parts of Wynaed and southern Mysore. Stray Feathers 10: 329–419.
Dawson, D. G. 1969. Article on Spanish Sparrow. In: Gooders, J. (Ed.) 1969.
Dawson, D. G. 1972. The breeding ecology of house sparrows. PhD Thesis, University of Oxford. (Quoted by Dyers *et al.* 1977.)
Dawson, D. G. 1973. House Sparrow, *Passer domesticus* (L.), breeding in New Zealand. In: Kendeigh & Pinowski (Eds.) 1973.
Dean, W. R. J. 1977. The moult of the Cape Sparrow. Ostrich Suppl. 12: 108–116.
Dean, W. R. J. 1978. Life expectancy of the Cape Sparrow. Ostrich 49: 16–20.
Deckert, G. 1962. Zur Ethologie des Feldsperlings (*Passer m. montanus* L.). J. Ornith. 103: 427–486.
Deckert, G. 1968. *Der Feldsperling*. Wittenburg Lutherstadt: Die Neue Brehm-Bücherei.
Deckert, G. 1969. Zur Ethologie und Ökologie des Haussperlings (*Passer d. domesticus* (L.)) Beitr. Vogelk. 15: 1–84.
De Greling, C. 1972. New records from northern Cameroun. Bull. Br. Orn. Cl. 92: 24–27.
Deignan, H. G. 1945. The birds of northern Thailand. U.S. Nat. Mus. Bull. No. 186: 1–616.
Delacour, J. 1930. On the birds collected during the fifth expedition to French Indo-China. Ibis (12) 6; 564–599.
Delacour, J. & Jabouille, P. 1931. *Les Oiseaux de l'Indochine Française*. Paris: Exposition Coloniale Internationale.
Delacour, J. & Mayr, E. 1946. *Birds of the Philippines*. New York: Macmillan.
Dementiev, G. P. & Gladkov, N. A. (Eds.) 1970. *Birds of the Soviet Union*, Vol. 5. Translated Gordon, E. D. Jerusalem: Israel Program for Scientific Translations. (Original publication 1954.)
de Naurois, R. 1969. Notes brève sur l'avifauna de l'Archipel de Cap Vert, faunistique, endemisme, ecologia. Bull. Inst. fond. d'Afr. noire 31A: 143–218.
Desfayes, M. & Praz, J. C. 1978. Notes on habitat and distribution of montane birds in southern Iran. Bonn. zool. Beitr. 29: 18–37.
Despott, G. 1917. Notes on the ornithology of Malta. Ibis (10) 5: 21–349.
Dexter, R. W. 1949. Banding studies on the English Sparrow. Bird Banding 20: 40–50.
Dexter, R. W. 1959. Two 13-year old age records for the House Sparrow. Bird Banding 30: 182.
Dohrn, H. 1871. Beiträge zur Ornithologie der Capverdischen Inseln. J. Ornith. 19: 1–10.
Doig, S. 1830. Birds' nesting in the 'Eastern Narra'. Stray Feathers 9: 277–282.
Dolnik, V. R. 1973. The winter storage by the migratory fat deposition in *Passer domesticus bactrianus* Zar. et Kud. – the arid zone migrant. In: Kendeigh & Pinowski (Eds.) 1973.
Dornbusch, M. 1973. Zur Siedlungsdichte und Ernährung des Feldsperlings in Kiefern-Dickungen. Der Falke 20: 193–195.
Dott, H. E. M. 1986. The spread of the House Sparrow *Passer domesticus* in Bolivia. Ibis 128: 132–137.
Dowsett, R. J. & Dowsett-Lamaire, F. 1980. The systematic status of some Zambian birds. Le Gerfaut 70: 151–199.
Du Pont, J. E. 1971. Philippine Birds. Greenville, Delaware: Delaware Mus. Nat. Hist.
Dyer, M. I., Pinowski, J. & Pinowska, B. 1977. Population dynamics. In: Pinowski & Kendeigh (Eds.). 1977.

Earlé, R. A. (in press). Reproductive isolation between urban and rural populations of Cape Sparrows and House Sparrows. Presented at 19th Int. Orn. Congr., Ottawa, June 1986.
Eisenhut, E. & Lutz, W. 1936. Beobachtungen über die Fortpflanzungsbiologie des Feldsperlings. Mitt. Vogelkw. Stuttgart 35: 1–14.
Elgood, J. L. 1982. *The Birds of Nigeria*. London: Brit. Orn. Union.
Eliseeva, V. I. 1961. [Nesting of Tree Sparrows in nest-boxes.] Zool. Zh. 40: 583–591. (In Russian) (Quoted by Dyers *et al.* 1977.)
Encke, F. W. 1965. Über Gelegen-, Schlupf- und Ausflugsstärken des Haussperlings (*Passer d. domesticus*) in Abhändigkeit von Biotop und Brutperiode. Beitr. Vogelk. 10: 268–287.
Enion, E. A. R. & Enion, D. 1962. Early breeding in Tenerife. Ibis 104: 158–166.

Erard, C. 1970. Short notes on the birds of Fezzan and Tripolitania. Bull. Br. Orn. Cl. 90: 107–111.
Erlanger, C. F. von. 1899. Beiträge zur Avifauna Tunisiens. J. Ornith. 47: 472–476.
Erlanger, C. F. von. 1907. Beiträge zur Vogelfauna Nordostafrikas. J. Ornith. 40: 1–58.
Escott, C. J. & Holmes, D. A. 1980. The avifauna of Sulawesi, Indonesia: faunistic notes and additions. Bull. Br. Orn. Cl. 100: 189–194.
Etchecopar, R. D. & Hüe, F. 1967. *The Birds of North Africa*. Translated Hollom, P. A. D. Edinburgh: Oliver & Boyd. (Original publication 1964.)
Every, B. 1976. Greyheaded Sparrows breeding in Cape Midlands. Bee-eater 27: 5–6.

Fallet, M. 1958. Zum Sozialverhalten des Haussperlings (*Passer domesticus* L.). Zool. Anzieger 161: 178–187.
Farkas, T. 1966. The birds of Barberspan. III. Some structural changes in the avifauna of the Barberspan Nature Reserve. Ostrich Suppl. 6: 463–491.
Feldmann, R. 1967. Haussperlinge an 'spatzensicheren' Meisenfütterungen Orn. Mitt. 19: 177–178.
Fisher, J. & Hinde, R. A. 1949. The opening of milk bottles by birds. Brit. Birds 42: 347–357.
Fleischer, R. C. 1982. Clutch size in Costa Rican House Sparrows. J. Field Nat. 53: 280–281.
Fleischer, R. C., Lowther, P. E. & Johnston, R. F. 1984. Natal dispersal in House Sparrows: possible causes and consequences. J. Field Orn. 55: 444–456.
Fleming, R. L. Sr., Fleming, R. L. Jr. & Bangdel, L. S. 1979. *Birds of Nepal*. Kathmandu: Avalok.
Flint, P. R. & Stewart, P. F. 1983. *The Birds of Cyprus*. London: Brit. Orn. Union.
Flower, S. S. 1925. Contribution to our knowledge of the duration of life in vertebrate animals. Proc. Zool. Soc. London, Ser. A, 95: 1365–1421.
Flower, S. S. 1938. Further notes on the duration of life in animals: Part IV, Birds. Proc. Zool. Soc. London, Ser. A, 108: 195–235.
Fogden, M. P. L. 1972. The seasonality and population dynamics of some equatorial forest birds in Sarawak. Ibis 114: 307–344.
Folk, C. & Novotny, I. 1970. Variation in body weight and wing length in the house sparrow, *Passer domesticus* (L.), in the course of a year. Zool. Listy 19: 333–342.
Frade, F. 1976. Aves do arquipélago de Cabo Verde. Garcia de Orta, Sér. Zool., Lisboa, 5: 47–58.
Friedmann, H. 1930. Birds collected by the Childe Frick expedition to Ethiopia and Kenya Colony. Part II: Passeres. U.S. Nat. Mus. Bull. No. 153: 386–439.
Friedmann, H. 1960. The parasitic weaverbirds. Bull. U.S. Nat. Mus. 223: 1–196.

Gallacher, M. D. & Rogers, T. D. 1978. On the breeding birds of Bahrain. Bonn. zool. Beitr. 29: 5–17.
Gauhl, F. 1984. Ein Beiträge zur Brutbiologie des Feldsperlings (*Passer montanus*). Die Vogelwelt 105: 176–187.
Gavrilov, E. I. 1962. A contribution to the biology of the Spanish Sparrow (*Passer hispaniolensis transcaspicus*). Ibis 104: 416–417.
Gavrilov, E. I. 1963. The biology of the Eastern Spanish Sparrow, *Passer hispaniolensis transcaspicus* Tschusi, in Kazakhstan. J. Bombay Nat. Hist. Soc. 60: 301–317.
Gavrilov, E. I. 1965. On the hybridisation of Indian and House Sparrows. Bull. Br. Orn. Cl. 85: 112–114.
Gavrilov, E. I. & Korelov, M. N. 1968. [The Indian Sparrow as a distinct good species] Bull. Mosc. Nat. Biol. Ser. 73: 115–122. (In Russian)
Gebhardt, E. 1959. Europäische Vögel in überseeischen Ländern. Bonn. zool. Beitr. 10: 510–542.
Geiler, H. 1959. Geschlechtsverhältnis, Körpergewicht und Flugellänge der Individuen einer mitteldeutschen Sperlingspopulation. Beitr. Vogelk. 6: 359–366.
Gee, J. P. 1984. The birds of Mauritania. Malimbus 6: 31–66.
Giglioli, H. 1865. Notes on the birds observed at Pisa and its neighbourhood. Ibis (2) 1: 50–63.
Gill, E. L. 1936. *A First Guide to South African Birds*. Cape Town: Maskow Millar.
Ginn, M. B. & Melville, D. S. 1983. *Moult in Birds*. Tring: British Trust for Ornithology.
Gistsov, A. P. & Gavrilov, E. I. 1984. Constancy in the dates and routes of spring migrations in the Spanish and Indian Sparrows in the foothills of the western Tian Shan. Int. Studies on Sparrows 11: 22–33.
Godfrey, W. E. 1966. *The Birds of Canada*. Ottawa: Nat. Mus. Canada, Bull. No. 203.
Gooders, J. (Ed.). 1969. *Birds of the World*, Vol. 9. London: IPC.

Gooding, C. D. & Walton, C. R. 1963. The sparrow invasion of 1962. J. Agric. Western Australia 4: 412–418.
Goodman, S. M. & Watson, G. E. 1983. Bird species records of some uncommon or previously unrecorded forms in Egypt. Bull. Br. Orn. Cl. 103: 101–106.
Gore, M. E. J. 1964. A new Borneo bird. Sabah Soc. J. 2; 109.
Gore, M. E. J. 1968. A check-list of the birds of Sabah, Borneo. Ibis 110: 165–196.
Gore, M. E. J. 1981. *Birds of the Gambia*. London: Brit. Orn. Union.
Gore, M. E. J. & Won Pyong-Oh. 1971. *The Birds of Korea*. Seoul: Royal Asiatic Society.
Gould, J. 1837. Exhibition of Mr. Darwin's birds, *etc.*, July 25th 1837. Proc. Zool. Soc.: 77.
Graczyk, R. 1970. The nesting density of tree sparrow (*Passer montanus* (L.)) populations in different biotopes in Poland. Int. Studies on Sparrows 4: 83–87.
Gramet, P. 1973. Structure et dynamique d'une population de Moineaux domestiques, *Passer domesticus* (L.). Resultats preliminaires. In Kendeigh & Pinowski (Eds.) 1973.
Gribbon, J. 1985. The drying of the Sahel. New Scientist 105: 8–9.
Grote, H. 1935. Die Nistweise des Feldsperlings. Beitr. Fortpfl.-biol. Vög. 1: 4–6.
Grun, G. 1975. Die Ernährung des Sperlinge *Passer domesticus* (L.) und *Passer montanus* (L.) unter verschiedenen Unwaltbedingungen. Int. Studies on Sparrows 8: 24–103.
Guichard, K. M. 1947. Birds of the innundation zone of the river Niger, French Sudan. Ibis 89: 450–489.
Guichard, K. M. 1955. The birds of Fezzan and Tibesti. Ibis 97: 393–424.
Guillory, H. D. & Deshotels, J. H. 1981. House Sparrow flushing prey from trees and shrubs. Wilson Bull. 93: 554.
Gurney, J. H. 1886. Letter. Ibis (11) 2: 419–423.

Hailbrun, L. H. & Arbib, R. 1974. The 74th Christmas bird count. American Birds 28: 135–540.
Hall, B. P. & Moreau, R. E. 1970. *An Atlas of Speciation in African Passerine Birds*. London: Brit. Mus. (Nat. Hist.).
Haller, W. 1936. Ein Beitrag zur Kenntnis der Verbreitung und Nistweise von Haus- und Feldsperling im schweizerischen Hugelland. Arch. Suisse d'Orn. 1: 350–357.
Hamilton, J. E. 1944. The House Sparrow in the Falkland Islands. Ibis 86: 553–554.
Hamilton, S. & Johnston, R. F. 1978. Evolution in the House Sparrow, VI. Variability and niche width. Auk 95: 313–323.
Hardy, J. W. & Morton, M. L. (Eds.). 1973. A symposium on the House Sparrow (*Passer domesticus*) and European Tree Sparrow (*P. montanus*) in North America. Orn. Monograph No. 14. Am. Orn. Union.
Harington, H. 1909–10. Some Rangoon birds. J. Bombay Nat. Hist. Soc. 19: 358–366.
Harington, H. H. 1914. Notes on the nidification of some birds from Burma. Ibis (10) 2: 1–26.
Harrison, C. 1975. *A Field Guide to the Nests, Eggs and Nestlings of European Birds*. London: Collins.
Harrison, C. J. O. 1965. The nest advertisement display as a *Passer/Ploceidae* link. Bull. Br. Orn. Cl. 85: 26–30.
Harrison, J. M. 1961. The significance of some plumage phases of the House-Sparrow, *Passer domesticus* (Linnaeus) and Spanish Sparrow *Passer hispaniolensis* Temminck. Bull. Br. Orn. Cl. 81: 96–103 & 119–124.
Harrisson, T. 1970. Birds from the rest house verandah, Brunei. Brunei Mus. J. 2; 269–278.
Harrisson, T. 1974. The Tree Sparrow in Borneo (East Malaysia and Brunei): A population explosion? Malay Nat. J. 27: 171.
Hartert, E. 1904–1922. *Die Vögel der paläarktischen Fauna*. Berlin: Friedländer und Sohn.
Hartert, E. 1921. Capt. A. Buchanan's Aïr expedition IV. The birds collected by Capt. A. Buchanan during journey from Kano or Aïr or Aslen. Novit. Zool. 28: 78–141.
Hartert, E. 1923. On the birds of Cyrenaica. Novit. Zool. 30: 1–32.
Harwin, R. M. & Irwin, M. P. S. 1966. The spread of the House Sparrow, *Passer domesticus*, in south-central Africa. Arnoldia 2: 1–17.
Haukioja, E. & Reponen, T. 1968. Varpusen (*Passer domesticus*) sulkasadosta eripainos porin. Lintutieteellinen Yhdistys ry:n vuosikirjasat: 49–51.
Haviland, M. D. 1918. Notes on the birds of the Bessarabian Steppe. Ibis (10) 2: 288–291.
Haviland, M. D. 1926. *Forest, Steppe and Tundra: Studies in Animal Environment*. Cambridge: University Press.
Hazevoet, C. J. 1986. Sirius West Africa expedition: Cape Verde Islands. Amsterdam: private publication.
Heij, C. J. 1985. Comparative ecology of the House Sparrow *Passer domesticus* in rural, suburban and urban situations. Thesis: Vrije Universiteit te Amsterdam.

Heim de Balsac, H. 1929. Remarques sur l'éthologie de *Passer simplex*. Alauda 1: 68–77.
Heim de Balsac, H. & Mayaud, N. 1962. *Les oiseaux du Nord-Ouest de l'Afrique*. Paris: Editions Paul Lechevalier.
Hill, A. 1977. House Sparrows feeding at suspended feeders. Brit. Birds 70: 84–85.
Hobbs, J. N. 1955a. House Sparrows taking dead grasshoppers from car radiators. Emu 55: 202.
Hobbs, J. N. 1955b. House Sparrow breeding away from man. Emu 55: 302.
Hoffmann, L. (Ed.). 1955. Premier compte rendu, 1950–1954, et recueil de travaux. Station Biologique de la Tour du Valat.
Hollom, P. A. D. 1959. Notes from Jordan, Lebanon, Syria and Antioch. Ibis 101: 183–200.
Holmes, D. A. & Wright, J. O. 1969. The birds of Sind: A Review, Part 2. J. Bombay Nat. Hist. Soc. 66: 8–30.
Howells, V. 1956. *A Naturalist in Palestine*. London: Andrew Melrose.
Hüe, F. & Etchecopar, R. D. 1970. *Les Oiseaux du Proche et du Moyen Orient*. Paris: N. Boubee.
Hume, A. O. 1873. Contributions to the ornithology of India: Sindh. No. II. Stray Feathers 1: 91–289.
Hume, A. O. 1880. *Passer pyrrhonotus* Blyth. Stray Feathers 9: 442–445.
Hume, R. A. 1963. Hybrid Tree × House Sparrow paired with House Sparrow. Brit. Birds 76: 234–235.

Ilyashenko, V. V. 1979. [On the biology of the Saxaul Sparrow.]. Ornitologiya 14: 213–214. (In Russian)
Immelmann, K. 1970. Brutbiologische Beobachtungen am Kapsperling (*Passer melanurus*) im südlichen Afrika. Beitr. Vogelk. 16: 195–203.
Inskipp, C. & T. P. 1984. Additions to the bird species recorded from Nepal. J. Bombay Nat. Hist. Soc. 81: 703–706.
Ion, I. 1973. [Biostatistic and ecological researches on the populations of two sparrow species (*Passer d. domesticus* (L.) and *Passer m. montanus* (L.)) in Moldavia.] PhD Thesis, Iaşi University. (In Roumanian). (Quoted by Dyers *et al.* 1977.)
Irwin, M. P. S. 1981. *The Birds of Zimbabwe*. Harare: Quest Publications.
Irwin, M. P. S. & Benson, C. W. 1967. Notes on the birds of Zambia: Part 4. Arnoldia 3: 1–27.

Janssen, R. R. 1983. House Sparrows build roost nests. Loon 55: 64–65.
Jacob, J.-P. & Schaetzen, de R. 1984. Decouverte du Moineau doré *Passer luteus* dans l'extreme sud de l'Algerie en relation avec le peuplement du nord du Niger. Malimbus 6: 73–74.
Jenni, L. & Schaffner, U. 1984. Herbstbewegungen von Haus- und Feldsperling *Passer domesticus domesticus* und *P. montanus* in der Schweiz. Orn. Beob. 81: 61–67.
Jennings, M. C. 1981a. *The Birds of Saudi Arabia*: A Check List. Private publication.
Jennings, M. C. 1981b. *Birds of the Arabian Gulf*. London: Geo. Allen & Unwin.
Jensen, J. V. & Kirkeby, J. 1980. *The Birds of the Gambia*. Arhus.
Johnston, A. W., Millie, W. R. & Moffat, G. 1970. Notes on the birds of Easter Island. Ibis 112: 532–538.
Johnston, R. F. 1967a. Sexual dimorphism in juvenile House Sparrow. Auk 84: 275–277.
Johnston, R. F. 1967b. Some observations on natural mass mortality of House Sparrows. Kansas Orn. Soc. Bull. 18: 9–12.
Johnston, R. F. 1969a. Taxonomy of House Sparrows and their allies in the Mediterranean basin. Condor 71: 129–139.
Johnston, R. F. 1969b. Aggressive foraging behaviour in House Sparrow. Auk 86: 558–559.
Johnston, R. F. 1969c. Character variation and adaptation in European sparrows. Syst. Zool. 18: 206–231.
Johnston, R. F. 1973. Evolution in the house sparrow. IV. Replicate studies in plumage covariation. Syst. Zool. 22: 219–226.
Johnston, R. F. 1976. Evolution in the house sparrow. V. Covariation of skull and hindlimb size. Occ. Papers Mus. Nat. Hist. Univ. Kansas. 56: 1–8.
Johnston, R. F. & Klitz, W. J. 1977. Variation and evolution in a granivorous bird: the house sparrow. In: Pinowski & Kendeigh (Eds.). 1977.
Johnston, R. F. & Selander, R. K. 1964. House Sparrows: rapid evolution of races in North America. Science 140: 548–550.
Johnston, R. F. & Selander, R. K. 1971. Evolution in the House Sparrow. II. Adaptive differentiation in North American populations. Evolution 25: 1–28.

Johnston, R. F. & Selander, R. K. 1973a. Evolution in the House Sparrow. III. Variation in size and sexual dimorphism in Europe, North America and South America. Am. Nat. 107: 373–390.
Johnston, R. F. & Selander, R. K. 1973b. Variation, adaptation and evolution in the North American House Sparrows. In: Kendeigh & Pinowski (Eds.). 1973.
Jones, A. E. 1912. Notes on birds from Lahore. J. Bombay Nat. Hist. Soc. 21: 1073–1074.
Jones, P. J. 1976. The tale of the Golden Sparrow. Spectrum 146: 7–9.
Jourdain, F. C. R. 1936. The birds of southern Spain. Part I – Passeres. Ibis (13) 6: 725–736.
Jourdain, F. C. R. & Witherby, H. F. 1918/19. The effect of the winter of 1916–17 on our resident birds. Brit. Birds 12: 26–35.

Kaatz, C. & Olburg, S. 1975. Investigations on the breeding biology of *Passer montanus* (L.). Int. Studies on Sparrow 8: 107–116.
Kalinoski, R. 1975. Intra- and interspecific aggression in House Finch and House Sparrow. Condor 77: 375–384.
Kalitsch, L. von. 1927. Freistehendes Nist des Feldsperlings. Beitr. Fortpfl.-biol. Vög. 3: 209.
Kalmbach, E. R. 1940. Economic status of the English Sparrow in the United States. U.S. Dept. Agric. Tech. Bull. No. 711.
Kashkarov, D. 1926. [Observations on the biology of the sparrows in Turkestan and their role in the deterioration of the crops.] Byull. Sred.-Asiat. Gosudar Univ. 13: 61–80. (In Russian)
Keay, R. W. J. (Ed.). 1959. *Vegetation map of Africa*. London: Oxford Univ. Press.
Keck, W. N. 1934. The control of the secondary sex characters in the English Sparrow *Passer domesticus* (Linnaeus). J. Exptl. Zool. 67: 315–347.
Kendeigh, S. C. & Pinowski, J. (Eds.). 1973. *Productivity, Population Dynamics and Systematics of Granivorous Birds*. Warszawa: PWN.
Kendrew, W. G. 1961. *The Climate of the Continents*. Oxford: Clarendon Press.
Keulemans, J. G. 1866. Opmerkingen over de vogels van de Kaap-Verdische Eilanden en van Prins-Eiland. Nederland Tijds. Dierk. 3: 363–374.
Keve, A. 1976a. Taxonomical position of the Tree Sparrow on Malta. Il-Merill (17): 13–14.
Keve, A. 1976b. Some remarks on the taxonomic position of the Tree Sparrow introduced into Australia. Emu 76: 152–153.
King, B. F. 1962. Guam field notes. Elepaio 23: 29–31.
King, B. F., Dickinson, E. C. & Woodcock, M. W. 1975. *A Field Guide to the Birds of South-East Asia*. London: Collins.
Kinnear, N. B. 1922. On the birds collected by Mr. A. F. Wollaston during the first Mount Everest expedition. Ibis (11) 4: 496–526.
Kinnear, N. B. 1929. On the birds collected by Mr. H. Stevens in northern Tonkin in 1923–24, Part II. Ibis (12) 5: 292–344.
Korzyukov, A. I. 1979. [On the mass migration of the Tree Sparrow over the north west part of the Black Sea]. Ornitologiya 14: 216. (In Russian)
Kozlova, E. V. 1932. The birds of south-west Transbaicalica, northern Mongolia and central Gobi. Ibis (13) 2: 316–347.
Kozlova, E. V. 1933. The birds of south-west Transbaicalica, northern Mongolia and central Gobi. Part IV. Ibis (13) 3: 59–87.
Kumerloeve, H. 1965a. Der Moabsperling, *Passer moabiticus* Tristram, Brutvogel in der Türkei. J. Ornith. 106: 112.
Kumerloeve, H. 1965b. Le moineau moabite *Passer moabiticus* près Bircek sur l'Euphrate. Alauda 33: 257–264.
Kumerloeve, H. 1969. The Dead Sea Sparrow: a second breeding place on Turkish and the first known breeding place on Syrian territory. Ibis 111: 617–618.
Kumerloeve, H. 1970. Zur Vogelwelt in Raume Ceylânpinar. Beitr. Vogelk. 16: 239–249.
Kumerloeve, H. 1978. Situation des Moineaux moabites nicheurs en Turquie. Alauda 46: 181–182.
Kunkel, P. 1961. Allgemeine und soziales Verhalten des Braunrücken-goldsperlings (*Passer* [*Auripasser*] *luteus* Licht.). Z. Tierpsychologie 18: 471–489.
Kuroda, N. L. 1966. Analysis of banding data (1924–43) of the Tree Sparrow in Japan. Misc. Rep. Yamashina Inst. for Ornithology. 4: 129–134.

Lack, D. 1976. *Island Biology*. Oxford: Blackwell.
Lack, P. (Compiler). 1986. *The Atlas of Wintering Birds in Britain and Ireland*. Calton: Poyser.
Lafresnaye, F. de 1850. Sur la nidification de quelque especes d'oiseaux de la famille ou sous-famille de Tisserins (Ploceinae). Rev. Mag. Zool. Pure et Applique, (2) 2: 315–320.

Lamarche, B. 1981. Liste commentée des oiseaux du Mali. 2eme partie: passeraux. Malimbus 3: 73–102.
La Touche, J. D. de 1922. A list of birds collected and observed in the island of Shaweisan. Bull. Br. Orn. Cl. 29: 124–160.
La Touche, J. D. de 1923. On the birds of south-east Yunnan. Ibis (11) 5: 629–645.
Leakey, R. E. 1981. *The Making of Mankind*. London: Michael Joseph.
Lees-Smith, D. T. 1986. Composition and origins of the southwest Arabian avifauna: a preliminary analysis. Sandgrouse 7: 71–93.
Le Grand, G. 1977. Apparition du moineau domestique aux Açores. Alauda 45: 339–340.
Le Grand, G. 1983. Le moineau domestique (*Passer domesticus*) aux Açores. Arquipélago 4: 85–116.
Lekagul, B. & Cronin, E. W. Jr. 1974. *Bird Guide to Thailand*, 2nd Edn. Bangkok: Kurusapa Ladprao Press.
Lever, Sir C. 1987. *Naturalised Birds of the World*. London: Longman.
Lewis, L. R. 1984. Observations on the status of the House Sparrow in the Newbury district. Newbury District Orn. Cl. Ann. Rep.: 34–39.
Lewis, M. D. 1981. The Somali Sparrow *Passer castanopterus*: a breeding record for East Africa. Scopus 5: 83.
Lippens, L. & Wille, H. 1976. *Les Oiseaux du Zaire*. Tielt: Lannoo.
Lobb, M. G. 1981. Dead Sea Sparrow *Passer moabiticus*: A new breeding species for Cyprus. RAF Orn. Soc. J. (12): 25–27.
Löhrl, H. 1963. Zur Hohenverbreitung einiger Vogel der Alpen. J. Ornith. 104: 62–68.
Löhrl, H. 1978. Hohlenkonkurrenz und Herbst-Nestbau beim Feldsperling (*Passer montanus*). Vogelwelt 99: 121–131.
Löhrl, H. & Böhringer, R. 1957. Untersuchungen an einer südwestdeutschen Population des Haussperlings (*Passer d. domesticus*). J. Ornith. 98: 229–240.
Long, J. L. 1981. *Introduced Birds of the World*. Newton Abbot & London: David Charles.
Louette, M. 1981. The Birds of Cameroun. An annotated Checklist. Brussels: Palais de Academien Klasse der Wettenschoppen 43: No. 163.
Loustau-Lalanne, P. 1962. The birds of the Chagos Archipelago, Indian Ocean. Ibis 104: 67–73.
Lowther, P. E. 1979a, Overlap of House Sparrow broods in the same nest. Bird Banding 50: 160–161.
Lowther, P. E. 1979b. The nesting biology of House Sparrows in Kansas. Kansas. Orn. Soc. Bull. 30: 23–26.
Lowther, P. E. 1979c. Growth and dispersal of nestling House Sparrows: sexual differences. Inland Bird banding 51: 23–29.
Ludlow, F. 1928. Birds of the Gyantse neighbourhood, southern Tibet, Part II. Ibis (12) 4; 51–73.
Ludlow, F. 1937. The birds of Bhutan and adjacent territories of Sikkim and Tibet, Part III. Ibis (14) 1: 467–504.
Ludlow, F. 1944. The birds of south east Tibet. Ibis 86: 348–389.
Ludlow, F. 1951. The birds of Kongbo and Pome, south Tibet. Ibis 93: 547–578.
Ludlow, F. & Kinnear, N. B. 1933. A contribution to the ornithology of Chinese Turkestan, Part II. Ibis (13) 3: 658–694.
Lulav, S. 1967. Northward extension of range of the Dead Sea Sparrow. IUCN Bull. N.S. 2: 11.
Lund, Hj. M.-K. 1956. Graspurven (*Passer domesticus* L.) i Nord-Norge. Dansk. Orn. Forenings Tids. 50: 67–76.
Lynes, H. 1924. On the birds of north and central Darfur, with notes on the western central Kordofan and north Nubia Provinces of British Sudan. Ibis (11) 6: 648–719.
Lynes, H. 1926. On the birds of North and Central Darfur. Taxonomic Appendix, Part I. Ibis (12) 2: 346–405.

McAtee, R. L. 1940. A study in song bird management. Auk 57: 338–348.
McClure, H. E. & Puntipa, K. 1973. The avifauna complex of an Open-billed Stork colony (*Anastomus oscitans*) in Thailand. Nat. Hist. Bull. Siam Soc. 25: 133–155.
Macdonald, J. D. 1957. Contribution to the ornithology of western South Africa. London: Brit. Mus.
McGillivray, W. B. 1980. Communal nesting in the House Sparrow. J. Field Orn. 51: 371–372.
McGillivray, W. B. 1981. Climatic influences on productivity in the House Sparrow. Wilson Bull. 93: 19–206.

McGillivray, W. B. 1983. Intraseasonal reproductive costs for the House Sparrow. (*Passer domesticus*). Auk 100: 25–32.

McGillivray, W. B. 1984. Nestling feeding rates and body size of adult House Sparrows. Can. J. Zool. 62: 381–385.

Mackinowitz, R., Pinowski, J. & Wieloch, M. 1970. Biomass production by house sparrow (*Passer d. domesticus* (L.)) and tree sparrow (*Passer m. montanus*. (L.)) populations in Poland. Ekologia Polska 18: 465–501.

Mackworth-Praed, C. W. & Grant, C. H. B. 2nd Edn. 1960. *African Handbook of Birds.* Series I, Vol. II. Birds of Eastern and North Eastern Africa. London: Longman.

Mackworth-Praed, C. W. & Grant, C. H. B. 1963. *African Handbook of Birds.* Series II, Vol. II. Birds of the Southern Third of Africa. London: Longman.

Mackworth-Praed, C. W. & Grant, C. H. B. 1973. *African Handbook of Birds.* Series III, Vol. II. Birds of West Central and Western Africa. London: Longman.

McLachlan, G. R. & Liversidge, R. 1978. *Roberts' Birds of South Africa.* 4th Edn. Cape Town: John Voelcker Bird Book Fund.

McLean, G. L. 1984. *Robert's Birds of Southern Africa.* 5th Edn. Cape Town: John Voelcker Bird Book Fund.

Macpherson, A. H. 1919. Wild hybrid between House- and Tree-Sparrow. Brit. Birds 13: 199.

Makatsch, W. 1955. Beitrag zur Biologie des Weidensperlings. Aquila 59–62: 347–350.

Makatsch, W. 1957. Observations on spring passage through Algeria. Vogelwelt 78: 19–31.

Mansfeld, K. 1939. Die Ernährung der Haussperlinge und eine Bekämpfung. Deutsche Vogelw. 64: 81–85.

Marchant, S. 1963. The breeding of some Iraqi birds. Ibis 105: 516–557.

Marples, B. J. & Gurr, L. 1943. A mechanism for recording automatically the nesting habits of birds. Emu 43: 67–71.

Mathew, K. L. & Naik, R. M. 1986. Interrelation between moulting and breeding in a tropical population of the House Sparrow *Passer domesticus.* Ibis 128: 260–265.

Mayes, W. E. 1926–27. House-Sparrow winter nest building. Brit, Birds 20: 273–274.

Mayr, E. 1927. Die Schneefinken (Gattungen *Montifringilla* und *Leucosticta*). J. Ornith. 75: 596–619.

Mayr, E. 1949. Enigmatic sparrows. Ibis 91: 304–306.

Mayr, E. 1963. *Animal Species and Evolution.* Cambridge, Mass.: Belknap Press.

Mayr, E. & Amadon, D. 1951. A classification of recent birds. Am. Mus. Novit. No. 1496.

Mayr, E. & Greenway, J. C. Jr. (Eds.). 1962. Check-list of the Birds of the World. Vol. XV. Cambridge, Mass., USA: Mus. Comp. Biology.

Medway, Lord & Wells, D. R. 1976. *The Birds of the Malay Peninsula.* Vol. V. London: Witherby.

Meinertzhagen, R. 1919. A preliminary study of the relations between geographical distribution and migration with special reference to the Palaearctic region. Ibis (11) 1: 379–392.

Meinertzhagen, R. 1920. Notes on the birds of southern Palestine. Ibis (11) 2: 195–259.

Meinertzhagen, R. 1921. Notes on some birds from the Near East and from tropical East Africa. Ibis (11) 3: 621–671.

Meinertzhagen, R. 1934. The biogeographical status of the Ahaggar Plateau in the central Sahara, with special reference to birds. Ibis (13) 4: 528–571.

Meinertzhagen, R. 1940. Autumn in central Morocco. Ibis (14): 4: 106–136.

Meinertzhagen, R. 1949. Notes on Saudi Arabian birds. Ibis 91: 465–482.

Meinertzhagen, R. 1951. Some relationships between African, Oriental and Palaearctic genera and species. Ibis 93: 443–459.

Meinertzhagen, R. 1954. *The Birds of Arabia.* Edinburgh & London: Oliver & Boyd.

Meinertzhagen, R. 1958. *Pirates and Predators.* Edinburgh & London: Oliver & Boyd.

Meise, W. 1934. Über Artbastarde bei paläarktischen Sperlingen. Orn. Monatsb. 42: 9–15.

Meise, W. 1936. Zur Systematik und Verbreitungsgeschichte der Haus- und Weidensperlinge, *Passer domesticus* (L.) und *hispaniolensis* (T.). J. Ornith. 84: 631–672.

Meise, W. 1951. Hampes Mischzucht von Haus- und Feldsperling *Passer d. domesticus* (L.) × *P. m. montanus* (L.). Bonn. zool. Beitr. 2: 85–98.

Melliss, J. C. 1870. Letter. Ibis (3) 1: 367–370.

Mendelsshon, H. 1955. [Biology and ethology of the Dead Sea Sparrow, *Passer moabiticus* Tristram 1864.] Sal'it (2): 27–36. (In Hebrew)

Menegaux, A. 1919, 1920 & 1921. Ênquete sur la disparition du moineau dans la Midi. Rev. franç. d'Orn. 11: 65–71 & 129–131, 12: 32–36, 52–55 & 77–78, 13: 41–42 & 127–128.

Merikallio, E. 1958. Finnish birds, their distribution and numbers. Helsinki (Societas pro Fauna et Flora. Fuana Fennica 5.).

Milstein, P. le S. 1975. The biology of Barberspan with special reference to the avifauna. Ostrich Suppl. 10 1: 74.
Minock, M. E. 1969. Salinity tolerance and discrimination in House Sparrow (*Passer domesticus*). Condor 71: 79–80.
Mirza, Z. B. 1973. Study on the fecundity, mortality, numbers, biomass and food of a population of House Sparrows in Lahore, Pakistan. In: Kendeigh & Pinowski (Eds.) 1973.
Mirza, Z. B. 1974. A preliminary study of the breeding, food, sexual dimorphism and distribution of the Spanish Sparrow, *Passer hispaniolensis* Temm. in Libya. Int. Studies on Sparrows 7: 76–87.
Mitchell, C. J., Hayes, R. O., Holden, P. & Hughes, T. B. Jr. 1973. Nesting activity of the House Sparrow in Hale County, Texas, during 1968. In: Hardy & Morton 1973.
Moltoni, E. 1929. Su alcuni uccelli della Sardegna. Atti Soc. Ital. Sci. Nat. 62: 121–128.
Momiyama. 1927. Annot. Ornith. Orient 1: 121.
Moore, N. C. 1962. Tree Sparrows (*Passer montanus*): 'Copy preening' behaviour. Northants Nat. Hist. Soc. J. 34: 130–132.
Moreau, R. E. 1931. An Egyptian sparrow roost. Ibis (13) 1: 204–208.
Moreau, R. E. 1940. Contributions to the ornithology of the East African islands. Ibis (14) 4: 48–91.
Moreau, R. E. 1950. The breeding season of African birds. I. Land Birds. Ibis 92: 223–267.
Moreau, R. E. 1962. Family Ploceidae. In: Mayr & Greenway (Eds.) 1962.
Moreau, R. E. 1967. Dust bathing in Ploceidae? Ibis 109: 445.
Morel, M.-Y. (in prep.). Successful introduction of *Passer domesticus indicus* Jardine & Selby in Senegambia. (Presented at 5th Pan African Orn. Congr., Malawi 1980).
Morel, M.-Y. & Morel, G. J. 1973a. Premieres observations sur la reproduction du Moineau doré, *Passer luteus* (Licht.) en zone semi-aride de l'ouest africain. Oiseau Rev. fr. Orn. 43: 97–118.
Morel, M.-Y. & Morel, G. J. 1973b. Elements de comparaison du comportement reproducteur colonial de trois especes de Ploceides: *Passer luteus, Ploceus cucullatus* et *Quelea quelea* en zone semi-aride de l'ouest africain. Oiseau Rev. fr. Orn. 43: 314–329.
Morel, G. J. & Morel, M.-Y. 1976. Nouvelles observations sur la reproduction du Moineau doré *Passer luteus* en zone semi-aride de l'ouest africain. Terre et Vie 30; 493–520.
Morel, G. J. & Morel, M.-Y. 1978. Elements du comparaison entre *Quelea qu. quelea* (L.) et *Passer luteus* (Lichtenstein) dans les savanes tropicales de l'ouest africain. Cah. O.R.S.T.O.M. ser. Biol. 13: 347–358.
Morel, G. J. & Morel, M.-Y. 1980. Has the Golden Sparrow replaced the Black-faced Dioch in West Africa? Symposium on genus *Passer*. Proc. 17th I.O.C. 1978: 1150–1154. Deutsch. Orn. Gesell. R. Nohrung (Ed.). Berlin.
Morel, G. J. & Morel, M.-Y. 1982. Dates de reproduction des oiseaux de Senegambia. Bonn. Orn. Beitr. 33: 249–268.
Moritz, D. 1981. Abnahme des Feldsperlings, *Passer montanus*, auch als Durchzugler auf Helgoland. Vogelwelt 102: 215–219.
Mountfort, G. 1965. *Portrait of a Desert*. London: Collins.
Müller, P. 1967. Zur Verbreitung von *Passer domesticus* in Brasilien. J. Ornith. 108: 497–499.
Murphy, E. C. 1978. Breeding ecology of House Sparrows: spatial variation. Condor 80: 180–193.
Murphy, R. C. 1924. The marine ornithology of the Cape Verde Islands, with a list of all the birds in the archipelago. Bull. Am. Mus. Nat. Hist. No. 50: 211–278.
Myrcha, A., Pinowski, J. & Tomek, T. 1973. Energy balance of nestlings of Tree Sparrows, *Passer m. montanus* (L.), and House Sparrows, *Passer d. domesticus* (L.). In: Kendeigh & Pinowski (Eds.). 1973.

Naik, R. M. 1974. Recent studies on the granivorous birds in India. Intl. Studies on Sparrows 7: 21–25.
Naik, R. M. & Mistry, L. 1980. Breeding season in a tropical population of the House Sparrow. J. Bombay Nat. Hist. Soc. (Suppl.): 1118–1142.
Nankinov, D. N. 1984. Nesting habits of Tree Sparrows, *Passer montanus* (L.) in Bulgaria. Int. Studies on Sparrows 11: 47–70.
Ndao, B. 1980. Le Moineau domestique (*Passer domesticus*) espèce nouvelle pour le Senegal. Bull. IFAN 40: 422–424.
Nelson, T. H. 1907. *The Birds of Yorkshire*. Vol. 1. London: A. Brown & Sons.
Nero, R. W. 1951. Pattern and rate of cranial ossification in the House Sparrow. Wilson Bull. 63: 84–88.

Newman, K. 1980. *Birds of Southern Africa*, 1: Kruger National Park. Johannesburg: Macmillan.
Nice, M. M. 1957. Nesting success in altricial birds. Auk 74: 305–321.
Nichols, J. B. 1919. Wild hybrid between House Sparrow and Tree Sparrow. Brit. Birds 13: 136.
Nichols, J. T. 1934. Distribution and seasonal movements of the House-Sparrow. Bird Banding 5: 20–23.
Nicoll, M. J. 1922. On a collection of birds made in the Sudan. Ibis (11) 4: 699–701.
Niethammer, G. 1937. *Handbuch der deutschen Vogelkunde*. Vol. I. Leipzig. Akademische Verlagsgesellschaft.
Niethammer, G. 1953. Gewicht und Flügellänge beim Haussperling. J. Ornith. 94: 282–289.
Niethammer, G. 1955. Zur Vogelwelt des Ennedi-Gebirges. Bonn. zool. Beitr. 6: 29–50.
Niethammer, G. 1958. Das Mischgebiet zwischen *Passer d. domesticus* and *Passer d. italiae* in Süd-Tirol. J. Ornith. 99: 431–437.
Niethammer, G. & Bauer, K. 1960. Das Mischgebiet zwischen *Passer d. domesticus* und *Passer d. italiae* im Tessim. Orn. Beob. 57: 241–242.
Nørrevang, A. N. & Den Hartog, J. C. 1984. Bird observations in the Cape Verde Islands. Courier Forsch.-Inst. Senckenberg. 69: 107–134.
North, C. A. 1968. A study of House Sparrow populations and their movements in the vicinity of Stillwater, Oklahoma. PhD Thesis, Oklahoma State Univ.
North, C. A. 1973. Population dynamics of the House Sparrow *Passer domesticus* (L.) in Wisconsin, USA. In: Kendeigh & Pinowski (Eds.). 1973.
North, C. A. 1980. Attentiveness and nesting behaviour of the male and female House Sparrow (*Passer domesticus*) in Wisconsin. Symposium on the genus *Passer*: Proc. 17th I.O.C. 1978: 1122–1128. Deutsch. Orn. Gesell. R. Nohrung (Ed.). Berlin.
Novotny, I. 1970. Breeding bionomy, growth and development of young House Sparrow (*Passer domesticus* Linne 1758). Acta Sc. Nat. Brno. 4: 1–57.

Oates, E. W. 1882. A list of the birds of Pegu. Stray Feathers 10: 175–248.
Oatley, T. B. & Skead, D. M. 1972. Nectar feeding by South African birds. Lammergeyer 15: 65–74.
O'Connor, R. J. 1973. Patterns of weight changes in the House Sparrow, *Passer domesticus* (L.). In: Kendeigh & Pinowski (Eds.). 1973.
Ogilvie-Grant, W. R. 1912. On the birds of Ngamiland. Ibis (9) 6: 355–404.
Ortiz-Crespo, F. I. 1977. La presencia del gorrión europeo, *Passer domesticus* L, en el Ecuador. Rev. Univ. Catolica, Quito 16: 193–197.
Oubron, G. 1967. Quelques observations sur *Passer luteus*. O.C.L.A.L.A.V. (Dakar), Report No. 707/403.33.21/LAV.
Oustalet, M. E. 1883. Description d'espèces nouvelles d'oiseaux provement des Iles du Cap Vert. Ann. Sci. Nat. Zool. (6) 16: 1–2.

Packard, G. C. 1967a. Seasonal variation in the bill length of House Sparrows, Wilson Bull. 79: 345–346.
Packard, G. C. 1967b. House Sparrows: evolution of populations from the Great Plains and Colorado Rockies. Syst. Zool. 16: 73–89.
Pakenham, R. H. W. 1943. Field-notes on the birds of Zanzibar and Pemba. Ibis 85: 165–189.
Pakenham, R. H. W. 1979. *The Birds of Zanzibar and Pemba*. London: Brit. Orn. Union.
Pantuwatana, S., Imlarp, S. & Marshall, J. T. 1969. Vertebrate ecology of Bang Phra. Nat. Hist. Bull. Siam Soc. 23: 133–183.
Pardo, R. 1980. Contribución al conocimiento del gorrión commún, *Passer domesticus*, en el naranjal de Sagunto (Valencia). Misc. Zool. Barcelona 6: 85–94.
Parkin, D. T. (in press). Genetic variation in the House Sparrow *Passer domesticus*. Presented at 19th I.O.C., Ottawa, June 1986.
Parkin, D. T. & Cole, S. R. 1984. Genetic variation in the house sparrow, *Passer domesticus*, in the East Midlands of England. Biol. J. Linnean Soc. 23: 287–301.
Parkin, D. T. & Cole, S. R. 1985. Genetic differentiation and rates of evolution in some introduced populations of the House Sparrow *Passer domesticus* in Australia and New Zealand. Heredity 54: 15–23.
Payne, R. B. 1969. Nest parasitism and display of Chestnut Sparrow in a colony of Grey-headed Social Weavers. Ibis 111: 300–307.
Phillips, E. L. 1898. Narrative of a visit to Somaliland in 1897, with field notes on the birds obtained during this expedition. Ibis (7) 1: 382–425.
Piechocki, R. 1954. Statistische Feststellungen an 20,000 Sperlingen (*Passer d. domesticus*). J. Ornith. 95: 297–305.

Pielowski, Z. & Pinowski, J. 1962. Autumn sexual behaviour of the Tree Sparrow. Bird Study 9: 116–122.
Piggot, S. 1965. *Ancient Europe from the Beginnings of Agriculture to Classical Antiquity: A Survey.* Chicago: Aldous Pub. Co.
Pihl, S. 1977. Birds in Eilat, spring 1977. Eilat Field Study Centre, Society for the Protection of Nature in Israel.
Pinowska, B. 1975. Foods of female house sparrows (*Passer domesticus* (L.)) in relation to stages of the nesting cycle. Polish Ecol. Studies 1: 211–225.
Pinowska, B. 1979. The effect of energy and building resources of females on the production of House Sparrows (*Passer domesticus* (L.)) populations. Ekol. Polska 27: 363–376.
Pinowska, B. & Pinowski, J. 1977. Fecundity, mortality, numbers and biomass dynamics of a population of the house sparrow *Passer domesticus* (L.). Int. Studies on Sparrows 10: 26–41.
Pinowski, J. 1965a. Overcrowding as one of the causes of dispersal of young Tree Sparrows. Bird Study 12: 27–33.
Pinowski, J. 1965b. Dispersal of young Tree Sparrows (*Passer m. montanus* L.). Bull. Acad. Pol. Sci. Cl. II 13: 509–514.
Pinowski, J. 1966. Der Jahreszyklus der Brutkolonie beim Feldsperling (*Passer m. montanus* L.). Ekol. Pol. Ser. A 14: 145–174.
Pinowski, J. 1967a. Die Auswahl des Brutbiotope beim Feldsperling (*Passer m. montanus* L.). Ekol. Pol. Ser. A 15: 1–30.
Pinowski, J. 1967b. Estimation of the biomass production by a Tree Sparrow (*Passer m. montanus* L.) population during the breeding season. In: Petrusewicz, K. (Ed.). 1967. *Secondary Production of Terrestrial Ecosystems.* Vol. 1. Warszawa: PWN.
Pinowski, J. 1968. Fecundity, mortality, numbers and biomass dynamics of a population of the Tree Sparrow (*Passer m. montanus* L.). Ekol. Pol. Ser. A 16: 1–58.
Pinowski, J. 1971. Dispersal, habitat preferences and the regulation of population numbers in Tree Sparrows, *Passer m. montanus* (L.). Int. Studies on Sparrows. 5: 21–39.
Pinowski, J. & Kendeigh, S. C. (Eds.). 1977. *Granivorous Birds in Ecosystems.* London: Cambridge Univ. Press.
Pinowski, J. & Pinowska, B. 1985. The effect of the snow cover on the Tree Sparrow (*Passer montanus*) survival. The Ring 124–125: 51–56.
Pinowski, J. & Wieloch, M. 1973. Energy flow through nestlings and biomass production of House Sparrow *Passer d. domesticus* (L.) and Tree Sparrow *Passer m. montanus* (L.) populations in Poland. In: Kendeigh & Pinowski (Eds.). 1973.
Pinto, A. A. Da Rosa. 1959. Alguns novos records de aves para o Sul do Save e Moçambique, incluindo de genera novo para a sub-região de Africa do Sul, com a descrição de nova sub-especies. Bol. Soc. Estudos Prov. Moçambique 118: 15–25.
Pocock, T. N. 1966. Contributions to the osteology of African birds. Proc. 2nd Pan African Orn. Cong. Ostrich Suppl. 6: 83–94.
Poltz, J. & Jacob, J. 1974. Bürzeldrüsensekrete bei Ammern (Emberizidae), Finken (Fringillidae) und Webern (Ploceidae). J. Ornith. 115: 119–127.
Preiser, F. 1957. Untersuchungen über die Ortsstetigkeit und Wanderung der Sperlinge (*Passer domesticus domesticus* L.) als Grundlage für die Bekämpfung. Diss. Landwirts. Hochschule, Hohenheim.
Price, T. D. 1979. The seasonality and occurrence of birds in the Eastern Ghats of Andra Pradesh. J. Bombay Nat. Hist. Soc. 76: 379–422.
Prozesky, O. P. M. 1970. *A Field Guide to the Birds of Southern Africa.* London: Collins.

Rademacher, B. 1951. Beringungsversuche über die Ortstreue der Sperlinge. Z. Pflanzenkrankheiten und Pflanzenschutz 48: 416–426.
Raju, K. S. P. & Price, J. D. 1973. Tree Sparrow *Passer montanus* (L.) in the Eastern Ghats. J. Bombay Nat. Hist. Soc. 70; 557–558.
Ralph, C. & Sakai, H. F. 1979. Forest bird and fruit bat populations and their conservation in Micronesia; notes and survey. Elepaio 40: 20–26.
Rassi (unpub.). Quoted by Pinowski & Kendeigh (Eds.). 1977.
Reynolds, J. & Stiles, F. G. 1982. Distribución y densidad de poblaciones del gorrión común (*Passer domesticus*: Aves: Ploceidae) en Costa Rica. Rev. Biol. Trop. 30: 65–77.
Richardson, R. A. 1957. Hybrid Tree × House Sparrow in Norfolk. Brit. Birds 50: 80–81.
Riley, J. H. 1938. Birds from Siam and the Malay Peninsula in the U.S. National Museum. U.S. Nat. Mus. Bull. No. 172: 532–535.

Ripley, S. D. 1982. A Synopsis of the Birds of India and Pakistan. Bombay Natural History Society.
Ris, H. 1957. Zum Vorkommen des Haussperlings im Kanton Wallis. Orn. Beob. 54: 195–197.
Rising, J. D. 1973. Age and seasonal variation in dimensions of House Sparrows, *Passer domesticus* (L.) from a single population in Kansas. In: Kendeigh & Pinowski (Eds.). 1973.
Robbins, C. S. 1973. Introduction, spread and present abundance of the House Sparrow in North America. In: Hardy & Morton (Eds.). 1973.
Rooke, K. B. 1957. Hybrid Tree × House Sparrow in Dorset. Brit. Birds 50: 79–80.
Rosetti, K. 1983. House Sparrow taking insects from spiders' webs. Brit. Birds 76: 412.
Rothschild, W. & Hartert, E. 1911. Ornithological explorations in Algeria. Novit. Zool. 18: 479–482.
Rowan, M. K. 1964. An analysis of the results of a South African ringing station. Ostrich 35: 160–187.
Rowan, M. K. 1966. Some observations of reproduction and mortality in the Cape Sparrow *Passer melanurus*. Ostrich Suppl. 6: 425–434.
Ruelle, P. J. & Semaille, R. 1982. Note sur l'envahissement du nord du Senegal par le moineau doré *Passer luteus* (Lichtenstein) en period de reproduction. Malimbus 4: 27–32.
Ruprecht, A. L. 1967. A hybrid House Sparrow × Tree Sparrow. Bull. Br. Orn. Cl. 87: 78–81.
Ruthke, P. 1930. Kommen Mischpaare von *Passer domesticus* und *montanus* vor?. Beitr. Fortpfl.-biol. Vög. 6: 29.
Ruthke, P. 1955. Feldsperling als Freibruter. Vogelwelt 76: 108.

Sacarrão, G. F. & Soares, A. A. 1975. Algumas observaçõs sobre a biologia de *Passer hispaniolensis* (Temm.) em Portugal. Est. Fauna Port. 8: 1–20.
Sage, B. L. 1956. Remarks on the racial status, history and distribution of the Tree Sparrow introduced into Australia. Emu 56: 137–40.
Sage, B. L. 1962. The breeding distribution of the Tree Sparrow. London Bird Report (27): 56–65.
Salamonsen, F. 1935. Aves. In: *Zoology of the Faroes*. Jenson, Ad. S., Lundbeck, W. & Mortmen, T. M. (Eds.). Copenhagen.
Savaltori, T. 1899. Collezioni ornitologiche fatte nette isola del Cabo Verde da Leonardi Fea. Ann. Mus. Civ. Genoa 20: 283–312.
Sánchez-Aguado, F. J. 1984. Fenologia de la reproduccion y tamaño de la puesta en el gorión molinero, '*Passer montanus* L.'. Ardeola 31: 33–45.
Sappington, J. N. 1977. Breeding biology of House Sparrows in north Mississippi. Wilson Bull. 89: 300–309.
Scherner, E. R. 1972a. Untersuchungen zur Ökologie des Feldsperlings *Passer montanus*. Vogelwelt 93: 41–68.
Scherner, E. R. 1972b. Dichte, Produktion und Umsatzrate bei 3 Höhlenbrüter-Populationen (*Parus caerulus, Parus major, Passer montanus*) in südöstlichen Niedersachsen. Angew. Ornith. 4: 35–42.
Schifferli, L. 1978. Die Rolle des Männchens während der Bebrütung der Eier beim Haussperling *Passer domesticus*. Orn. Beob. 75: 44–47.
Schifferli, L. & Schifferli, A. 1980. Die Verbreitung des Haussperlings *Passer domesticus domesticus* und des Italiensperlings *Passer domesticus italiae* im Tessin und im Misox. Orn. Beob. 77: 21–26.
Schmidl, D. 1982. *The Birds of the Serengeti National Park*. London: Brit. Orn. Union.
Schnurre, O. 1930. Bemerkenswerte zur Nistweise von Haus- und Feldsperling. Beitr. Fortpfl.-biol. Vög. 6: 67–76.
Schöll, R. W. 1959. Über des Vorkommen von Sperlingen am Brenner-Pass (Tirol). J. Ornith. 100: 439–440.
Schöll, R. W. 1960. Die Sperlingsbesiedlung des Pustertales/Südtirol und seiner nördlichen Seitentäler. Anz. Ornith. Ges. Bayern. 5: 591–596.
Schönfeld, M. & Brauer, P. 1972. Ergebnisse der achtjährigen Untersuchungen an der Höhlenbrüterpopulationen eines Eichen-Hainbuchen-Linden-Waldes in der 'Alten Göhle' bei Freyburg/Unstrut. Hercynia N.F. 9: 40–68.
Schweiger, H. 1959. Über das Vorkommen des Rotkopfsperlings (*Passer domesticus italiae* (Vieill.), in Südkärnten. J. Ornith. 100: 350–351.
Sclater, W. L. 1930. *Systema Avium Aethiopicarum*. Vol. 2: 305–992. London: Taylor & Francis.
Scully, J. 1881. A contribution to the ornithology of Gilgit. Ibis (4) 5: 567–594.
Seel, D. 1960. The behaviour of a pair of House Sparrows while rearing young. Brit. Birds 53: 303–310.

Seel, D. C. 1964. An analysis of the nest record cards of the Tree Sparrow. Bird Study 11: 265–271.
Seel, D. C. 1968a. Breeding seasons of the House Sparrow and Tree Sparrow. Ibis 110: 129–144.
Seel, D. C. 1968b. Clutch size, incubation and hatching success in the House Sparrow and Tree Sparrow *Passer* spp. at Oxford. Ibis 110: 270–282.
Seel, D. C. 1969. Food, feeding rates and body temperature in the nestling House Sparrow *Passer domesticus* at Oxford. Ibis 111: 36–47.
Seel, D. C. 1970. Nestling survival and nestling weight in the House Sparrow and Tree Sparrow *Passer* spp. at Oxford. Ibis 112: 1–14.
Selander, R. K. & Johnston, R. F. 1967. Evolution in the house sparrow. 1. Intrapopulation variation in North America. Condor 69: 217–258.
Serle, W. 1940. Field observations on some northern Nigerian birds. Part II. Ibis (14) 4: 1–47.
Serle, W. 1950. A contribution to the ornithology of the British Cameroons. Ibis 92: 602–638.
Serle, W. 1981. The breeding season of birds in the lowland rainforest and in the montane forest of West Cameroon. Ibis 123: 62–74.
Seth-Smith, L. M. 1913. Notes on some birds around Mpumu, Uganda. Ibis (10), 1: 485–508.
Severinghaus, S. R. & Blackshaw, K. T. 1976. *A New Guide to the Birds of Taiwan.* Taipei: Mai Ya Publ. Inc.
Sharpe, R. B. 1888. *Catalogue of the birds of the British Museum.* Vol. XII. London: Brit. Mus. (Nat. Hist.).
Sharpe, R. B. 1891. On the birds collected by Mr R. B. Jackson, F.Z.S., during his recent expedition to Uganda through the territory of the Imperial British East Africa Company. Ibis (5) 3: 233–260.
Shnitnikov, V. N. 1949. [*Birds of Semireche*] Moscow: Akad. Nauk. (In Russian)
Short, L. L. 1969. Taxonomic aspects of avian hybridisation. Auk 86: 84–105.
Shulpin. 1956. Trans. Zool. Inst. Sci. Akad. Kaz. SSR: 158–193. (Quoted by Gavrilov 1962.).
Sibley, C. G. 1970. A comparative study of the egg-white proteins of passerine birds. Bull. Peabody Mus. Nat. Hist. Yale Univ. 32: 1–131.
Sibley, C. G. & Ahlquist, J. E. 1981. The relationships of the Accentors (*Prunella*) as indicated by DNA × DNA hybridisation. J. Ornith. 122: 369–378.
Sibley, C. G. & Ahlquist, J. E. 1983. Phylogeny and classification of birds on the data of DNA × DNA hybridisation. In: Johnston, R. F. (Ed.). *Current Ornithology*: 245–292. New York: Plenum Press.
Sibley, C. G. & Ahlquist, J. E. 1985. The relationships of some groups of African birds, based on comparisons of the genetic material, DNA. Proc. Int. Symp. on African Vertebrates 1984: 115–161. K.-L. Schuchmann (Ed.). Bonn. zool. Forschungsinst. und Museum Alexander Koenig.
Sibley, C. G. & Ahlquist, J. E. 1986. Reconstructing bird phylogeny by comparing DNAs. Sci. Amer. 254 (2): 68–78.
Sibley, C. G., Ahlquist, J. E. & Monroe, B. L. Jr. (in press). An avian phylogeny. Poster presentation at 19th I.O.C., Ottawa, June 1986.
Sick, H. 1968. Über in Südamerika eingeführte Vogelarten. Bonn. zool. Beitr. 19: 293–306.
Sick, H. 1979. Notes on some Brazilian birds. Bull. Br. Orn. Cl. 99: 115–120.
Siegfried, W. R. 1973. Breeding success and reproductive potential in the Cape Sparrow, *Passer melanurus* (Müller). In: Kendeigh & Pinowski (Eds.) 1973.
Simon, P. 1965. Synthèse de l'avifauna du massif montagneux de Tibesti et distribution géographique de ces espèces en Afrique du Nord et environs. Gerfaut 55: 26–71.
Simpson, G. G. 1961. *Principles of Animal Taxonomy*. New York: Columbia Univ. Press.
Simwat, G. S. 1977. Studies on the feeding habits of House Sparrow *Passer domesticus* (L.) and its nestlings in Punjab, J. Bombay Nat. Hist. Soc. 74: 175–179.
Skead, D. M. 1977. Weights of birds handled at Barberspan. Ostrich Suppl. 12: 117–131.
Skead, D. W. & Dean, W. R. J. 1977. Status of the Barberspan avifauna, 1971–1975. Ostrich Suppl. 12: 3–42.
Sladen, A. G. 1919. Notes on birds observed in Palestine. Ibis (11) 1: 222–250.
Slater, P. 1975. *A Field Guide to Australian Birds: Passerines.* Edinburgh: Scottish Academic Press.
Smith, F. & Borg, S. 1976. New hybrid: *Passer montanus* × *Passer hispaniolensis.* Il-Merill (17): 25–26.
Smith, K. D. 1955a. Recent records from Eritrea. Ibis 97: 65–80.
Smith, K. D. 1955b. The winter breeding season of land-birds in eastern Eritrea. Ibis 97: 480–507.

Smith, K. D. 1957. An annotated check list of the birds of Eritrea. Ibis 99: 307–337.
Smith, N. H. J. 1973. House Sparrows (*Passer domesticus*) in the Amazon. Condor 75: 242–243.
Smith, N. H. J. 1980. Further advances of House Sparrows into the Brazilian Amazon. Condor 82: 109–111.
Smithers, R. H. N. 1964. A check list of the birds of the Bechuanaland Protectorate. Nat. Mus. S. Rhodesia.
Smithers, R. H. N., Irwin, M. P. S. & Paterson, M. L. 1957. A Check List of the Birds of Southern Rhodesia. Rhodesian Orn. Soc.
Smythies, B. E. 1953. *The Birds of Burma*. Edinburgh: Oliver & Boyd.
Sopyev, O. 1965. [The Desert Sparrow in the Kara Kum.] Ornitolgiya 7: 134–141. (In Russian)
Spennemann, A. 1934. Zur Nistweise des Feldsperlings auf Java. Beitr. Fortpfl. Vog. 10; 147–148.
Stanford, J. K. 1941. The Vernay-Cutting expedition to northern Burma. Part IV. Ibis (14) 5: 353–378.
Stanford, J. K. 1954. A survey of the ornithology of northern Libya. Ibis 96: 449–473 & 606–624.
Stanford, J. K. & Ticehurst, C. B. 1935. Notes on some new or rarely recorded Burmese birds. Part II. Ibis (13) 5: 249–279.
Steinbacher, J. 1952a. Jahrzeitliche Veränderung am Schnabel des Haussperlings (*Passer domesticus* L.). Bonn. zool. Beitr. 3: 23–30.
Steinbacher, J. 1952b. Zur Verbreitung und Biologie der Vögel Sardiniens. Vogelwelt 73: 197–208.
Steinbacher, J. 1954. Über die Sperlins-Formen von Sardinien und Sizilien. Senckenbergiana 34: 307–310.
Steinbacher, J. 1956. Zur Variation des Gefieders und Verhalten bei den Sperlingen Sardiniens und Siziliens. Senckenbergiana 37: 213–219.
Stenhouse, J. H. 1928. Remarkable decrease of the house sparrow on Fair Isle and Shetland. Scot. Nat.: 162–163.
Stephan, B. 1965. Beiträge zur Ethologie des Feldsperlings, *Passer montanus*. Beitr. Vogelk. 10: 380–385.
Stevens, H. 1925. Notes on the birds of the Sikkim Himalayas. Part V. J. Bombay Nat. Hist. Soc. 30; 352–379.
Strawinski, S. & Wieloch, M. 1972. [The characteristics of the populations of *Passer domesticus* (L.) and *P. montanus* (L.) as a background to activity studies.] Zeszyty Naukowe, Instytut Ekologii PAN. 5: 329–340. (In Russian). Quoted by Dyers et al. 1977.
Stresemann, E. 1936. A nominal list of the birds of Celebes. Ibis (13) 6: 351–368.
Stresemann, E. 1939. Zum neue Rassen aus Süd-Tibet und Nord Sikkim. Orn. Monatsb. 47: 176–179.
Stresemann, E. & Stresemann, V. 1966. Die Mauser der Vogel. J. Ornith. 107 (Sonderheft): 1–445.
Styan, F. W. 1891. On the birds of the lower Yangtse basin, Part I. Ibis (5) 3: 316–359.
Sultana, J. 1969. The Tree Sparrow *Passer montanus* breeding in the Maltese Islands. Bull. Br. Orn. Cl. 89: 29–31.
Sultana, J. & Gauchi, C. 1982. *A New Guide to the Birds of Malta*. Valetta, Malta: The Ornithological Society.
Summers-Smith, D. 1956. Movements of House Sparrows. Brit. Birds 49: 465–488.
Summers-Smith, D. 1958. Nest-site selection, pair formation and territory in the House-Sparrow *Passer domesticus*. Ibis 100: 190–203.
Summers-Smith, D. 1959. The House Sparrow *Passer domesticus*: population problems. Ibis 101: 449–454.
Summers-Smith, D. 1963. *The House Sparrow*. London: Collins.
Summers-Smith, D. 1978. The Spanish Sparrow on Malta. Il-Merill (19): 9–10.
Summers-Smith, D. 1979. *Passer* species on Sardinia. Il-Merill (20): 18–19.
Summers-Smith, D. 1980a. House Sparrows down coal mines. Brit. Birds 73: 325–327.
Summers-Smith, D. 1980b. Sparrows on Crete. Il-Merill (21): 17–18.
Summers-Smith, D. 1981. Notes on the Pegu Sparrow, *Passer flaveolus*. Nat. Hist. Bull. Siam. Soc. 29: 73–84.
Summers-Smith, D. 1984a. The rufous sparrows of the Cape Verde Islands. Bull. Br. Orn. Cl. 104: 138–142.
Summers-Smith, D. 1984b. The sparrows of the Cape Verde Islands. Ostrich 55: 141–146.
Summers-Smith, D. 1986. Articles on 'House Sparrow' and 'Tree Sparrow'. In: Lack, P. (Compiler). *The Atlas of Wintering Birds in Britain and Ireland*. Calton: Poyser.

Summers-Smith, D. & Vernon, J. D. R. 1972. The distribution of *Passer* in north-west Africa. Ibis 114: 259–262.
Sushkin, P. P. 1927. On the anatomy and classification of weaver-birds. Bull. Am. Mus. Nat. Hist. 57: 1–32.
Suyin, H. 1959. The sparrow shall fall. New Yorker 10th Oct. 1959.
Szlivka, L. 1983. Data on the biology of the Tree Sparrow (*Passer montanus montanus*). Larus 33–35: 141–159.

Tarboton, W. R. 1968. Check List of the Birds of South Central Transvaal. Witwatersrand Bird Club.
Tatschl, J. L. 1968. Unusual nesting site for House Sparrows. Auk 85: 514.
Tchernov, E. 1962. Paleolithic avifauna in Palestine. Bull. Res. Council Israel 11B3: 95–131.
Tchernov, E. 1984. Commensal animals and human sedentism in the Middle East. In: *Animals and Archaeology*. 3. Early Herders and their Flocks. Clutton-Brock, J. & Grigson, C. (Eds.). BAR International Series 202: 91–115.
Temme, M. 1985. First records of Wood Sandpiper, Ruff and Eurasian Tree Sparrow from the Marshall Islands. Atoll Res. Bull. 292: 23–28.
Thiollay, J. M. 1985. The birds of the Ivory Coast: status and distribution. Malimbus 7: 1–59.
Thomsen, P. & Jacobsen, P. 1979. *The Birds of Tunisia*. Copenhagen: Nature-Travels I/S.
Ticehurst, C. B. 1922. The birds of Sind. Ibis (11) 4: 605–662.
Ticehurst, C. B., Buxton, P. A. & Cheesman, R. E. 1923. The birds of Mesopotamia. J. Bombay Nat. Hist. Soc. 28: 210–250.
Ticehurst, C. B., Cox, P. & Cheesman, R. E. 1926. Additional notes on the avifauna of Iraq. J. Bombay Nat. Hist. Soc. 31: 91–119.
Ticehurst, C. B. & Whistler, H. 1927. On the summer avifauna of the Pyrenees Orientales. Ibis (12) 3: 284–313.
Tinbergen, L. 1946. De Sperwer als Roofvijand van Zangvogels. Ardea 34: 1–213.
Traylor, M. A. 1963. *Check List of Angolan Birds*. Lisbon: Museo do Dundo.
Tree, A. J. 1972. An ornithological comparison between differing dry seasons at a pan in Botswana. Ostrich 43: 363–396.
Tristram, H. B. 1859. On the ornithology of North Africa. Part II. The Sahara. Ibis (1) 1: 277–301.
Tristram, C. B. 1884. *The Survey of West Palestine*. London: Committee of the Palestine Exploration Fund.

Urban, E. K. & Brown, L. H. 1971. *A Checklist of the Birds of Ethiopia*. Addis Adaba: Haile Sellasie I Univ. Press.
Urban, E. K. & Boswall, J. 1969. Bird observations from the Dahlak Archipelago, Ethiopia. Bull. Br. Orn. Cl. 89: 121–129.

Van Someren, V. G. L. 1922. Notes on the birds of east Africa. Novit. Zool. 29: 1–246.
Vaughan, R. E. & Jones, K. H. 1913. The birds of Hong Kong, Macao, and the West River or Si Kiang in south-eastern China, with special reference to their nidification and seasonal movements. Part III. Ibis (10) 1: 163–201.
Vaurie, C. 1949. Notes on some Ploceidae from western Asia. Am. Mus. Novit. No. 1406.
Vaurie, C. 1956. Systematic notes on Palearctic birds, No. 24. Ploceidae; the genera *Passer, Petronia* and *Montifringilla*. Am. Mus. Novit. No. 1814.
Vaurie, C. 1958. The Rufous-backed Sparrows of the Cape Verde Islands. Ibis 100: 275–276.
Vaurie, C. 1959. *The Birds of the Palearctic Fauna, Passeriformes*. London: Witherby.
Vaurie, C. 1972. *Tibet and its Birds*. London: Witherby.
Verheyen, C. 1957. Over de verplaatsingen van de Boommus, *Passer montanus* (L.), in en door Belgie. Gerfaut 47: 161–170.
Vierke, J. 1970. Die Beseidlung Südafrikas durch dem Haussperling (*Passer domesticus*). J. Ornith. 111: 94–103.
Vik, R. 1962. Bird observations in the North Atlantic. Sterna 5: 15–23.
Voous, K. H. 1960. *Atlas of European Birds*. London: Nelson.
Voous, K. H. 1977. List of recent holarctic bird species. Ibis 119: 376–406.

Wade, V. E. & Rylander, M. K. 1982. Unrecorded foraging behaviour in the House Sparrow. Bird Study 29: 166.
Wagner, H. O. 1959. Die Einwanderung des Haussperlings in Mexico. Z. Tierpsychol. 16: 584–592.

Waller, C. S. 1981. Spanish Sparrow: new to Britain and Ireland. Brit. Birds 74: 109–110.
Wallis, H. M. 1887. Notes upon the northern limit of the Italian Sparrow (*Passer italiae*). Ibis (5) 2: 454–455.
Walsberg, H. 1975. Digestive adaptation of *Phainopepla nitans* associated with the eating of mistletoe berries. Condor 77: 169–174.
Ward, P. 1968. Origin of the avifauna of urban and suburban Singapore. Ibis 110: 239–255.
Ward, P. & Poh, G. E. 1968. Seasonal breeding in an equatorial population of the Tree Sparrow *Passer montanus*. Ibis 110: 359–363.
Wardlaw Ramsay, R. G. 1923. *Guide to the Birds of Europe and North Africa*. London & Edinburgh: Gurney & Jackson.
Waterston, G. (Ed.). 1978. Fair Isle Bird Observatory Report, 1977 (30): 47.
Weaver, R. L. 1939. Winter movements and a study of the nesting of English Sparrows. Bird Banding 10: 73–79.
Weaver, R. L. 1942. Growth and development of English Sparrows. Wilson Bull. 54: 183–191.
Webster, N. & Phillipps, K. 1976. *A New Guide to the Birds of Hong Kong*. Hong Kong: Sino-American Pub. Co.
Wettstein, O. Von. 1938. Die Vogelwelt der Ägais. J. Ornith. 86: 9–53.
Wettstein, O. Von. 1941. Biologische Notizen über einiges Vogelarten des Gschnitztales. Beitr. z. Fortpflz. Biol. Vog. 17: 169–170.
Wettstein, O. Von. 1959. Ergänzende Nachrichten uber das süd-alpine Mischgebiet der Haussperlinge. J. Ornith. 100: 103–104.
Whistler, H. 1910–11. The Rufous-backed Sparrow (*Passer domesticus pyrrhonotus* Blyth). J. Bombay Nat. Hist. Soc. 20: 1151.
Whistler, H. 1913–14. The Rufous-backed Sparrow, *Passer domesticus pyrrhonotus*, Blyth. J. Bombay Nat. Hist. Soc. 22: 392.
Whistler, H. 1922. The birds of the Jhang district, S. W. Punjab. Part I. Passerine birds. Ibis (11) 4: 259–309.
Whistler, H. 1926. The birds of the Kangra district, Punjab. Part II. Ibis (12) 2: 724–783.
Whistler, H. 1930. The birds of the Rawalpindi district, N. W. India. Ibis (12) 6: 67–119.
Whistler, H. 1949. *Popular Handbook of Indian Birds*. 4th Edn. revised by Kinnear, N. B. Edinburgh: Gurney & Jackson.
Whitaker, J. 1907. *Notes on the Birds of Nottinghamshire*. Nottingham: Walter Black & Co.
White, C. M. N. 1963. *A Revised Check List of African Flycatchers, Weavers and Waxbills*. Lusaka: Govt. Printers.
White, C. M. N. & Moreau, R. E. 1958. Taxonomic notes on the Ploceidae. Bull. Br. Orn. Cl. 78: 140–145 & 157–163.
Whitehead, C. H. T. 1899. Field-notes on birds collected in the Philippine Islands in 1893–96. Part II. Ibis (7) 5: 210–246.
Whitehead, C. H. T. 1909. On the birds of Kohat and Kurram, northern India. Ibis (9) 3: 214–284.
Widmann, O. 1889. History of the House Sparrow, *Passer domesticus*, and the European Tree Sparrow, *Passer montanus*, at St. Louis, Mo. In: Barrows 1889.
Wiens, J. A. & Dyer, M. I. 1977. Assessing the potential impact of granivorous birds in ecosystems. In Pinowski & Kendeigh (Eds.) 1977.
Wiens, J. A. & Johnston, R. F. 1977. Adaptive correlates of granivory in birds. In: Pinowski & Kendeigh (Eds.) 1977.
Wildash, P. 1968. *Birds of South Vietnam*. Rutland, Vermont & Tokyo: Tuttle.
Will, R. L. 1969. Fecundity, density and movements of a house sparrow population in southern Wisconsin. PhD Thesis, Univ. of Kansas. (Quoted by Dyers *et al*. 1977.)
Will, R. L. 1973. Breeding success, numbers and movements of House Sparrows at McLeansboro, Illinois. In: Hardy & Morton (Eds.) 1973.
Williams, G. R. 1953. The dispersal from New Zealand and Australia of some introduced European passerines. Ibis 95: 676–692.
Williams, J. G. & Arlott, N. 1980. *A Field Guide to the Birds of East Africa*. London: Collins.
Williamson, K. 1945. Some new and scarce breeding species in the Faeroe Islands. Ibis 87: 550–558.
Williamson, K. & Spencer, R. 1958. Ringing recoveries and the interpretation of bird-movements. Bird Migration 1: 176–181.
Winterbottom. J. M. 1936. Distributional and other notes on some Northern Rhodesian birds. Ibis (13) 6: 763–791.
Winterbottom, J. M. 1959. Expansion of the range of the House Sparrow. Cape Dept. Nat. Conserv. Rep. (16): 92–94.

Winterbottom, J. M. 1966. Systematic notes on birds of Cape Province. XXVII: the type locality of *Passer motitensis* (A. Smith). Ostrich 37: 138–139.
Winterbottom, J. M. 1971. A Preliminary Check-List of the Birds of South West Africa. Windhoek: S. W. A. Scientific Society.
Witherby, H. F. 1903. An ornithological journey in Fars, S. W. Persia. Ibis (8) 3: 501–571.
Witherby, H. F. 1905. On a collection of birds from Somaliland. Ibis (8) 5: 509–524.
Wodzicki, K. 1956. Breeding of the House Sparrow away from man in New Zealand. Emu 56: 146–147.
Woods, R. W. 1975. *The Birds of the Falkland Islands*. Oswestry: Nelson.
Woods, R. W. 1982. *Falkland Island Birds*. Oswestry: Nelson.
Wong, M. 1983. The effect of unlimited food availability on the breeding biology of wild Eurasian Tree Sparrows in west Malaysia. Wilson Bull. 95: 287–294.
Wynne-Edwards, V. C. 1927–28. House-Sparrow roosting in a lamp. Brit. Birds 21: 229–230.

Yakobi, V. W. 1979. [On the species independence of the Indian Sparrow *Passer indicus*.] Zool. Zh. 58: 136–137. (In Russian).
Yamashina, Y. 1961. *Birds in Japan: A Field Guide*. Tokyo: Tokyo News Service Ltd.
Yom-Tov, Y. 1980. Intraspecific nest parasitism among Dead Sea Sparrows *Passer moabiticus*. Ibis 121: 234–237.
Yom-Tov, Y. & Ar, A. 1980. On the breeding ecology of the Dead Sea Sparrow, *Passer moabiticus*. Israel J. Zoology. 29: 171–187.
Yom-Tov, Y., Ar, A. & Mendelssohn, H. 1978. Incubation behaviour of the Dead Sea Sparrow. Condor 80: 340–343.
Yom-Tov, Y., Mendelssohn, H. & Ar, A. 1976. Extension of range of the Dead Sea Sparrow *Passer moabiticus* and its effect on breeding success. Israel J. Zool. 25: 202–203.
Young, J. G. 1962. Unseasonable breeding of House Sparrows. Scot. Birds 2: 102.

Zarudny, N. 1904. *Passer mesopotamicus spec. nov.* Orn. Jahrb. 15: 108.
Zarudny, N. & Härms, M. 1902. Neue Vogelarten. Orn. Monatsb. 10: 49–54.
Zedlitz, O. Graf. 1913. Ornithologische Ergebnisse der Reise von Paul Spatz in der Algerische Sahara in Sommer 1912. Novit. Zool. 200: 167–170.
Zheng, Zuo-Xin. 1976. [*A Distributional List of Chinese Birds*.] Peiping: Science Publishing House. (In Chinese)
Zheng, Zuo-Xin & Tan, Yao Kuang *et al.* 1963. Taxonomic studies on birds from southeastern Szechwan and northeastern Yunnan, Part III: Passeriformes (cont.). Acta zool. Sinica 15: 295–316.
Ziedler, K. 1966. Untersuchungen über Flügelsbefiederung und Mauser des Haussperlings (*Passer domesticus* L.). J. Ornith. 107: 113–153.
Ziswiler, V. 1965. Zur Kenntnis des Samenoffens der Struktur des horneren Gaumens bei kornfressenden Oscines. J. Ornith. 106: 1–48.
Ziswiler, V. 1967a. Der Verdaaungstrakt kornerfressender Singvogel also taxonomischer Merkmalskomplex. Rev. Suisse Zool. 74: 620–628.
Ziswiler, V. 1967b. Die taxonomische Stellung des Schneefinken *Montifringilla nivalis* (L.). Orn. Beob. 64: 105–110.
Ziswiler, V. 1968. Die taxonomische Stellung der Gattung *Sporopipes cabanis*. Bonn. zool. Beitr. 19: 260–279.

Index

Accipiter nisus see Sparrowhawk
Adult (defined) 15
Aegypius tracheliotus see Vulture, Lappet-faced
Age 26, 59, 76, 155, 177, 240
Allen's rule 294, 295
Allospecies 17
Altitude, effect of 118, 146, 207, 218, 274
  occurrence at high 11, 118, 124, 128, 137, 172, 207, 209–210, 223, 226
Amblyospiza 298
Anting 140
Apus affinis see Swift, House
Auripasser 46, 301, 302
auripasser superspecies 304–305
Autumn sexual activity 145, 175–176, 230, 231, 242, 253
Avadavat 196

Barbet 24, 42
Bathing, dust 98, 140, 232, 300
  sand 52
  sun 140, 232
  water 52, 140, 232
Beagle, voyage of 79, 93
Bergmann's rule 92, 93, 114, 261, 294, 295
Bigamy 142
Bill, seasonal change in colour 20, 27, 32, 35, 39, 47, 62, 68, 80, 94, 102, 107, 116, 117, 164, 181, 195, 199, 206, 217, 242, 253–254, 302
  seasonal change in length 15, 120–121, 182, 218
  wiping 140, 142, 189, 232
Biometric data 15–16, see also under Sparrow species
  seasonal change in linear dimensions 15, 182
Bishop 70
Blossom expedition 96, 135, 167–168
Breeding, density 22, 55, 83, 156–158, 174, 212, 240–241
  opportunistic 48, 53–54, 56, 72–73, 76, 265
  precocious 58, 143
Brood patch 56, 149, 191
Bubalornis 298
Bubalornithinae 298, 300, 303
Buildings, occurrence in 120, 133, 138, 141, 231, 271, 274
Bulbul, Arabian 192
Bunting 231, 297
  Rock 196

Carduelis 298
Carpospiza 301, 303.
  brachydactyla 303

Chrysococcyx caprius see Cuckoo, Didric
Colonisation 167, 168–169, 183, 186, 189, 219, 223, 271–275
Competition, intrageneric 41–42, 42–43, 71, 72, 192, 221, 222, 255–261, 264, 266, 268, 275, 287
Control 178–179, 244
Corvids 193, 233
Corvinella melanoleuco see Shrike, Long-tailed
Cuckoo, Didric 76
Cytology 300

Delichon urbica see Martin, House
Dinemellia 298
DNA analysis 300
domesticus-lineage 284, 287
domesticus sub-group 114–116, 304, 305
  superspecies 107, 288, 304, 305
Dunnock 13

Ecographical rules 39, 92, 93, 294, 295
Egg-dumping 149, 190
Emberiza cia see Bunting, Rock
Emberizidae 13, 297, 299, 300
Emberezines 300
Enzyme analysis 280, 281, 295
Estrilda amadava see Avadavat
Estrildidae 13, 297, 302
Estrildinae, 300, 301
Estrildines 298
Euplectes 70
Evolution 114, 129, 214, 266, 275, 276–296
Evolutionary centres 289–290
Extinctions 107–108, 109, 167, 169, 220, 222, 223, 264, 275, 283, 288

Feeding, mechanisms 298
  techniques 203, 243, 270–271
Fertile Crescent 281, 287
Ficedula hypoleuca see Flycatcher, Pied
Fighting 42, 72
Finch 231, 297, 299, 300, 301
  Quail 83
  Snow 297, 301
Flycatcher, Pied 235
Flycatching 243, 270
Fossil remains 181, 278, 288, 305
'Founder effect' 295
Fringillidae 297, 298, 299, 300, 301, 302, 303
Fringilloidea 301

Gloger's rule 39, 114, 294, 295
griseus superspecies 17, 21, 269, 290, 304
Gymnoris 299, 301

Helpers 58, 150
Heron 173, 233, 276
Hirundo aethiopica see Swallow, Ethiopian
  daurica see Swallow, Red-rumped
  semirufa 24
  striata see Swallow, Striated
hispaniolensis-lineage 284, 287
Histurgops 297, 298, 301, 302, 303
  ruficauda 303
Homoiothermy 150, 237
Horse, association with 23
Hybridisation 105, 122–123, 126, 163, 166, 168, 170, 259, 260, 264, 267, 280, 283, 287, 288, 293, 305

iagoensis superspecies 304
Immature (defined) 15
indicus-lineage 285, 287
indicus sub-group 114–116, 304, 305
Intersex 115
Irrigation, effect of on breeding 24, 34, 49, 188, 202, 275

Juvenile (defined) 15

Latitude, effect of 74, 118–120, 143, 144, 146–151, 156, 198, 210, 218, 227, 236, 237, 274
Lonchura punctualata see Munia, Spotted
luteus superspecies 47, 61

Martin, House 235
  Sand 235, 266
Melospiza melodia see Sparrow, Song
Mines, occurrence in 138–139, 271
Montifringilla 297, 298, 299, 300, 301, 302, 303
  adamsi 302
  blanfordi 302
  davidiani 302
  nivalis 302
  ruficollis 302
  taczanowski 302
  theresae 302
Mossie 67
Moult see under Sparrow species
  arrested 158
  post-juvenile 15, 43, 59, 76, 143, 158, 198, 241, 252, 277, 299, 302
Movements, dispersal 24, 51, 83, 123–124, 141, 172, 184, 187, 221, 228, 230, 272
  migration 128, 141, 142, 164, 167, 169, 170–171, 184, 185, 186, 187, 207–208, 209, 210, 221, 222, 223, 228–230, 273
  nomadism 28, 51, 53, 56, 70, 83, 109, 169, 172, 201, 230, 263, 270
Munia, Spotted 202
Musculature see Myology
Myology 298

Nest size 55, 87, 89, 110, 145, 174, 190, 197, 202, 235
Nomenclature 16, see also under Sparrow species

Numbers 135, 156–158, 220, 222, 241
  decrease 82, 157–158, 166, 220, 241
  increase 157, 184, 186, 241

Onychostruthus 299, 302
Ortygospiza fuscocrissa see Finch, Quail

Padda orzivora see Sparrow, Java
Pair faithfulness 71–72, 142–143, 202, 232, 271
Parasitism, brood 65, 76, 190, 275
  nest 63–64, 65, 190, 263
Parus see Tit
  melanocephalus see Tit, Black-crested
  caeruleus see Tit, Blue
Passer 13, 46, 278, 297, 298, 299, 300, 301, 302, 303
  origin of genus 276–278
  ammodendri 269, 277, 304 see also Sparrow, Saxaul
  castanopterus 269, 277, 304, 305 see also Sparrow, Somali
  diffusus 269, 277, 304 see also Sparrow, Southern Grey-headed
  domesticus 269, 277, 296, 298, 303, 304, 305 see also Sparrow, House
  eminibey 269, 277, 300, 304, 305 see also Sparrow, Chestnut
  flaveolus 269, 277, 304 see also Sparrow, Pegu
  griseus 269, 277, 298, 304 see also Sparrow, Grey-headed
  gongonensis 269, 277, 304 see also Sparrow, Parrot-billed
  hispaniolensis 269, 277, 304, 305 see also Sparrow, Willow
  iagoensis 269, 277, 304 see also Sparrow, Iago
  indicus 304 see also Sparrow, Indian
  italiae 282, 304, 305 see also Sparrow, Italian
  luteus 269, 277, 300, 304, 305 see also Sparrow, Golden
  melanurus 269, 277, 304 see also Sparrow, Cape
  moabiticus 269, 277, 300, 304, 305 see also Sparrow, Dead Sea
  montanus 269, 277, 304 see also Sparrow, Tree
  motitensis 269, 277, 304 see also Sparrow, Rufous
  predomesticus 269, 277, 304, 305
  pyrrhonotus 269, 277, 304, 305 see also Sparrow, Sind Jungle
  rutilans 269, 277, 304, 305 see also Sparrow, Cinnamon
  simplex 269, 277, 304, 305 see also Sparrow, Desert
  suahelicus 269, 277, 304 see also Sparrow, Swahili
  swainsonii 269, 277, 304 see also Sparrow, Swainson's
Passeridae 298, 300, 301, 302, 303
Passerinae 298, 300, 301, 302

Passerine finches 297, 298, 302
Passeroidea 301
Pest of agriculture 30, 50, 60, 70, 77, 100, 112, 139, 171, 172, 178, 243–244
Pests, control of 129, 244
Petronia 13, 297, 298, 299, 300, 301, 302, 303
   brachydactyla 302
   dentata 302
   petronia 302
   pyrgita 302
   superciliosus 302
   xanthicollis 302, 303
   xanthosterna see *Petronia pyrgita*
Philetairus 297, 298, 300, 301, 302, 303
   socius 303
Physiological adaptation 133, 142, 200, 274
Pleistocene 279, 280, 281, 282, 283, 284, 290, 291
Pliocene 279
Ploceidae 268, 297, 298, 299, 300, 301, 302, 303
Ploceinae 298, 300, 301, 302, 303
Ploceines 298, 300
Plocepasser 297, 298, 299, 301, 302, 303
   donaldsonii 303
   mahili 303 see also Sparrow-weaver, White-browed
   rufoscapulatus 303
   superciliosus 303
Plocepasserinae 298, 300, 301, 303
Plocepasserine 302
Ploceus 47, 300
   capensis see Weaver, Cape
   intermedius see Weaver, Masked
   philippinus see Weaver, Baya
   rubiginosus 64
   spekei see Weaver, Skeke's
   velatus 64
Plumage, aberrant 116, 123, 283–284
Population dynamics 154–155, 272
Promiscuity 143
Prunella modularis see Dunnock
Prunellidae 13
Pseudonigrita 297, 298, 299, 301, 302, 303
   arnaudi 303 see also Weaver, Grey-capped Social
   cabanis 303 see also Weaver, Black-capped Social
Pterology 299
Pyconotatus capensis see Bulbul, Arabian
Pyrgolauda 299, 300, 302

Quaternary 280, 291
Quelea 62

Range extension 22, 40, 49, 54, 102, 121–122, 129–137, 166–167, 170, 182–184, 185, 185–186, 200, 219–220, 223, 224, 258, 261, 271–275, 287, 291
Reproductive isolation 94, 261, 267, 287–288, 291
Riparia riparia see Martin, Sand

Roller feeding 52, 211–212, 231
Roosting 24, 26, 52, 70–71, 110, 140–141, 149, 173, 196, 202, 212, 230–231
Roost nest 70, 141, 230

Semi-species 304, 305, 306
Serology 300
Sesquimorphic plumage 14, 67, 80, 245, 276, 278
Ship-assisted passage 105, 131, 132, 133, 135, 137, 220, 223, 224, 274
Shrike, Long-tailed 83
Social behaviour 24, 28, 33, 42, 51–52, 62, 70, 71, 98, 103–104, 110, 139–141, 159, 172, 173–174, 189, 193, 196, 201–202, 211–212, 228, 231, 243, 251
Sorella 61, 299, 301, 302
Sparrow, American 13
   American Tree 13, 217
   'black-bibbed' 14, 80, 94, 101, 107, 116, 163, 174, 181, 199, 206, 217, 245, 254, 267, 278, 279, 287, 291, 292
   Bush 13, 297
   Cape 41, 42, 67–77, 82, 263, 265, 268, 269, 275, 276, 277, 292
      behaviour 70–71
      biometrics 67, 68
      breeding biology 71–76
      description 67–68
      distribution 68–69
      food 77
      habitat 70
      moult 76
      nomenclature 67
      survival 76
      voice 76–77
   Cape Verde Island 94
   Chestnut 61–66, 254, 263, 268, 269, 275, 276, 277, 290
      behaviour 62
      biometrics 61, 62
      breeding biology 63–65
      description 61–62
      distribution 62, 63
      food 66
      habitat 62
      nomenclature 61
      voice 66
   Cinnamon 71, 101, 205–215, 227, 268, 269, 277, 288, 290, 292
      behaviour 211–212
      biometrics 206–207
      breeding biology 212–214
      description 206
      distribution 207–210
      habitat 210–211
      food 214
      moult 214
      nomenclature 205–206
      voice 214
   Cinnamon Tree 206
   Cisalpine 164

Sparrow—*contd*
  Dead Sea 15, 180–193, 218, 244, 254, 263, 268, 269, 274, 277, 278, 281, 288, 290
    behaviour 189
    biometrics 181–182
    breeding biology 189–192
    description 181
    distribution 182–187
    food 193
    habitat 187–188
    moult 193
    nomenclature 180–181
    voice 193
  Desert 106–113, 173, 268, 269, 271, 275, 277, 288, 290
    behaviour 110
    biometrics 107, 108
    breeding biology 110–112
    description 107
    distribution 107–110
    food 113
    habitat 110
    moult 112
    nomenclature 106–107
    voice 113
  Emin Bey's 61
  Eurasian Tree 13, 217
  European Tree 217
  Golden 14, 42, 46–59, 61, 62, 73, 104, 254, 263, 265, 269, 270, 275, 276, 277, 278, 290
    behaviour 51–52
    biometrics 47–48
    breeding biology 52–58
    description 47
    distribution 48–50
    food 59–60
    habitat 50–51
    moult 59
    nomenclature 46–47
    survival 58
    voice 59
  Great 80, 82–85, 100
    behaviour 83
    biometrics 80, 81–82
    breeding biology 83–85
    description 80
    distribution 79, 82–83
    food 85
    habitat 83
    moult 85
    voice 85
  grey-headed 14, 19–45, 254, 255–263, 269, 275, 290, 291, 304
  Grey-headed 17, 19–26, 27, 28, 31, 35, 37, 40, 42, 44, 91, 254, 255–261, 262, 269, 277, 291
    behaviour 23–24
    biometrics 20–21, 256
    breeding biology 24–26
    description 19–20
    distribution 21–22, 23
    food 26
    habitat 22–23
    moult 26
    nomenclature 19
    survival 26
    voice 26
  Hedge see Dunnock
  House 15, 23, 41, 42, 56, 71, 73, 97, 98, 102, 104, 105, 114–161, 167, 172, 173, 174, 175, 177, 189, 192, 193, 196, 210, 214, 218, 221, 224, 226, 227, 228, 233, 234, 235, 237, 243, 244, 251, 253, 254, 261, 264, 265, 266, 267, 268, 269, 270, 271, 272, 273, 274, 275, 277, 278, 280, 281, 282, 283, 285, 286, 287, 288, 293, 294, 295, 296, 303, 304, 305
    behaviour 139–142
    biometrics 116, 117–121
    breeding biology 142–154
    breeding density 156–158
    description 116–117
    distribution 121–137
    food 159–161
    habitat 137–139
    introductions 129–137
    moult 158
    nomenclature 114–116
    numbers 156–158
    survival 154–156
    voice 158–159
  Iago 93–100, 172, 263–265, 269, 276, 277, 292
    behaviour 98
    biometrics 94, 95–96
    breeding biology 98–100
    description 93–95
    distribution 79, 95–97
    food 100
    habitat 97–98
    moult 100
    nomenclature 93–94
    voice 100
  Indian 115, 304, 305, 306
  Italian 122–126, 162–163, 164, 169–170, 174, 175, 282, 284, 285, 286, 303
  Java 13
  Kenya Rufous 86–87, 89, 265
    behaviour 87
    biometrics 81
    breeding biology 87
    description 80
    distribution 79, 86
    food 87
    habitat 86
    voice 87
  Kordofan Rufous 90–91
    behaviour 90–91
    biometrics 81
    breeding biology 91
    description 80
    distribution 79, 90
    habitat 90
  Moab 180
  New World 13

Old World 13, 301
Old World Tree 217
Parrot-billed 17, 22, 31–34, 35, 45, 82, 255–257, 261–263, 265, 269, 277, 291
  behaviour 33
  biometrics 31, 32, 256
  breeding biology 33–34
  description 31–32
  distribution 32, 33, 257, 261, 262
  food 34
  habitat 32
  nomenclature 31
  voice 34
Pegu 199–204, 268, 269, 274, 275, 277, 288
  behaviour 201–202
  biometrics 199, 200
  breeding biology 202–203
  description 199
  distribution 200, 201
  food 204
  habitat 200–201
  moult 203
  nomenclature 199
  voice 204
Pegu House 199, 200
'plain' 14, 19, 290
Plain-backed 199
Rock 13, 297, 301
rufous 263–264, 291
Rufous 78–92, 93, 94, 95, 194, 263, 264, 269, 275, 277, 288, 292
  biometrics 81
  habitat 92
  nomenclature 78–80
  description 80, 92
  distribution 79–80
Rufous-backed 94, 194
Russet 206
Saxaul 14, 245–252, 268, 269, 277, 288, 290
  behaviour 251
  biometrics 245, 246
  breeding biology 251–252
  description 245 246
  distribution 246–250, 251
  food 252
  habitat 250–251
  moult 252
  nomenclature 245
  voice 252
Scrub 180
Sind Jungle 194–198, 268, 269, 277, 288
  behaviour 196
  biometrics 194, 195
  breeding biology 197–198
  description 194–195
  distribution 195, 196
  food 198
  habitat 195–196
  moult 198
  nomenclature 194
  voice 198

Socotra 92
  biometrics 81
  breeding biology 92
  description 80, 92
  distribution 79, 92
  habitat 92
Somali 73, 101–105, 265, 267, 269, 270, 277, 292, 293
  behaviour 103–104
  biometrics 101, 102
  breeding biology 104
  description 101–102
  distribution 102, 103
  food 105
  habitat 102
  nomenclature 101
  voice 104
Song 13
Southern Grey-headed 17, 38–43, 45, 71, 72, 256, 259–261, 263, 265, 268, 269, 274, 275, 277, 291
  behaviour 42
  biometrics 38, 39–40, 256
  breeding biology 42–43
  description 38–39
  distribution 40–41, 259, 260
  food 43
  habitat 41–42
  moult 43
  nomenclature 38
  voice 43
Spanish 163
Swahili 17, 22, 35–37, 45, 256, 257–258, 262–263, 269, 277
  behaviour 36
  biometrics 35, 256
  breeding biology 36–37
  description 35
  distribution 35–36, 258, 262
  food 37
  habitat 36
  nomenclature 35
  voice 37
Swainson's 17, 27–30, 31, 35, 44–45, 255, 256, 261–262, 269, 277, 291
  behaviour 28
  biometrics 27, 28, 256
  breeding biology 28–29
  description 27
  distribution 28, 29, 256, 261
  food 30
  habitat 28
  nomenclature 27
  voice 30
Tree 13, 15, 71, 163, 167, 210, 216–244, 251, 255, 264, 265, 266, 267, 268, 269, 271, 272, 274, 275, 277, 287, 288, 295, 296
  behaviour 228–232
  biometrics 217, 218, 219
  breeding biology 232–239
  breeding density 240–241
  description 217–218

Sparrow—*contd*
  Tree—*contd*
    distribution 219–224, 225
    food 243–244
    habitat 224–228
    introductions 218, 221–222, 223, 224
    moult 241–242
    nomenclature 216–217
    numbers 240–241
    survival 240
    voice 242–243
  'True' 13, 14, 303
  White Nile Rufous 88–89, 90
    behaviour 89
    biometrics 81
    breeding biology 89
    description 80
    distribution 79, 88
    habitat 89
  Willow 96, 97, 98, 99, 112, 162–179, 189, 192, 234, 244, 251, 264, 265, 266, 267, 269, 271, 275, 277, 278, 280, 281, 283, 285, 286, 288, 290, 295, 303
    behaviour 172–173
    biometrics 163, 164–165
    breeding biology 173–177
    description 163–164
    distribution 165–171
    food 178–179
    introductions 166–167
    habitat 171–172
    moult 177
    nomenclature 162–163
    survival 177
    voice 177
Sparrowhawk 158, 184, 296
Sparrow-weaver 297, 298, 302, 303
  White-browed 83
Spizella arborea see Sparrow, American Tree
Sporopipes 29, 298, 299, 302, 303
  frontalis 303
  squamifrons 303

Sporopipinae 30, 302, 303
Starling 234, 235
Stork 174
Street lamp, used for nesting 98–99, 144
Sturnus vulgaris see Starling
Superpecies 17, 47, 61, 254, 290, 304, 305
Swallow 24, 42, 268
  Ethiopian 29
  Red-rumped 212, 235
  Rufous-chested 24
  Striated 235
Swift 24, 42
Systematics 297–306

Tail flicking 52, 98, 142, 189, 193, 232
Time-differentiated species 177, 305
Tit 235, 269, 270, 271
  Black-crested 213
  Blue 270

Uropygial gland 300

Vulture, Lappet-faced 111

Walking 42, 52
Waxbill 297, 301
Weaver 268, 297, 299, 300, 301
  Baya 197
  Black-capped Social 63
  Cape 90
  Grey-capped Social 63
  Masked 63
  Speke's 63
Weight, diurnal variation 48, 120
  seasonal variation 40, 48, 120, 142, 147, 164–165, 242
Wing-loading 119, 155–156
Woodpecker 24, 42, 233

Young, out of nest 100, 237